Metallic Glasses

Special Issue Editors

Keith K.C. Chan

Jordi Sort Viñas

MDPI • Basel • Beijing • Wuhan • Barcelona • Belgrade

MDPI

Special Issue Editors
Keith K.C. Chan
The Hong Kong Polytechnic University
China

Jordi Sort Viñas
ICREA
Universitat Autònoma de Barcelona
Spain

Editorial Office
MDPI AG
St. Alban-Anlage 66
Basel, Switzerland

This edition is a reprint of the Special Issue published online in the open access journal *Metals* (ISSN 2075-4701) in 2015 (available at: http://www.mdpi.com/journal/metals/special_issues/metallic-glasses).

For citation purposes, cite each article independently as indicated on the article page online and as indicated below:

Author 1, Author 2. Article title. *Journal Name*. **Year**. Article number/page range.

First Edition 2017

ISBN 978-3-03842-506-9 (Pbk)
ISBN 978-3-03842-507-6 (PDF)

Table of Contents

About the Special Issue Editors

Keith K.C. Chan is currently the Director of Advanced Manufacturing Technology Research Center, and Head of the Department of Industrial and Systems Engineering of The Hong Kong Polytechnic University. He is very active in research. He has published more than 230 international journal papers and secured more than fifteen external research grants and 2 US patents. His research areas include advanced manufacturing technology, processing of advanced materials and bulk metallic glasses. He was a member of the Engineering panel of the Research Grants Council of Hong Kong between 2005 and 2010. He is currently a member of the Innovation & Technology Support Programme Assessment Panel under the Innovation and Technology Fund. He received the President's Award for Achievement in the Category of Research and Scholarly Activities (1996), and a number of Faculty or Departmental awards in different categories including the Research Grant Achievement Award. Professor Chan was the Editor-in-Chief of HKIE Transactions, and a guest editor of a special issue of two international journals. Currently, he is a member of the Editorial Board of various international journals such as Scientific Reports and Metals.

Jordi Sort Viñas (aged 40) is currently working as an ICREA Research Professor at the Universitat Autònoma de Barcelona (UAB), where he leads the Group of Smart Nanoengineered Materials, Nanomechanics and Nanomagnetism (Gnm3) (2014-SGR-1015), now consisting of 19 researchers. The research activities of the group focus on the synthesis of functional materials (electrodeposited films, lithographed structures, porous materials, bulk metallic glasses, nanocomposites, etc.) and the study of their structural, mechanical, catalytic and magnetic properties. The group develops new types of materials, whose microstructure can be precisely controlled at the nanoscale, leading to enhanced performance and functionalities, that go beyond the state-of-the-art. For further details on our group, see: http://jsort-icrea.uab.cat/index.htm.

Prof. Sort got his PhD degree in Materials Science at UAB in 2002. His research experience abroad includes a stay of 5 months at the Grenoble High Magnetic Fields Laboratory (France), 4 months at Los Alamos National Laboratory (USA), 2 years as postdoc at the laboratory SPINTEC in the Commissariat à l'Energie Atomique (Grenoble) and 6 months at Argonne National Laboratory (USA). Prof. Sort received the PhD UAB Extraordinary Award in 2004 and his work was also awarded by the Catalan Physical Society (Jordi Porta i Jué's Prize, 2000), the Spanish Royal Physical Society (Young Researcher Award in Experimental Physics, 2003) and the Federation of European Materials Societies (FEMS Materials Science & Technology Prize, 2015).

So far, Prof. Sort has supervised 7 PhD students (7 more are in progress) and has published around 250 articles which have received more than 5400 citations (h = 36 in ISI Web of Science. The work of the Team has been presented as invited talks in 160 international conferences. Prof. Sort has filled 4 patents and has managed 20 national/international research projects. His research is being funded both from private contracts with industries, as well as from Public Grants from the Catalan Government, the Spanish Ministry of Economy and Competitiveness and the European Commission (Horizon 2020). Remarkably, he was awarded an ERC 2014-Consolidator Grant from the European Research Council. His project entitled "Merging Nanoporous Materials with Energy-Efficient Spintronics (SPIN-PORICS)" aims to integrate engineered nanoporous materials into novel spintronic applications. Also, he is the Coordinator of a Marie Sklodowska-Curie Innovative Training Network (ITN-ETN) entitled "Smart electrodeposited alloys for environmentally sustainable applications: from advanced protective coatings to micro/nano-robotic platforms (SELECTA)" (http://selecta-etn.eu/), which gathers a Consortium of 15 Partners (including 3 companies) from 12 different countries.

Preface to "Metallic Glasses"

Metallic glasses, a family of metallic materials with metastable glassy states, are obtained by rapid cooling of liquid alloys. Because of their amorphous atomic structures, metallic glasses exhibit unique mechanical, physical, and chemical properties, which are superior to conventional metals' and alloys', within a wide range of potential applications.

With the combination of high strength, elasticity, and fracture resistance, bulk metallic glasses (BMGs) have attracted tremendous research interest over the past few decades. However, low tensile ductility is still one of the key issues of BMGs in structural applications. Due to its high accuracy in replicating complicated mold shapes, the die casting technique has been successfully applied to produce BMG products. On the other hand, metallic glasses can also be formed into complex shapes, like plastics, in the supercooled liquid region, due to their high thermal-plastic forming abilities.

In addition to their unique mechanical properties, metallic glasses have also demonstrated interesting physical and chemical properties. For example, some metallic glasses have been found to have good magnetocaloric effects and display catalytic behavior, with potential for magnetic refrigeration or catalytic applications. In recent years, metallic glass thin films and coatings have also drawn much attention due to their potential engineering and biomedical applications.

Up to now, despite the encouraging achievements, the wide application of metallic glasses is still somewhat hindered by the mysterious physical origins of the unique properties, and by the difficulties in exploiting the alloys with benchmark performances. The wide range of studies will continue to further uncover the underlying physical meanings, and to expand the application potential of metallic glasses. Papers on recent advances, and review articles, particularly in regard to fundamental properties and the structural and functional applications of metallic glasses, are invited for inclusion in this Special Issue on "Metallic Glasses".

Keith K.C. Chan and Jordi Sort Viñas
Special Issue Editors

![metals logo] *metals*

MDPI

Article

A New Ni-Based Metallic Glass with High Thermal Stability and Hardness

Aytekin Hitit *, Hakan Şahin, Pelin Öztürk and Ahmet Malik Aşgın

Department of Materials Science and Engineering, Afyon Kocatepe University, Afyonkarahisar 03200, Turkey; hakansahin@aku.edu.tr (H.S.); pelinsuozturk@gmail.com (P.O.); amalikasgin@gmail.com (A.M.A.)
* Author to whom correspondence should be addressed; hitit@aku.edu.tr; Tel./Fax: +90-272-228-1441.

Academic Editors: Jordi Sort Viñas and Hugo F. Lopez
Received: 20 October 2014; Accepted: 27 January 2015; Published: 2 February 2015

Abstract: Glass forming ability (GFA), thermal stability and microhardness of $Ni_{51-x}Cu_xW_{31.6}B_{17.4}$ ($x = 0, 5$) metallic glasses have been investigated. For each alloy, thin sheets of samples having thickness of 20 μm and 100 μm were synthesized by piston and anvil method in a vacuum arc furnace. Also, 400 μm thick samples of the alloys were synthesized by suction casting method. The samples were investigated by X-ray diffractometry (XRD) and differential scanning calorimetry (DSC). Crystallization temperature of the base alloy, $Ni_{51}W_{31.6}B_{17.4}$, is found to be 996 K and 5 at.% copper substitution for nickel increases the crystallization temperature to 1063 K, which is the highest value reported for Ni-based metallic glasses up to the present. In addition, critical casting thickness of alloy $Ni_{51}W_{31.6}B_{17.4}$ is 100 μm and copper substitution does not have any effect on critical casting thickness of the alloys. Also, microhardness of the alloys are found to be around 1200 Hv, which is one of the highest microhardness values reported for a Ni-based metallic glass until now.

Keywords: metallic glass; refractory metal; glass forming ability; crystallization temperature; microhardness

1. Introduction

For the last two decades, multicomponent bulk metallic glasses (BMGs) have attracted great attention because of their unusual physical, chemical and mechanical properties [1]. The major factor which restricts the utilization of the bulk metallic glasses at high temperatures is their crystallization temperatures, above which they crystallize and lose their excellent properties. In general, if a metallic glass alloy contains elements having high melting point, it is expected to have high crystallization temperature. For this reason, number of metallic glass alloys containing high amount of refractory metals, such as tungsten, ruthenium, rhenium, iridium, osmium, tantalum and niobium, have been studied in order to develop metallic glasses having high crystallization temperatures [2–10]. In these studies, metallic glasses which have crystallization temperatures higher than 1100–1200 K have been developed. In addition, microhardnesses of the refractory based metallic glasses determined to be between 1200–2000 Hv, which are much higher than almost all of the non-refractory metal based metallic glasses. Examinations of the compositions of the refractory metal based metallic glasses show that most of these alloys contain high amount of boron in addition to refractory metals [2–10]. This indicates that these alloys owe their attractive mechanical and thermal properties to strong bonds form between refractory metals and boron.

Although refractory metal based metallic glasses are superior to other metallic glasses in terms of mechanical properties and thermal stability, their critical casting thicknesses are quite low, which are less than 30 μm [2–10]. Such low critical casting thickness values prevent them from being used in industrial applications. The reason of these low critical casting values is that their liquidus temperatures are quite high because of high refractory metal contents. As known, increasing the liquidus temperature of an alloy for a constant glass transition temperature increases the cooling rate required for glass formation without crystallization. As a result, critical casting thickness decreases. In fact, most of the refractory metal based metallic glasses have such high liquidus temperatures that they can not be measured with thermal analysis equipment.

Therefore, in order to develop refractory metal based metallic glasses which have high critical casting thickness, alloy compositions having sufficiently low liquidus temperatures must be found. Unfortunately, there is very limited information about RM-B-X (X: other elements) systems in the literature. In a theoretical study on Ni-W-B system, it has been reported that one of the eutectic compositions is $Ni_{51}W_{31.6}B_{17.4}$ and its euctectic temperature is 1622 K [11]. This composition contains significantly high amount of tungsten and boron and has a quite low liquidus temperature with respect to other refractory metal based metallic glasses. Therefore, we believe that it deserves an investigation to determine its glass forming ability (GFA), thermal stability and microhardness.

In this study, GFA, thermal stability and microhardness of $Ni_{51-x}Cu_xW_{31.6}B_{17.4}$ ($x = 0, 5$) alloys are investigated. The base alloy, $Ni_{51}W_{31.6}B_{17.4}$, is modified with copper due to the fact that copper has such a low solubility with boron and tungsten that it is supposed not to form any phase with them. Therefore, copper substitution for nickel is expected to improve GFA of the alloy by hindering formation of precipitating phases during solidification.

2. Results

2.1. XRD Results

XRD patterns of 100 μm thick samples of the alloys are given in Figure 1a. In both of the XRD patterns, a broad diffraction peak, which is typical of amorphous structure, is observed. Also, no crystalline peak is visible in the patterns, which indicates that 100 μm thick samples of the alloys are fully amorphous. However, for the casting thickness of 400 μm, alloys almost completely crystallize during solidification (Figure 1b). Also, it should be noted that 400 μm thick sample of $Ni_{51}W_{31.6}B_{17.4}$ alloy contains some amount of amorphous phase. Examination of the XRD patterns revealed that for both of the alloys, Ni (space group Fm3m) and W_2B (space group I4/mcm) phases form in the samples having casting thickness of 400 μm. Lattice parameter of nickel is found to be 3.59 Å, which is higher than the lattice parameter of pure nickel, 3.5238 Å (JCPDS-PDF-4-850). This result clearly shows that this phase contains some amount of tungsten, so it is actually a Ni-W solid solution rather than pure nickel. In addition, lattice parameters of W_2B phase are also determined. They are found to be $a = 5.58$ Å and $c = 4.73$ Å. These values are very close to the tabulated cell parameters, which are $a = 5.568$ Å and $c = 4.744$ Å (JCPDS-PDF-25-990).

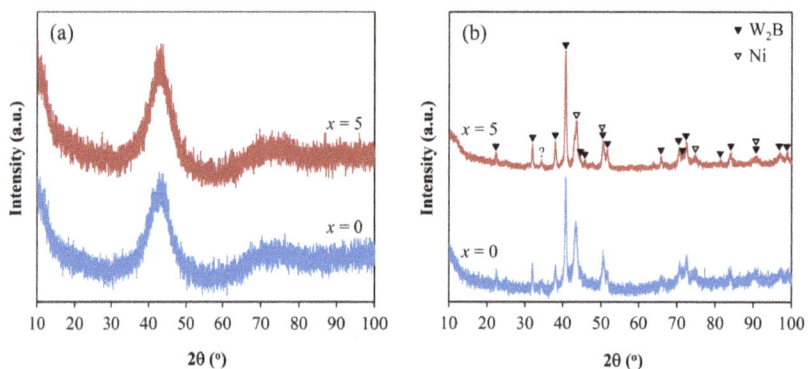

Figure 1. X-ray diffractometry (XRD) patterns of $Ni_{51-x}Cu_xW_{31.6}B_{17.4}$ alloys (a) 100 μm thick samples; (b) 400 μm thick samples.

2.2. Thermal Stability

DSC scans of the alloys are shown in Figure 2. Each DSC scan exhibits exothermic reactions corresponding to crystallization of the undercooled liquid. However, glass transition temperatures of the alloys cannot be well defined. Only tentatively determined T_g values of the alloys are shown in Figure 2a. Glass transition temperatures of $Ni_{51}W_{31.6}B_{17.4}$ and $Ni_{46}Cu_5W_{31.6}B_{17.4}$ alloys are determined as 929 and 971 K, respectively. Also, crystallization temperatures of $Ni_{51}W_{31.6}B_{17.4}$ and $Ni_{46}Cu_5W_{31.6}B_{17.4}$ alloys are found to be 996 and 1063 K, respectively. In addition, solidus temperatures of $Ni_{51}W_{31.6}B_{17.4}$, and $Ni_{46}Cu_5W_{31.6}B_{17.4}$ alloys are measured as 1610 and 1620 K, respectively. Moreover, liquidus temperatures of $Ni_{51}W_{31.6}B_{17.4}$ and $Ni_{46}Cu_5W_{31.6}B_{17.4}$ alloys are found to be 1784 and 1780 K, respectively. Thermal properties of the alloys are summarized in Table 1. In addition, XRD patterns of the samples of $Ni_{46}Cu_5W_{31.6}B_{17.4}$ alloy which were annealed at 950 and 1025 K are given in Figure 3. XRD patterns show that after annealing at 950 K small amount of nickel precipitates. Annealing at 1025 K results in significant increase in the amount of nickel which forms in the structure. These results show that T_g values of the alloys are lower than those determined tentatively.

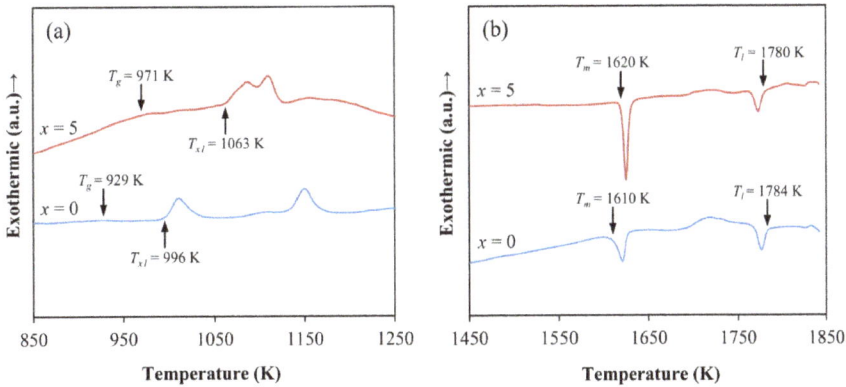

Figure 2. Differential scanning calorimetry (DSC) curves of $Ni_{51-x}Cu_xW_{31.6}B_{17.4}$ alloys (**a**) low temperature measurements; (**b**) high temperature measurements.

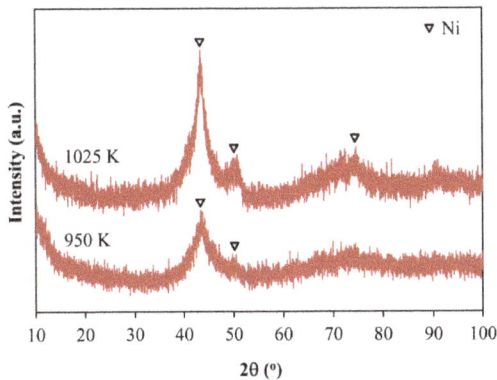

Figure 3. XRD patterns of annealed samples of $Ni_{46}Cu_5W_{31.6}B_{17.4}$ alloy.

Metals **2015**, *5*, 162–171

2.3. Microhardness

Vickers hardness values of $Ni_{51}W_{31.6}B_{17.4}$ and $Ni_{46}Cu_5W_{31.6}B_{17.4}$ alloys are found to be 1265 and 1213 Hv, respectively. Also, standart deviations of hardness measurements for $Ni_{51}W_{31.6}B_{17.4}$ and $Ni_{46}Cu_5W_{31.6}B_{17.4}$ alloys are 49 and 31 Hv, respectively. SEM image of an indent obtained from 100 µm thick sample of $Ni_{46}Cu_5W_{31.6}B_{17.4}$ alloy is shown in Figure 4.

Table 1. Thermal properties (T_g, T_x, T_m, T_l) and microhardnesses of $Ni_{51-x}Cu_xW_{31.6}B_{17.4}$ alloys.

Alloy	T_g (K)	T_x (K)	T_m (K)	T_l (K)	H_v
$Ni_{51}W_{31.6}B_{17.4}$	929	996	1610	1784	1265
$Ni_{46}Cu_5W_{31.6}B_{17.4}$	971	1063	1620	1780	1213

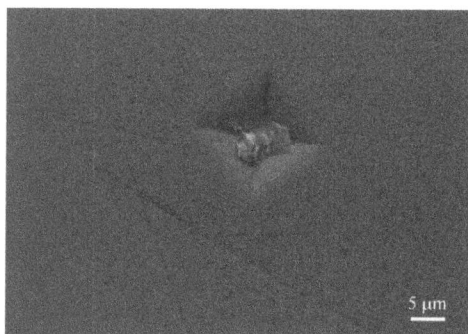

Figure 4. Scanning electron microscopy (SEM) image of an indent obtained from 100 µm thick sample of $Ni_{46}Cu_5W_{31.6}B_{17.4}$ alloy.

3. Discussion

XRD results clearly show that critical casting thicknesses of the alloys are at least 100 µm but less than 400 µm. This critical casting thickness value is higher than critical casting thicknesses of most of the refractory metal based metallic glasses [2–10]. This is due to the fact that liquidus temperatures of the alloys investigated in this study are much lower than those of other refractory metal based metallic glasses. Because of lower liquidus temperatures, $Ni_{51}W_{31.6}B_{17.4}$ and $Ni_{46}Cu_5W_{31.6}B_{17.4}$ alloys have lower reduced glass transition temperatures, T_g/T_l; as a result, they have higher GFA values.

Glass transition and crystallization temperatures of the alloys are increasing with copper addition. Copper is not soluble with tungsten in even liquid state [12], but completely soluble with nickel. Also, nickel and tungsten are completely soluble in liquid state and tungsten is partially soluble in nickel in solid state. Therefore, copper substitution for nickel must be lowering solubility of tungsten in nickel in liquid state. For this reason, in $Ni_{46}Cu_5W_{31.6}B_{17.4}$ alloy, the number of Ni-W neighbors is decreasing and the number of W-W and W-B neighbors is increasing. Because of the reduction in the number of Ni-W pairs, formation of the first phase, nickel-tungsten solid solution, becomes more difficult during heating. In other words, crystallization of nickel takes places at a higher temperature (Figure 5). On the other hand, due to the increase in number of W-B neighbors, formation of the second phase, W_2B, becomes easier. In fact, the second crystallization temperature of $Ni_{46}Cu_5W_{31.6}B_{17.4}$ alloy is lower than that of $Ni_{51}W_{31.6}B_{17.4}$ alloy. Crystallization temperature of $Ni_{46}Cu_5W_{31.6}B_{17.4}$ alloy, 1063 K, is higher than those of many of the refractory metal based metallic glasses [2–10] and the highest value reported for a Ni-based metallic glass until now (Table 2).

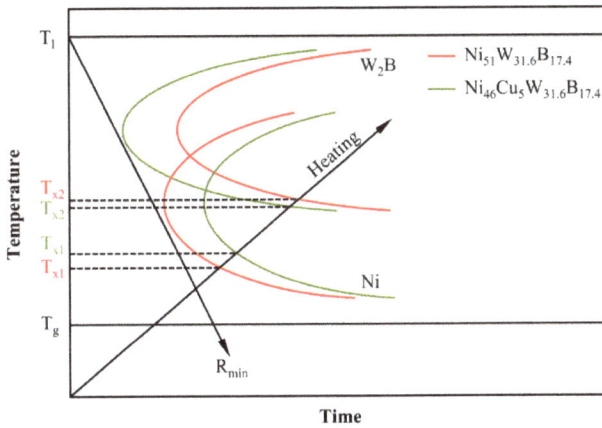

Figure 5. Schematic time-temperature-transformations (TTT) diagram of $Ni_{51}W_{31.6}B_{17.4}$ and $Ni_{46}Cu_5W_{31.6}B_{17.4}$ alloys.

Table 2. Thermal properties (T_g, T_x) and microhardnesses of various Ni-based metallic glasses.

Alloy	D (mm)	T_g (K)	T_x (K)	Ref.
$Ni_{40}Ta_{42}Co_{18}$	2	993	1032	[13]
$Ni_{62}Ta_{36}Sn_2$	1	977	1035	[14]
$Ni_{64}(Nb_{0.85}Zr_{0.15})_{36}$	2	899	935	[15]
$Ni_{60}Nb_{40}$	1	891	924	[16]
$Ni_{60}Zr_{20}Ta_{10}Nb_5Al_5$	3	870	938	[17]
$Ni_{45}Ti_{20}Zr_{23}Al_{12}$	<0.5	783	832	[18]

For a constant T_l, increasing T_g and T_x is expected to result in improved critical casting thickness. However, as mentioned, since formation of W_2B phase becomes easier because of copper addition, minimum cooling rate, R_{min}, required for the formation of glass phase is still high (Figure 5). As a result no change in critical casting thickness is observed. In order to improve GFA of Ni-W-B metallic glasses, R_{min} must be reduced, which can be accomplished by decreasing the liquidus temperature of the alloy and hindering the formation of precipitating phases, W_2B and nickel. These can be achieved by decreasing the tungsten and boron contents of the alloy. However, it should be considered that decreasing tungsten and boron contents of the alloy also decreases T_g and T_x, which causes deterioration of thermal stability. Therefore, reduction of tungsten and boron contents of the alloy should be carried out systematically to obtain optimum critical casting thickness-thermal stability combination. In addition, if element(s) which will be substituted for tungsten and boron have low solubility with nickel in solid state, formation of nickel solid solution is hindered. As a result, R_{min} decreases further, which causes more improvement in GFA.

Microhardness values of the alloys are found to be around 1200 Hv (11.7 GPa), which is one of the highest values reported for a Ni-based metallic glass (Table 3). Also, tensile yield strengths of the alloys can be estimated by using the equation $\sigma_y = Hv/3$ [19]. Based on the microhardness values, the tensile yield strengths of the alloys are determined to be about 3.9 GPa, which is higher than yield strengths of other Ni-based metallic glasses [1]. In addition to mechanical properties, one of the advantages of Ni-W-B metallic glasses over those reported in Table 3 is that Ni-W-B metallic glasses have higher thermal stability. In fact, crystallization temperature of $Ni_{59.35}Nb_{34.45}Sn_{6.2}$ alloy is 930 K, which is about 130 K less than that of $Ni_{46}Cu_5W_{31.6}B_{17.4}$ alloy. The other advantage of Ni-W-B metallic glasses is their lower cost with respect to costs of those containing high amount of Nb, Ta and Hf.

Table 3. Microhardnesses of various Ni-based metallic glasses.

Alloy	H_v	Ref.
$Ni_{59.35}Nb_{34.45}Sn_{6.2}$	1280	[20]
$Ni_{60}(Nb_{60}Ta_{40})_{34}Sn_6$	1192	[21]
$Ni_{59.5}Nb_{40.5}$	1173	[22]
$Ni_{58}Ta_{36}Sn_6$	1066	[15]
$Ni_{50}Nb_{28}Zr_{22}$	1020	[23]
$Ni_{62}Ta_{36}Sn_2$	1002	[14]
$Ni_{40}Ta_{42}Co_{18}$	993	[13]
$Ni_{59}Ta_{41}$	969	[24]
$Ni_{62}Nb_{33}Zr_5$	965	[25]
$Ni_{40}Cu_5Ti_{17}Zr_{28}Al_{10}$	862	[18]
$Ni_{60}Nb_{20}Hf_{15}Ti_5$	847	[26]
$Ni_{53}Nb_{20}Ti_{10}Zr_8Co_6Cu_3$	759	[27]
$Ni_{60}Zr_{17}Al_6Hf_7Nb_{10}$	713	[28]
$Ni_{58.5}Nb_{20.25}Y_{21.25}$	663	[22]

Although the alloys synthesized in this study have low critical casting thicknesses, 100 µm, there are still some potential applications where they can be utilized because of their high thermal stabilities and microhardnesses. Also, since the alloys contain significantly high amount of nickel, they are expected to have high corrosion resistance. For example, alloys can be coated on substrates by thermal spraying techniques to obtain hard, wear resistant and corrosion resistant coatings [29,30]. Besides, powders of the alloys having diameter less than 100 µm can be produced by gas atomization, and then they can be sintered by spark plasma sintering to manufacture components with desired shape and dimensions [31]. In addition, considering the fact that nickel and W_2B phases precipitates, the alloys can be used as precursors to develop Ni-W_2B metal matrix composite coatings having high hardness and high toughness.

4. Experimental Section

Ni-Cu-W-B alloy ingots with compositions of $Ni_{51-x}Cu_xW_{31.6}B_{17.4}$ (x = 0, 5) were prepared by arc melting the mixtures of Ni (99.9 mass%), W (99.9 mass%) and Cu (99.9 mass%) metals and crystalline B (98 mass%) in a Ti-gettered high purity argon atmosphere. In order to obtain homogeneous master alloys, samples were melted three times. Compositions of the alloys represent nominal atomic percentages. 20 µm and 100 µm thick thin foils of the alloys were produced by piston and anvil method in an arc furnace. For the production of the thin foils, a molten sample of each alloy was squeezed between two copper plates pushed by pneumatic pistons. Velocity of each piston was about 400 mm/s and 50 mm/s for 20 µm and 100 µm thick samples, respectively. Thicknesses of the foils were determined by optical microscopy. Also, samples having thickness of 400 µm were produced by suction casting method. In addition, samples of copper containing alloy, $Ni_{46}Cu_5W_{31.6}B_{17.4}$, were annealed at 950 and 1025 K for 5 min. under high purity argon atmosphere to determine the thermal stability of the alloy. The structures of the samples were examined by X-ray diffraction (XRD) (Bruker Karlsruhe, Germany, D8 Advance equipped with Vantec-1 detector) with Cu-K_α radiation. The glass transition temperature (T_g), crystallization temperature (T_x), solidus temperature (T_m) and liquidus temperatures (T_l) of the alloys were measured by differential scanning calorimetry (DSC) (Netzsch Selb, Germany, STA 449 F3) at a heating rate of 0.33 K/s. Vickers hardnesses of the samples were measured with a Vickers hardness tester (Shimadzu Kyoto, Japan, HMV 2L) under a load of 1.96 N. Images of the indents obtained after microhardness measurements were acquired with scanning electron microscopy (SEM) (Leo Cambridge, UK, 1430 VP) under secondary electron (SE) imaging mode.

5. Conclusions

GFA, thermal stabilities and microhardnesses of $Ni_{51-x}Cu_xW_{31.6}B_{17.4}$ (x = 0, 5) metallic glass alloys are investigated. It is found that copper substitutions for nickel improves the glass transition and crystallization temperatures, which are comparable to those of other refractory metal based metallic glasses and higher than those of other Ni-based metallic glasses. However, copper substitutions for nickel do not have any effect on critical casting thicknesses of the alloys studied. Critical casting thicknesses of the alloys are found to be 100 µm. Although this value is quite low, it is still higher than critical casting thicknesses of most of the refractory metal based metallic glasses. Also, microhardness of the alloys are found to be around 1200 Hv and this value is higher than microhardness of other Ni-based metallic glasses. Furthermore, alloys can be used as a precursor to manufacture $Ni-W_2B$ metal matrix composites.

Acknowledgments: The authors would like to acknowledge the support provided by Afyon Kocatepe University (AKU) for funding this work through project 13.MUH.10.

Author Contributions: Aytekin Hitit wrote and edited the paper, and contributed in all activities. Pelin Öztürk and Ahmet Malik Aşgın synthesized the alloys and cast the samples. Hakan Şahin performed XRD and DSC experiments and analyzed the results.

Conflicts of Interest: The authors declare no conflict of interest.

References

1. Inoue, A.; Takeuchi, A. Recent developments and application products of bulk glassy alloys. *Acta. Mater.* **2011**, *59*, 2243–2267. [CrossRef]
2. Yoshitake, T.; Kubo, Y.; Igarashi, H. Preparation of refractory transition metal-metalloid amorphous alloys and their thermal stability. *Mater. Sci. Eng.* **1988**, *97*, 269–271. [CrossRef]
3. Inoue, A.; Sakai, S.; Kimura, H.; Masumoto, T. Crystallization temperature and hardness of new chromium-based amorphous alloys. *Trans. JIM* **1979**, *20*, 255–262.
4. Koch, C.C.; Kroeger, D.M.; Scarbrough, J.O.; Giessen, B.C. Superconductivity in amorphous T5 T9 transition metal alloys (T5 = Nb, Ta; T9 = Rh, Ir). *Phys. Rev. B* **1980**, *22*, 5213–5224. [CrossRef]
5. Masumoto, T.; Inoue, A.; Sakai, S.; Kimura, H.; Hoshi, A. Superconductivity of ductile Nb-based amorphous alloys. *Trans. JIM* **1980**, *21*, 115–122.
6. Inoue, A.; Sakai, S.; Kimura, H.; Masumoto, T.; Hoshi, A. Superconductivity of Mo-Si-B and W-Si-B amorphous alloys obtained by liquid quenching. *Scr. Metall.* **1980**, *14*, 235–239. [CrossRef]
7. Mahan, M.K.; Jha, B.L. Relaxation time and molar free energy of activation for some rare-earth complexes of kaolinite. *J. Mater. Sci. Lett.* **1980**, *15*, 1594–1596. [CrossRef]
8. Mehra, M.; Schultz, R.; Johnson, W.L. Structural studies and relaxation behaviour of $(Mo_{0.6}Ru_{0.4})_{100-x}B_x$ metallic glasses. *J. Non-Cryst. Solids* **1984**, *61–62*, 859–864. [CrossRef]
9. Yoshimoto, R.; Nogi, Y.; Tamura, R.; Takeuchi, S. Fabrication of refractory metal based metallic glasses. *Mater. Sci. Eng. A* **2007**, *449*, 260–263. [CrossRef]
10. Suo, Z.Y.; Song, Y.L.; Yu, B.; Qiu, K.Q. Fabrication of tungsten-based metallic glasses by low purity industrial raw materials. *Mater. Sci. Eng. A* **2011**, *528*, 2912–2916. [CrossRef]
11. Morishita, M.; Koyama, K.; Maeda, K.; Zhang, G. Calculated phase diagram of the Ni-W-B ternary system. *Mater. Trans. JIM* **1999**, *40*, 600–605. [CrossRef]
12. Naidu, S.V.N.; Rao, P.R. *Phase Diagrams of Binary Tungsten Alloys*; Indian Institute of Metals: Calcutta, India, 1991.
13. Meng, D.; Yi, J.; Zhao, D.Q.; Ding, D.W.; Bai, H.Y.; Pan, M.X.; Wang, W.H. Tantalum based bulk metallic glasses. *J. Non-Cryst. Solids* **2011**, *357*, 1787–1790. [CrossRef]
14. Tien, H.Y.; Lin, C.Y.; Chin, T.S. New ternary Ni-Ta-Sn bulk metallic glasses. *Intermetallics* **2006**, *14*, 1075–1078. [CrossRef]
15. Zhu, Z.W.; Zhang, H.F.; Ding, B.Z.; Hu, Z.Q. Synthesis and properties of bulk metallic glasses in the ternary Ni-Nb-Zr alloy system. *Mater. Sci. Eng. A* **2008**, *492*, 221–229. [CrossRef]

16. Chang, H.J.; Park, E.S.; Jung, Y.S.; Kim, M.K.; Kim, D.H. The effect of Zr addition in glass forming ability of Ni-Nb alloy system. *J. Alloy Compd.* **2007**, *434–435*, 156–159. [CrossRef]
17. Qiang, J.B.; Zhang, W.; Inoue, A. Ni-(Zr/Hf)-(Nb/Ta)-Al bulk metallic glasses with high thermal stabilities. *Intermetallics* **2009**, *17*, 249–252. [CrossRef]
18. Xu, D.; Duan, G.; Johnson, W.L.; Garland, C. Formation and properties of new Ni-based amorphous alloys with critical casting thickness up to 5 mm. *Acta. Mater.* **2004**, *52*, 3493–3497. [CrossRef]
19. Zhang, P.; Li, S.X.; Zhang, Z.F. General relationship between strength and hardness. *Mater. Sci. Eng. A* **2011**, *529*, 62–73. [CrossRef]
20. Yim, H.C.; Xu, D.; Johnson, W.L. Ni-based bulk metallic glass formation in the Ni-Nb-Sn and Ni-Nb-Sn-X (X = B, Fe, Cu) alloy systems. *Appl. Phys. Lett.* **2003**, *82*, 1030–1032. [CrossRef]
21. Yim, H.C.; Tokarz, M.; Bilello, J.C.; Johnson, W.L. Structure and properties of $Ni_{60}(Nb_{100-x}Ta_x)_{34}Sn_6$ bulk metallic glass alloys. *J. Non-Cryst. Solids* **2006**, *352*, 747–755. [CrossRef]
22. Concustell, A.; Mattern, N.; Wendrock, H.; Kuehn, U.; Gebert, A.; Eckert, J.; Greer, A.L.; Sort, J.; Baró, M.D. Mechanical properties of a two-phase amorphous Ni-Nb-Y alloy studied by nanoindentation. *Scripta Mater.* **2007**, *56*, 85–88. [CrossRef]
23. Santos, F.S.; Sort, J.; Fornell, J.; Baró, M.D.; Suriñach, S.; Bolfarini, C.; Botta, W.J.; Kiminami, C.S. Mechanical behavior under nanoindentation of a new Ni-based glassy alloy produced by melt-spinning and copper mold casting. *J. Non-Cryst. Solids* **2010**, *356*, 2251–2257. [CrossRef]
24. Wang, Y.; Wang, Q.; Zhao, J.; Dong, C. Ni-Ta binary bulk metallic glasses. *Scripta Mater.* **2010**, *63*, 178–180. [CrossRef]
25. Hu, H.T.; Chen, L.Y.; Wang, X.D.; Cao, Q.P.; Jiang, J.Z. Formation of Ni-Nb-Zr-X (X = Ti, Ta, Fe, Cu, Co) bulk metallic glasses. *J. Alloy Compd.* **2008**, *460*, 714–718. [CrossRef]
26. Zhang, W.; Inoue, A. Formation and mechanical properties of Ni-based Ni-Nb-Ti-Hf bulk glassy alloys. *Scripta Mater.* **2003**, *48*, 641–645. [CrossRef]
27. Ishida, M.; Takeda, H.; Nishiyama, N.; Kita, K.; Shimizu, Y.; Saotome, Y.; Inoue, A. Wear resistivity of super-precision microgear made of Ni-based metallic glass. *Mater. Sci. Eng. A* **2007**, *449–451*, 149–154. [CrossRef]
28. Kim, W.B.; Ye, B.J.; Yi, S. Effect of Nb addition on the glass forming ability and mechanical properties in the Ni-Zr-Al-Hf-Nb alloys. *J. Mater. Sci.* **2006**, *41*, 3805–3809. [CrossRef]
29. Wang, A.P.; Chang, X.C.; Hou, W.L.; Wang, J.Q. Preparation and corrosion behaviour of amorphous Ni-based alloy coatings. *Mater. Sci. Eng. A* **2007**, *449–451*, 277–280. [CrossRef]
30. Yugeswaran, S.; Kobayashi, A. Metallic glass coatings fabricated by gas tunnel type plasma spraying. *Vacuum* **2014**. [CrossRef]
31. Xie, G.; Qin, F.; Zhu, S.; Inoue, A. Ni-free Ti based bulk metallic glass with potential for biomedical applications produced by spark plasma sintering. *Intermetallics* **2012**, *29*, 99–103. [CrossRef]

metals

MDPI

Article

Effect of Yttrium Addition on Glass-Forming Ability and Magnetic Properties of Fe–Co–B–Si–Nb Bulk Metallic Glass

Teruo Bitoh * and Dai Watanabe

Department of Machine Intelligence and Systems Engineering, Faculty of Systems Science and Technology, Akita Prefectural University, Yurihonjo 015-0055, Japan; b12a084@akita-pu.ac.jp
* Author to whom correspondence should be addressed; teruo_bitoh@akita-pu.ac.jp; Tel.: +81-184-27-2161; Fax: +81-184-27-2188.

Academic Editors: K. C. Chan and Jordi Sort Viñas
Received: 28 April 2015; Accepted: 23 June 2015; Published: 29 June 2015

Abstract: The glass-forming ability (GFA) and the magnetic properties of the $[(Fe_{0.5}Co_{0.5})_{0.75}B_{0.20}Si_{0.05}]_{96}Nb_{4-x}Y_x$ bulk metallic glasses (BMGs) have been studied. The partial replacement of Nb by Y improves the thermal stability of the glass against crystallization. The saturation mass magnetization (σ_s) exhibits a maximum around 2 at. % Y, and the value of σ_s of the alloy with 2 at. % Y is 6.5% larger than that of the Y-free alloy. The coercivity shows a tendency to decrease with increasing Y content. These results indicate that the partial replacement of Nb by Y in the Fe–Co–B–Si–Nb BMGs is useful to simultaneous achievement of high GFA, high σ_s, and good soft magnetic properties.

Keywords: soft magnetic material; bulk metallic glass; iron-based alloy; magnetization; coercivity

---

1. Introduction

The Fe-based bulk metallic glasses (BMGs) are expected as a new class of soft magnetic materials with extremely low core losses [1]. The BMGs have large glass-forming ability (GFA) and, therefore, they can be used to prepare amorphous alloys with thicknesses of few millimeters by casting. Furthermore, the soft magnetic properties of the Fe-based BMGs are better than those of ordinary amorphous alloys which require extremely high cooling rate, typically 10^5–10^6 K/s, for amorphous formation due to their low GFA [2–4].

One of the disadvantages of the Fe-based soft magnetic BMGs is the smaller saturation magnetization (typically 1.2 T or less) compared with the ordinary Fe-based amorphous alloys. The demands on soft magnetic materials include higher combined magnetization and permeability. In order to achieve high magnetization, it is necessary to reduce the contents of solute elements such as B, C, Si and P. However, the reduction of the solute elements content leads to a decrease of GFA.

Recently, we reported that the effect of replacement of Nb by Y on GFA and the magnetic properties of the $(Fe_{0.8}Co_{0.2})_{96-x}B_ySi_1Nb_{3-x}Y_x$ ($y = 15, 17$) alloys [5], which is close to the limit of the amorphous formation [6]. The results obtained in the study indicate that the partial replacement of Nb by Y in the Fe–Co–B–Si–Nb alloys is useful to simultaneous achievement of high GFA, high magnetization, and good soft magnetic properties. In the present study, we have investigated the effect of the replacement of Nb by Y on GFA and the magnetic properties of the $[(Fe_{0.5}Co_{0.5})_{0.75}B_{0.20}Si_{0.05}]_{96}Nb_4$ alloy. Although this alloy has the same Fe–Co–B–Si–Nb system as the previous ones, the GFA is quite different. The present alloy system has a large GFA which enables us to produce rod specimens with 5 mm in diameter by Cu-mold casting [7]. In addition, the alloy exhibits the rather high magnetization of 1.13 T as well as the good soft magnetic properties [8,9]. Therefore, this alloy has a possibility to be able to form magnetic cores into complicated shapes by casting or by superplastic deformation in supercooled liquid region.

2. Materials and Methods

The mother alloys with nominal composition of $[(Fe_{0.5}Co_{0.5})_{0.75}B_{0.20}Si_{0.05}]_{96}Nb_{4-x}Y_x$ were prepared as follows. First, the eutectic Fe-33.1 mass % Y alloy was prepared by arc-melting the mixture of pure Fe (99.99%) and Y (99.9%) metals in an Ar atmosphere. Subsequently, the mixtures of pure Fe, Co (99.9%), and Nb (99.9%) metals, pure B (99.5%) and Si (99.999%) crystals, and the eutectic Fe–Y alloy were melted by an arc furnace in an Ar atmosphere. The rapidly-solidified ribbons with approximately 1 mm in width and 30 μm in thickness were prepared by a single-roller melt-spinning apparatus with a Cu wheel in an Ar atmosphere.

The structure of the specimens was examined by X-ray diffractometry (XRD, PANalytical, Almelo, The Netherlands) with Cu K_α incident radiation. The thermal stability of the glass was investigated using a differential scanning calorimetry (DSC, NETZSCH-Gerätebau, Selb, Germany) during heating at various heating rates (β) between 0.167 and 0.667 K/s. The saturation mass magnetization (σ_s) was measured with a magnetic balance in an applied magnetic field (H) up to 800 kA/m at 296 ± 3 K. The hysteresis loops of the 70 mm long straight specimens were measured by a hysteresis loop tracer with a compensation coil under a maximum magnetic field of 10 kA/m at room temperature. The hysteresis loops and σ_s were measured for the five specimens cut from the same ribbons.

3. Results and Discussion

Figure 1 shows the XRD profiles of the as-quenched specimens (x = 0, 2, 4) taken from the free surface. All the profiles consist only of a halo which originates from an amorphous phase. The similar results were obtained by both the free and wheel-contacted surfaces of all the alloys. Figure 2 shows the glass-transition temperature (T_g) and the onset temperature of crystallization (T_x) together with the supercooled liquid region (ΔT_x) that is defined as the temperature interval between T_g and T_x as a function of Y content. All the alloys exhibit the distinct glass transition before crystallization. Both T_g and T_x increase with increasing Y content. The super-cooled liquid region also increases with replacing Nb by Y. The maximum value of ΔT_x is 47 K for x = 1, which is 6 K larger than that of the Y-free alloy. The alloys with x = 2–3 also show the larger ΔT_x than that of the Y-free alloy. However, ΔT_x remarkably decreased to 28 K when Nb is completely replaced by Y.

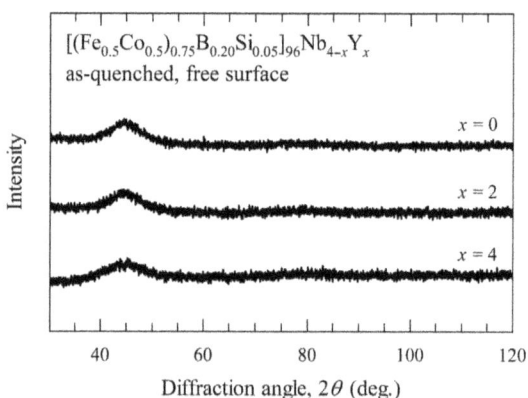

Figure 1. X-ray diffraction profiles of $[(Fe_{0.5}Co_{0.5})_{0.75}B_{0.20}Si_{0.05}]_{96}Nb_{4-x}Y_x$ alloys taken from free surface in an as-quenched state.

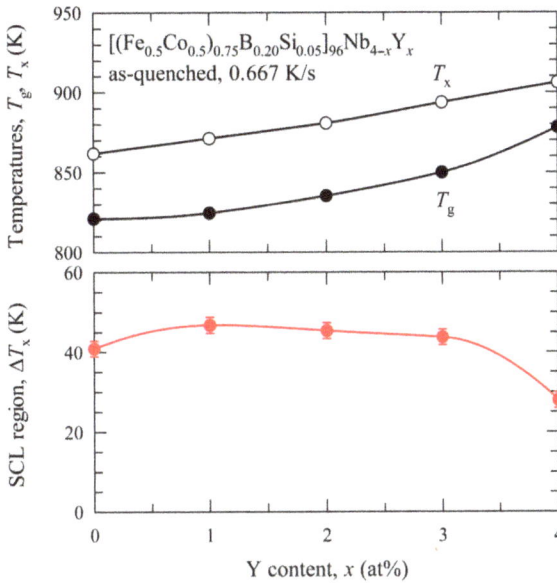

Figure 2. Glass-transition temperature (T_g), crystallization temperature (T_x) and supercooled liquid (SCL) region ($\Delta T_x = T_x - T_g$) of $[(Fe_{0.5}Co_{0.5})_{0.75}B_{0.20}Si_{0.05}]_{96}Nb_{4-x}Y_x$ alloys as a function of Y content.

In order to evaluate GFA of the alloys, the continuous heating transformation (CHT) curves have been derived by the Kissinger analysis in which $\ln(\beta/T_p^2)$ *vs.* $1/T_p$ plot shows a linear relationship as shown in the following equation [10–12]:

$$\ln\left(\frac{\beta}{T_p^2}\right) = -\frac{E_a}{RT_p} + \ln\left(\frac{E_a K_0}{R}\right) \tag{1}$$

where β is the heating rate, T_p is the peak temperature of the DSC curve (at which the transformation rate reaches maximum), E_a is the activation energy for nucleation and growth, R is the gas constant, K_0 is the frequency factor, respectively. The values of E_a and K_0 can be obtained by the linear fit of the $\ln(\beta/T_p^2)$ *vs.* $1/T_p$ plot. The CHT curves are derived by using the relationship between T_p and the heating time, $t_h = (T_p - 298)/\beta$, where

$$\beta = \frac{E_a K_0}{R} T_p^2 \exp\left(-\frac{E_a}{RT_p}\right) \tag{2}$$

In general, T_p is used in Kissinger analysis to investigate the maximum transformation rate during crystallization of glass. However, T_p can be replaced by T_x (the onset temperature of crystallization) to calculate a CHT curve for the crystallization of glass, which indicates as actual starting point for the transformation [11,12]. Figures 3 and 4 show the heating rate dependence of T_x and $\ln(\beta/T_x^2)$ *vs.* $1/T_x$ plot, respectively. The values of the kinetics parameters required for calculation of the CHT curves are listed in Table 1. All the coefficients of determination (R^2) for the linear regression of Figure 4 are larger than 0.9998. Figure 5 shows the CHT curves that show the relationship between T_x and corresponding heating time, t_h. It should be noted that the boundary between the glass and the crystalline phases moves to the longer time side with increasing Y content. This means that the incubation time for crystallization is postponed, *i.e.*, GFA is improved by the replacement of Nb by Y.

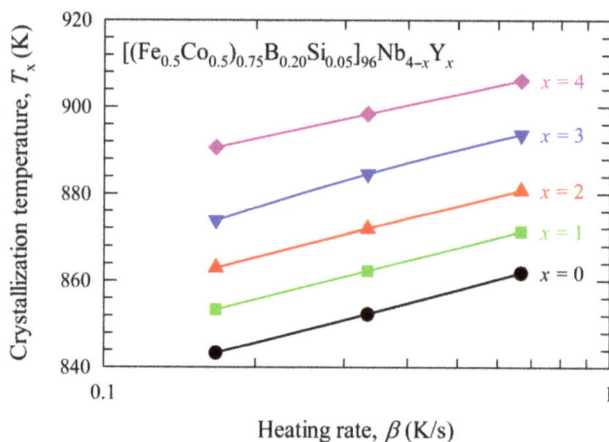

Figure 3. Heating rate (β) dependence of the crystallization temperature (T_x) of $[(Fe_{0.5}Co_{0.5})_{0.75}B_{0.20}Si_{0.05}]_{96}Nb_{4-x}Y_x$ alloy.

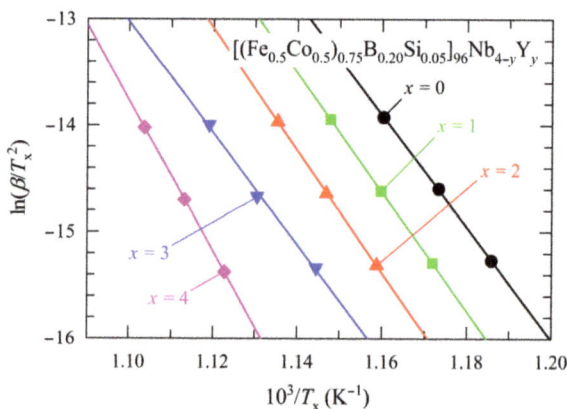

Figure 4. $\ln(\beta/T_x^2)$ vs. $1/T_x$ plot for $[(Fe_{0.5}Co_{0.5})_{0.75}B_{0.20}Si_{0.05}]_{96}Nb_{4-x}Y_x$ alloys.

Table 1. Kinetics parameters for onset crystallization temperatures for $[(Fe_{0.5}Co_{0.5})_{0.75}B_{0.20}Si_{0.05}]_{96}Nb_{4-x}Y_x$ alloys.

x	T_x* (K)	$-E_a/R$ (10^3 K)	$\ln(E_aK_0/R)$	E_a (kJ/mol)	K_0 (s^{-1})
0	862	52.8	47.3	439	6.6×10^{15}
1	871	55.9	50.2	464	1.1×10^{17}
2	881	57.4	51.3	478	3.1×10^{17}
3	894	58.8	51.9	489	5.6×10^{17}
4	906	70.9	64.2	589	1.0×10^{23}

* Measured at a heating rate of 0.667 K/s.

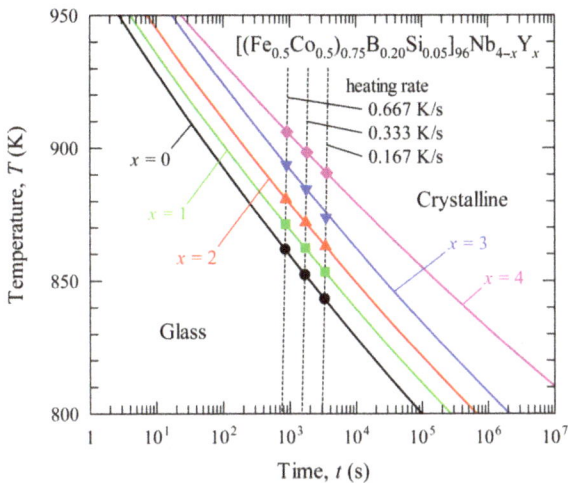

Figure 5. Continuous heating transformation (CHT) diagram of $[(Fe_{0.5}Co_{0.5})_{0.75}B_{0.20}Si_{0.05}]_{96}Nb_{4-x}Y_x$ alloys. The symbols represent the experimental values of the onset temperature of crystallization (T_x) obtained by the differential scanning calorimetry (DSC) in the continuous heating.

The stabilization of the amorphous phase can be achieved by increasing the atomic packing density in the amorphous phase. Yttrium has a large atomic radius of 181 pm, which is much larger than that of Nb (143 pm), Co (125 pm), Fe (124 pm), Si (117 pm), and B (83 pm) [13]. A large difference of the atomic radius between Y and Fe is favourable for increase the atomic packing density of the amorphous structure. It has been reported that Fe–TM–B (TM: transition metals) and Fe–Ln–B (Ln: lanthanides) type BMGs have unique network-like structure, in which triangular prisms consisting of Fe and B are connected to each other through glue atoms of TM or Ln [14,15]. If the atomic packing density increases, the atomic diffusion becomes more difficult. In addition, the much lower diffusivity of Y than Nb also contributes to the improvement of GFA. Therefore, it can be concluded that the improvement of GFA is brought by the replacement of Nb by Y with the larger radius than that of Nb.

Figure 6 shows the saturation mass magnetization (σ_s) in an as-quenched state as a function of the Y content. As previously reported [5], the changes of σ_s are interesting. The magnetization exhibits a maximum around 2 at. % Y. The cause of this phenomenon is unclear. However, we can point out a possibility of the influence of the nanoscale phase separation (NPS) or chemical short-range ordering (CSRO). It is known that the values of the magnetic moments depend on the local environment of Fe and Co atoms: the types of neighbors, the fluctuation of interatomic distance, and the average coordination number [16,17]. Therefore, the values of σ_s will change according to degree of NPS or CSOR even if the contents of Fe and Co are fixed. The heat of mixing of Y and Fe, Co, and Nb atomic pairs are −1, −22, and +30 kJ/mol, respectively [18,19]. Thus Fe–Y is nearly ideal solution. However, Y atoms attract Co ones and repulse Nb ones. These interatomic attractive/repulsive forces may promote the formation of NPS or CSRO.

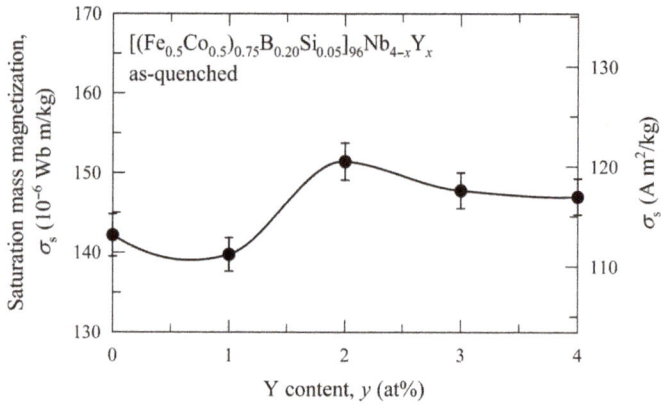

Figure 6. Saturation mass magnetization (σ_s) of $[(Fe_{0.5}Co_{0.5})_{0.75}B_{0.20}Si_{0.05}]_{96}Nb_{4-x}Y_x$ alloys in an as-quenched state as a function of Y content. The graph shows the mean value for the five specimens (closed circles). The error bars represent the 95% confidence limits ($\pm 2 \times$ standard errors).

Figure 7 shows the examples of the hysteresis loops of the alloy with $x = 0, 2$ in an as-quenched state. The hysteresis loops indicate that the alloys exhibit the good soft magnetic properties, *i.e.*, low coercivity (H_c) and high permeability. Figure 8 shows H_c in an as-quenched state as a function of the Y content. The coercivity gradually decreases with increasing the Y content. This result suggests that the soft magnetic properties are improved by replacing Nb with Y, which means the increase of GFA. This is considered to be due to the reduction of the free volume in an amorphous phase [2–4].

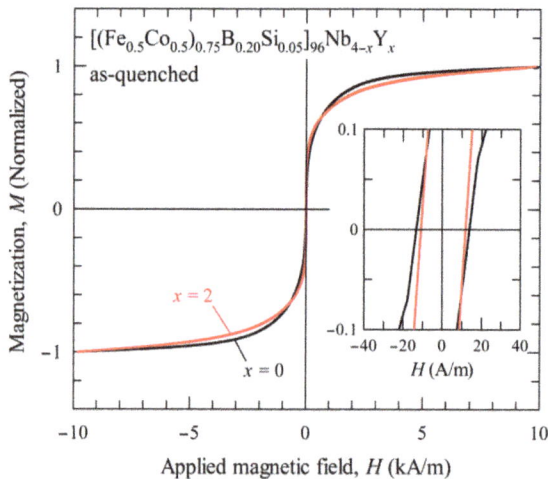

Figure 7. Hysteresis loops of $[(Fe_{0.5}Co_{0.5})_{0.75}B_{0.20}Si_{0.05}]_{96}Nb_{4-x}Y_x$ ($x = 0, 2$) alloys in an as-quenched state. Inset shows enlarged view near the origin.

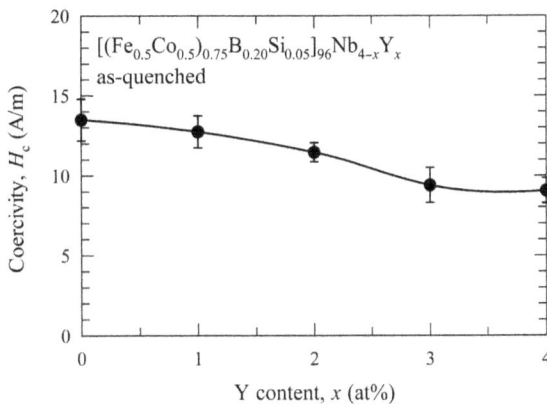

Figure 8. Coercivity (H_c) of $[(Fe_{0.5}Co_{0.5})_{0.75}B_{0.20}Si_{0.05}]_{96}Nb_{4-x}Y_x$ alloys in an as-quenched state as a function of Y content. The graph shows the mean value for the five specimens (closed circles). The error bars represent the 95% confidence limits ($\pm 2 \times$ standard errors).

4. Conclusions

It has been confirmed by the glass-forming ability (GFA) of the $[(Fe_{0.5}Co_{0.5})_{0.75}B_{0.20}Si_{0.05}]_{96}Nb_{4-x}Y_x$ bulk metallic glasses (BMGs) is improved by replacing Nb with Y by the continuous heating transformation (CHT) diagram. The improvement of GFA is brought by the replacement of Nb by Y with the larger radius than that of Nb.

The saturation mass magnetization (σ_s) exhibits a maximum around 2 at. % Y. The value of σ_s of the alloy with $x = 2$ in an as-quenched state is 151×10^{-6} Wb m/kg, which is 6.5% larger than that of the Y-free alloy. The coercivity (H_c) of the present alloys in an as-quenched state show a tendency to decrease with increasing Y content, which means an increase of GFA.

The results obtained in the present study indicate that the partial replacement of Nb by Y in the Fe–Co–B–Si–Nb BMGs is useful to simultaneous achievement of high GFA, high magnetization, and good soft magnetic properties. The good magnetic properties of the (Fe, Co)–B–Si–(Nb, Y) BMGs bring greater efficiency and miniaturization to magnetic devices.

Acknowledgments: This work was partly supported by Japan Society for the Promotion of Science (JSPS), Grant-in-Aid for Scientific Research (C) (KAKENHI), No. 24560806.

Author Contributions: Teruo Bitoh conceived and designed the experiments, and wrote the paper. Dai Watanabe performed the experiments and analyzed the data.

Conflicts of Interest: The authors declare no conflict of interest.

References

1. Inoue, A.; Takeuchi, A.; Shen, B. Formation and functional properties of Fe-based bulk glassy alloys. *Mater. Trans.* **2001**, *42*, 970–978. [CrossRef]
2. Bitoh, T.; Makino, A.; Inoue, A. Origin of low coercivity of Fe–(Al, Ga)–(P, C, B, Si, Ge) bulk glassy alloys. *Mater. Trans.* **2003**, *44*, 2020–2024. [CrossRef]
3. Bitoh, T.; Makino, A.; Inoue, A. Magnetization process and coercivity of Fe–(Al, Ga)–(P, C, B, Si) soft magnetic glassy alloys. *Mater. Trans.* **2004**, *45*, 1219–1227. [CrossRef]
4. Bitoh, T.; Makino, A.; Inoue, A. Origin of low coercivity of $(Fe_{0.75}B_{0.15}Si_{0.10})_{100-x}Nb_x$ ($x = 1$–4) glassy alloys. *J. Appl. Phys.* **2006**, *99*, 08F102. [CrossRef]
5. Bitoh, T.; Kikuchi, S. Glass-forming ability and magnetic properties of $(Fe_{0.80}Co_{0.20})_{96-x}B_xSi_1Nb_{3-y}Y_y$ ($x = 15$, 17) amorphous alloys. *IEEE Trans. Magn.* **2014**, *50*, 1–5. [CrossRef]

6. Ishikawa, T.; Tsubota, T.; Bitoh, T. Soft magnetic properties of ring-shaped Fe–Co–B–Si–Nb bulk metallic glasses. *J. Magn.* **2001**, *16*, 431–434. [CrossRef]

7. Inoue, A.; Shen, B.L.; Chang, C.T. Super-high strength of over 4000 MPa for Fe-based bulk glassy alloys in $[(Fe_{1-x}Co_x)_{0.75}B_{0.2}Si_{0.05}]_{96}Nb_4$ system. *Acta Mater.* **2004**, *52*, 4093–4099. [CrossRef]

8. Bitoh, T.; Makino, A.; Inoue, A.; Greer, A.L. Large bulk soft magnetic $[(Fe_{0.5}Co_{0.5})_{0.75}B_{0.20}Si_{0.05}]_{96}Nb_4$ glassy alloy prepared by B_2O_3 flux melting and water quenching. *Appl. Phys. Lett.* **2006**, *88*, 182510. [CrossRef]

9. Bitoh, T.; Shibata, D. Improvement of soft magnetic properties $[(Fe_{0.5}Co_{0.5})_{0.75}B_{0.20}Si_{0.05}]_{96}Nb_4$ bulk metallic glass by B_2O_3 flux melting. *J. Appl. Phys.* **2008**, *103*, 07E702. [CrossRef]

10. Kissinger, H.E. Variation of peak temperature with heating rate in differential thermal analysis. *J. Res. Natl. Bur. Stand.* **1956**, *57*, 217–221. [CrossRef]

11. Louzguine, D.V.; Inoue, A. Comparison of the long-term thermal stability of various metallic glasses under continuous heating. *Scr. Mater.* **2002**, *47*, 887–891. [CrossRef]

12. Kim, J.H.; Park, J.S.; Fleury, E.; Kim, W.T.; Kim, D.H. Effect of yttrium addition on thermal stability and glass forming ability in Fe–TM (Mn, Mo, Ni)–B ternary alloys. *Mater. Trans.* **2004**, *45*, 2770–2775. [CrossRef]

13. Emsley, J. *The Elements*, 3rd ed.; Oxford University Press: Oxford, UK, 1998.

14. Nakamura, T.; Matsubara, E.; Imafuku, M.; Koshiba, H.; Inoue, A.; Waseda, Y. Structural study of amorphous $Fe_{70}M_{10}B_{20}$ (M = Cr, W, Nb, Zr and Hf) alloys by X-ray diffraction. *Mater. Trans.* **2001**, *42*, 1530–1534. [CrossRef]

15. Nakamura, T.; Koshiba, H.; Imafuku, M.; Inoue, A.; Matsubara, E. Determination of atomic sites of Nb dissolved in metastable $Fe_{23}B_6$ phase. *Mater. Trans.* **2002**, *43*, 1918–1920. [CrossRef]

16. O'Handley, R.C. *Modern Magnetic Materials: Principles and Applications*; Wiley-Interscience: New York, NY, USA, 1999; pp. 391–431.

17. Kakehashi, Y. *Modern Theory of Magnetism in Metals and Alloys*; Springer: Berlin, Germany, 2013; pp. 253–299.

18. De Boer, F.R.; Boom, R.; Mattens, W.C.M.; Miedema, A.R.; Niessen, A.K. *Cohesion in Metals*; North-Holland: Amsterdam, The Netherlands, 1988.

19. Takeuchi, A.; Inoue, A. Classification of bulk metallic glasses by atomic size difference, heat of mixing and period of constitute elements and its application to characterization of the main alloying element. *Mater Trans.* **2005**, *46*, 2817–2829. [CrossRef]

metals

MDPI

Article

Mechanical and Structural Investigation of Porous Bulk Metallic Glasses

Baran Sarac [1,*], Daniel Sopu [1], Eunmi Park [1], Julia Kristin Hufenbach [1], Steffen Oswald [1], Mihai Stoica [1,2] and Jürgen Eckert [1,3]

[1] Institute for Complex Materials, Leibniz Institute for Solid State and Materials Research Helmholtzstrasse 20, D-01069 Dresden, Germany; d.sopu@ifw-dresden.de (D.S.); e.m.park@ifw-dresden.de (E.P.); j.k.hufenbach@ifw-dresden.de (J.K.H.); s.oswald@ifw-dresden.de (S.O.); m.stoica@ifw-dresden.de (M.S.); j.eckert@ifw-dresden.de (J.E.)

[2] Politehnica University of Timisoara, P-ta Victoriei 2, RO-300006 Timisoara, Romania

[3] TU Dresden, Institute of Materials Science, D-01062 Dresden, Germany

[*] Author to whom correspondence should be addressed; b.sarac@ifw-dresden.de; Tel.: +49-351-4659-1877; Fax: +49-351-4659-540.

Academic Editors: K. C. Chan and Jordi Sort Viñas

Received: 28 April 2015; Accepted: 25 May 2015; Published: 2 June 2015

Abstract: The intrinsic properties of advanced alloy systems can be altered by changing their microstructural features. Here, we present a highly efficient method to produce and characterize structures with systematically-designed pores embedded inside. The fabrication stage involves a combination of photolithography and deep reactive ion etching of a Si template replicated using the concept of thermoplastic forming. Pt- and Zr-based bulk metallic glasses (BMGs) were evaluated through uniaxial tensile test, followed by scanning electron microscope (SEM) fractographic and shear band analysis. Compositional investigation of the fracture surface performed via energy dispersive X-ray spectroscopy (EDX), as well as Auger spectroscopy (AES) shows a moderate amount of interdiffusion (5 at.% maximum) of the constituent elements between the deformed and undeformed regions. Furthermore, length-scale effects on the mechanical behavior of porous BMGs were explored through molecular dynamics (MD) simulations, where shear band formation is observed for a material width of 18 nm.

Keywords: bulk metallic glass; porous materials; Si lithography; thermoplastic forming; mechanical testing; shear band; microstructure; toughening mechanism; MD simulations; compositional analysis

1. Introduction

The effectiveness of a second phase in metallic glass structures, like foams and composites, to improve mechanical properties has been agreed upon widely. Unfortunately, the current methods used to fabricate such structures have insufficient control over the arrangement of microstructural features [1–4]. To overcome this challenge, a novel strategy to analyze microstructure-property relationships was postulated by Sarac *et al.* [5–8]. The strength and purpose of this method is the ability to vary microstructural features completely independently and to determine the individual effect of these features on the mechanical and morphological response, respectively. Bulk metallic glasses (BMGs) are excellent candidates for these investigations due to the intrinsic Newtonian flow behavior in their supercooled liquid region and the length-scale influence on properties [9]. In the micro-scale regime, the control mechanism of fracture in BMGs relies on the multiplication/deflection of the shear bands (SBs). These bands can be stabilized as long as the second phase spacing is less than the plastic zone size [10], creating a toughening mechanism in metallic glasses. Numerical simulations conducted by a non-local gradient-enhanced continuum mechanical model implemented by a finite element method [11,12] verify this proof of concept. The model sheds light on the scalability and versatility of this approach by using periodic boundary conditions along the external boundaries of the representative volume element [13] or by analyzing the spectrum of samples with stochastic pore designs [14]. In addition, the SB-mediated inhomogeneous plastic flow below the micron level is further regarded as the fundamental deformation mechanism for sub-micron-sized Zr-BMG pillars at no expense of yield strength [15].

The concept of thermoplastic forming has been broadly adopted for BMGs, which enabled a wide span of geometries on a ten orders of magnitude length scale [16]. This phenomenon has lately been utilized in a variety of micro- to nano-meter-scale applications, which can combine multiple length scale features in one component with easiness and precision via a thermoplastic forming (TPF) process [16,17]. The intrinsic properties in BMGs (*i.e.*, dramatic decrease in viscosity and increase in the formability at high temperatures) enhance the competitiveness of different TPF-based amorphous alloy processing within the commercial production routes [18].

The versatility in deformation behavior of BMGs is inherently related to the size effects, where the deformation is solely controlled by Newtonian flow behavior. Particularly interesting is the region where the deformation of the macro-sized samples, defined by global catastrophic fracture, turns into localized shear-induced deformation [6,7]. In this contribution, the toughening mechanism of the porous BMGs in terms of the mechanical and structural changes is investigated. The structural changes in the vicinity of shear bands and crack surfaces of the actual samples are sought to determine the influence of compositional alterations over the mechanical properties. Furthermore, molecular dynamics simulations will be carried out under uniaxial tension to understand the influence of porosity on the nanometer scale.

2. Materials and Methods

Intelligently-designed porous structures involve a unique fabrication strategy, which permits the arrangement of microstructural features within the BMG matrix with high precision and easiness. The layout drawings of these complex patterns was created by a CAD software (AutoCAD 2013, Autodesk Inc., San Rafael, CA, USA), which was subsequently transferred to a direct laser beam writer to fabricate the chromium glass photomask. Roughly, 50 samples with distinct pore designs were created on a 150-mm (or bigger) diameter template. Si wafers are utilized for this purpose, because of their low shrinkage, the precision in the directional etch, close tolerances to the original design and the cost-effectiveness of commercially-viable silicon. Figure 1a illustrates a small section of the Si wafer, where the entire wafer was coated uniformly by spinning PDMS photoresist. After the pre-baking process, the photoresist was patterned by UV exposure, where the selected regions of the photoresist were polymerized with the aid of the chromium mask. Figure 1b outlines the cured regions of the periodic circular pore design (in red), which become insoluble to the developer solution. The resist was then post-baked, and the uncured regions were stripped. Patterns with a depth of 400 μm were successively etched using deep reactive ion etching (Bosch process, Robert Bosch GmbH, Gerlingen, Germany), which allowed creating high-quality sidewall and top surface (roughness in hundreds of nanometers), as well as an anisotropic etch profile with a maximum span (pore diameter to etch length) ratio of 1:8. Figure 1c shows the sketch of the template with periodically-spaced pillars.

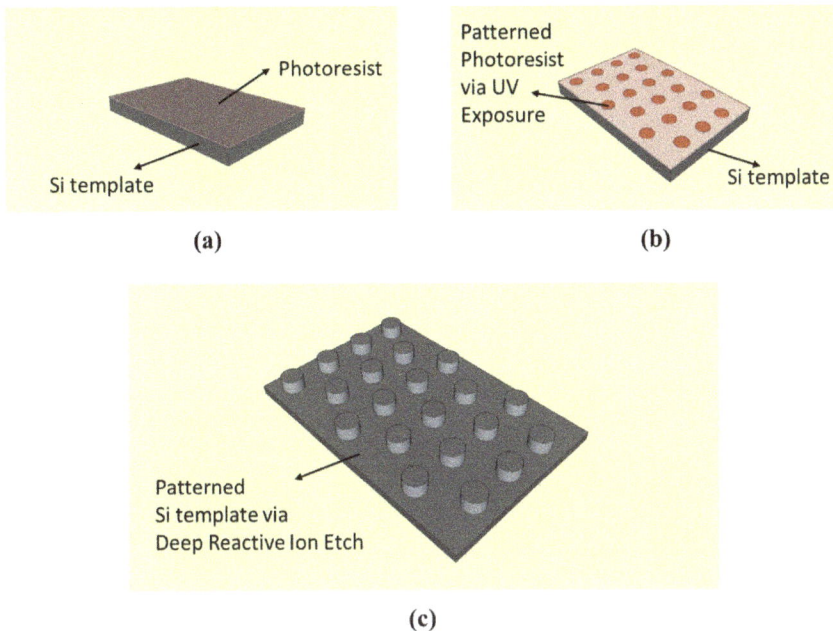

Figure 1. Patterning procedure of the periodic microfeatures. (**a**) Coating the Si template (wafer) surface with photoresist. (**b**) Patterning the photoresist using the chromium mask and UV light source. (**c**) Selective anisotropic etching of the entire Si template through deep reactive ion etching, which generates the negative pattern of the final structure.

For the manufacturing process, $Pt_{57.5}Cu_{14.7}Ni_{5.3}P_{22.5}$ was selected, because of its exclusively high fracture toughness $K_{1c} \approx 80$ MPam$^{1/2}$ and Poisson's ratio $v = 0.42$ in combination with low shear to bulk modulus ratio $G/B = 0.168$, placing this material into a different category, even among the BMG alloy families with proven records of extensive shear banding and plasticity [10,19]. In addition, a large processing window of ≈ 90 K and a large formability parameter S [20] enable this BMG alloy type to be used in pattern replication with high complexity and precision. The cast rods of this alloy were prepared in a quartz tube under a vacuum of 10^{-6} mbar, where the tube was purged with an inert gas to remove the residual oxygen. B_2O_3 is used to flux the ingot, which was confirmed to increase the glass forming ability by reducing the oxide content even to a higher extent [21]. The final diameter of the cast rod is 2 mm. The cast rod displays a broad peak in the X-ray diffraction (XRD) pattern, as well as a distinct glass transition and a sharp crystallization in the differential scanning calorimetry (DSC) curve. These findings are ascribed to the glassy nature of the as-cast state [22].

For comparison, $Zr_{35}Ti_{30}Cu_{7.5}Be_{27.5}$ metallic glass with relatively smaller fracture toughness K_{1c} of about 42.5 MPam$^{1/2}$ (averaged value from [5]), lower Poisson's ratio $v = 0.37$ and higher shear to bulk modulus ratio $G/B = 0.285$ was selected. This BMG type exhibits one of the widest supercooled regions of the known BMG types, rendering these materials to have excellent glass forming ability, as well as very low viscosity during hot forming (*viz.* 10^5 Pas) [23]. For this reason, it is suitable for rapid manufacturing of highly complex structures (including undercuts/multi-layered patterns) within very tight tolerances. The final rods of 10 mm in diameter, which were cast using copper mold suction casting in a Ti-gettered argon atmosphere, are fully amorphous [23]. The rods from both compositions were then sliced and hot-pressed to the desired thicknesses using an Instron compression device with a custom-built cartridge heating system. The pre-pressed BMG disc was situated on the patterned sample, and both were heated together to the TPF temperature of 550 K (Figure 2a).

The viscous BMG flows into the etched regions as the applied stress (σ_{app}) exceeds the flow stress defined by the Newtonian flow rule:

$$\sigma_{app} > \sigma_{flow} = 3\eta\dot{\varepsilon} \tag{1}$$

where η is the viscosity of the BMG at TPF temperature and $\dot{\varepsilon}$ is the strain rate of the TPF. For strain rates of $10^{-1}s^{-1}$ or lower, σ_{flow} for BMGs with high formability (e.g., Zr- and Pt-based BMGs [20]) is calculated to be 0.3 MPa, placing them in the deformation scale of the commercially available polymers [13]. The sample with embedded pores was then detached from the assembly by BMG surface polishing (the polished Si-BMG assembly is given in Figure 2b) and subsequent Si etching using a diluted KOH (35% by mass) solution (Figure 2c). The middle section of the actual porous BMG sample (left) and the entire tensile sample (middle) is depicted in Figure 2d. Both samples remain fully amorphous after thermoplastic forming (Figure 2d, right). The samples were pinned from the grip sections, and the tensile test was conducted under quasi-static conditions of $\dot{\varepsilon} = 5 \times 10^{-3}s^{-1}$ until rupture.

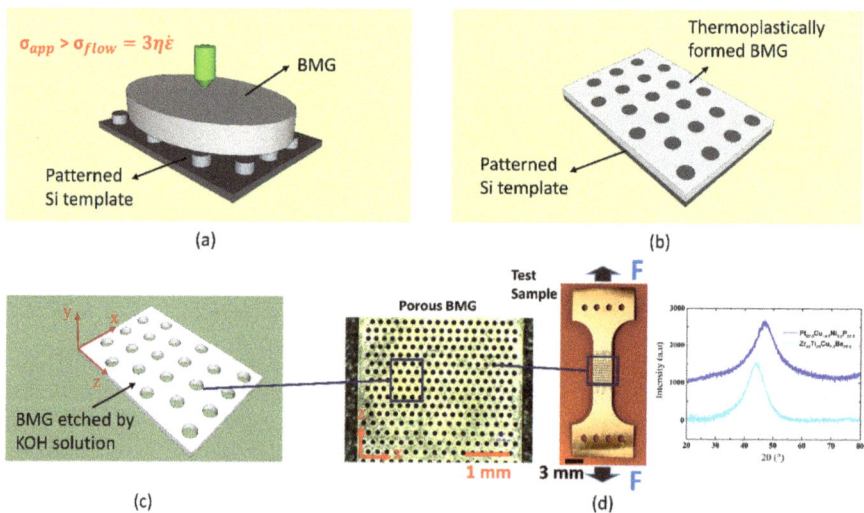

Figure 2. Thermoplastic forming of bulk metallic glasses (BMGs) into a complex geometry. (**a**) BMG disc placed on Si template with embedded negative pattern. (**b**) BMG is thermoplastically formed when the stress resisting the flow behavior σ_{flow} is exceeded. (**c**) Final sample shape is attained by metallographic treatment, and the sample is released from the Si template via KOH etching. (**d**) Actual sample geometry (middle) with the embedded periodic pores (left). The broad diffraction maximum after thermoplastic forming (right) reveals the fully amorphous nature of the alloys.

The chemical analysis of Pt-BMG is performed using a mixture of HCl (10 mL), HNO_3 (3 mL) and H_2O_2 (1 mL) added to deionized water. Four samples with approximately 50 mg are dissolved separately via microwave digestion at 513 K for 1 hour, where the total mass of these acidic solutions is 1000 g. The analysis of the dissolved elements is carried out through inductive coupled plasma optical emission spectroscopy (ICP-OES analysis), where the constituent elements are identified by their characteristic emission lines and quantified by the intensity of the same lines.

Metals **2015**, *5*, 920–933

3. Results and Discussion

3.1. Influence of Pore Design

Earlier results revealed the importance of the pore design within the matrix [5,13]; the mechanical properties are optimized, as the diameter-to-spacing ratio (d/s) of the embedded pores is chosen to be between 0.5 and two. For this ideal range, the deformation is mainly controlled by the periodically-spaced pores, and thereby, the global fracture of the sample is delayed by the redirection of SBs and redistribution of the resolved shear stress around the pores. The decomposition of deformation into elastic and plastic parts through cyclic loading has demonstrated that the majority of the deformation is attributed to the geometric lineup of the pores (AB-type pore lineup, where the center of apore is situated equidistant in x-direction to two pores in the neighboring row; see Figure 2d). The material remaining between the pores resembles a helical pattern, which creates additional elastic stretching compared to that of the sample with AA-type (center of the pores match between each row) pore stacking during tension [5]. The rest of the deformation is accumulated in the vicinity of the crack zone, where the SBs between the lateral pores perpendicular to the loading direction favor the additional macroscopic plasticity generation.

The quasi-static tensile behavior of BMG monolithic and porous structures for Pt- and Zr-BMGs is illustrated in a 3D stress-strain-absorbed energy per unit volume plot (Figure 3). The monolithic BMGs fabricated by thermoplastic forming exhibit an elastic limit of about 2% with no remarkable plastic deformation. On the other hand, the porous counterparts, including a periodic AB-type lineup with pore diameter and spacing of 100 and 50 μm (*i.e.*, d100s50), respectively, show almost 6% total strain at the expense of their fracture strength. The absorbed energy per unit volume is described by the area under the strain-stress curve until rupture, where the multiplication of these two components creates a second-degree parabolic profile. Further information pertaining to the mechanical and structural properties of these alloys can be found in Table 1.

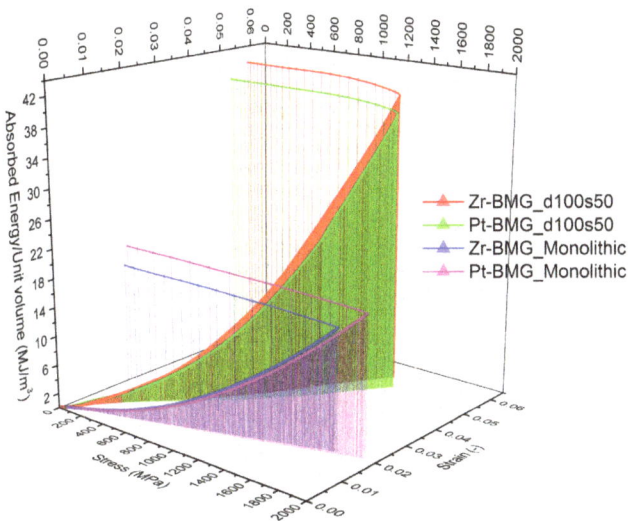

Figure 3. Representative graph of the deformation behavior of monolithic *vs.* porous Zr- and Pt-BMGs. Dimensions of the samples are within the same scale. $t \approx 300$ μm (sample thickness), $w = 3.75$ mm (width of porous BMG samples), $w = 3$ mm (width of Pt- and Zr-BMGs). The nominal fracture strength of the samples is calculated from the subtracted pore region where the shear bands (SBs) are connecting the pores perpendicular to the applied tensile load.

Table 1. Experimental parameters of the thermoplastically formed bulk (monolithic) and porous (*d* = 100 μm, *s* = 50 μm, *d*/*s* = 2) samples. The error percentages for σ$_y$ and ε$_y$ are within ±5%, whereas for *W*, the error is within ±10.5%.

Sample Name	σ$_y$ (MPa)	ε$_y$ (-)	E (GPa)	W (MJ/m^3)
Zr-BMG bulk	1741	0.019	87	16.8
Pt-BMG bulk	1924	0.020	95	19.4
Zr-BMG d100s50	1305	0.058	23	41.6
Pt-BMG d100s50	1399	0.053	28	39.6

The fracture analysis of Pt-BMG through scanning electron microscopy (LEO 1530 Gemini, Carl Zeiss, Oberkochen, Germany) reveals the characteristics of the multiple SB formation before fracture occurs. As an example, the network of pores via SB formation near the fracture surface is seen in Figure 4a. In this region, a number of around 20 SBs is observed (Figure 4b), where the local cleavage is ascribed to the high stress accumulation on the tip of an already existing SB (indicated with dark blue dotted region), which further develops and becomes irrepressible.

The phenomenon of hindering shear band propagation by an obstacle (e.g., *in situ* and *ex situ* second phases, heterogeneities, pores) is correlated with the intrinsic plastic zone size R_P defined by:

$$R_P = \frac{1}{\pi}\left(\frac{K_{1C}}{\sigma_y}\right)^2 \tag{2}$$

where K_{1c} is defined as the Mode 1 fracture toughness of the sample. Computing Equation (2) with the values for Pt-BMG and Zr-BMG results in R_P of 550 μm and 190 μm, respectively. Therefore, the second phase features become the prominent mechanism for the localized SB stabilization given that the spacing between the second phases are equal to or smaller than the intrinsic plastic zone size of each alloy type.

The fracture surface topology of the highly deformed Pt-BMG samples under uniaxial tension exhibit cores (indicated by white arrows) and radiating vein-like structures (Figure 4c). The cores represent the initial microcracks before fracture during tension [24]. The regions with a lava texture (pointed by black arrows in Figure 4d) appearing on the ridge of the veins confirm the molten phase of the Pt-BMG porous sample, which areattributed to localized melting generated by the elastic energy dissipated as heat on the fracture plane (as previously suggested in [25,26]), with an estimated temperature increase of 900 K [27]. Similar shear and fracture surface patterns are observed in Zr-BMG samples (Figure 5a–c). The smooth region accounts for the shear offset before global failure (which is also seen in Pt-BMGs; see the bottom part of Figure 4c) accompanied by vein-like deformation during fracture. The most significant difference between the fracture surfaces of two different alloys is that for the Zr-BMG, the molten spots are sparsely distributed within the vein-like patterns with a small and round shape, whereas for Pt-BMG, large molten regions scattered around veins are observed.

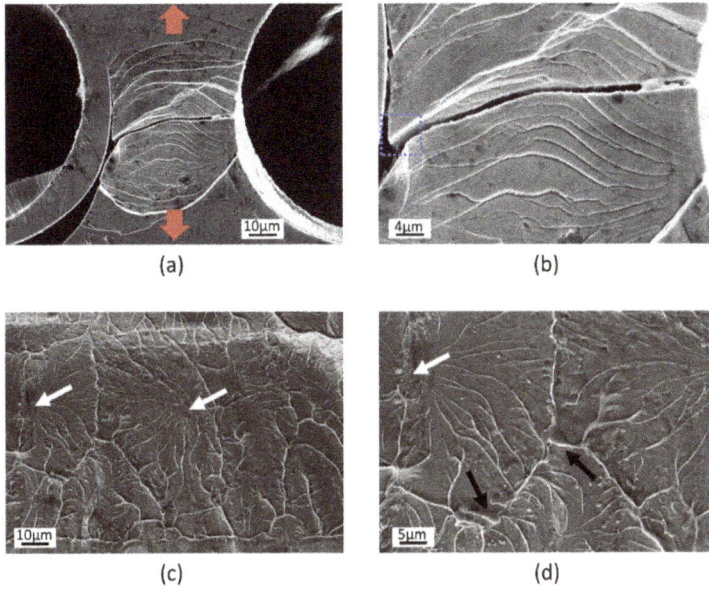

Figure 4. Morphological analysis of Pt-BMG pulled until rupture. (**a,b**) Deformation is localized to multiple SBs connecting adjacent pores perpendicular to the loading direction. (**c,d**) Fracture plane analysis shows the core regions (white arrows), as well as the vein-like deformation pattern containing molten regions (black arrows).

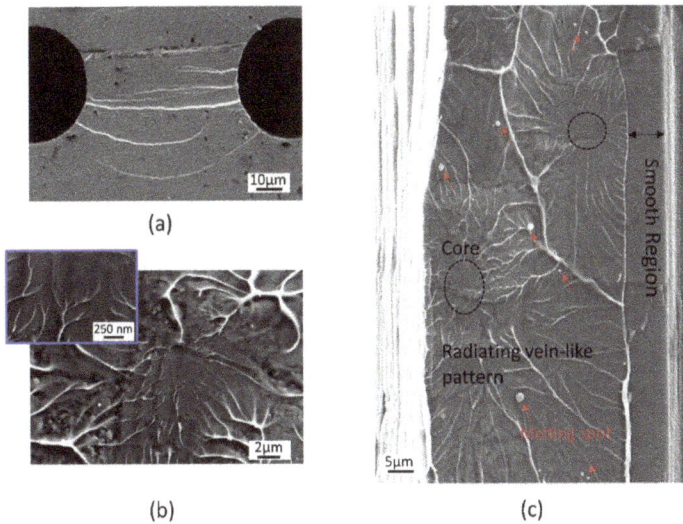

Figure 5. Morphological analysis of the Zr-BMG sample. (**a**) SBs are formed between pores perpendicular to loading, some of which further branch in multiple bands. (**b**) The ridges of the veins on the fracture surface radiate and point toward a core, where the tip of the veins has a width smaller than 50 nm (inset). (**c**) Fracture plane analysis shows various deformation morphologies, including melting spots.

3.2. Size Effects in Metallic Glasses

Next, it would be interesting to see how the tensile ductility of BMG changes when the characteristic length scale goes down to the nanometer regime. Since the critical crack length is in the order of micrometers, SB-mediated inhomogeneous flow is observed when approaching the nm scale [15]. We conducted molecular dynamics simulations in order to analyze the deformation behavior of bulk metallic glass heterostructures with pore sizes of 18 nanometers. The SB propagation and distribution across the pores were investigated for the ratio $d/s = 1$. As a prototype material, a CuZr BMG plate with dimensions of 36 nm × 7 nm × 85 nm, giving a total number of 1,172,450 atoms, was considered [28]. By using periodic boundary conditions, we excluded the presence of free surfaces that promote strain localization and formation of critical SBs [29]. Thus, the pores are the only heterogeneities in the BMG plate. The BMG heterostructure is deformed in tension along the z-direction with a strain rate of $\dot{\varepsilon} = 4 \times 10^7 \, s^{-1}$.

The stress-strain curve together with the local atomic shear strain calculated by OVITO (open visualization and analysis tool for atomistic simulations) [30] is presented in Figure 6.

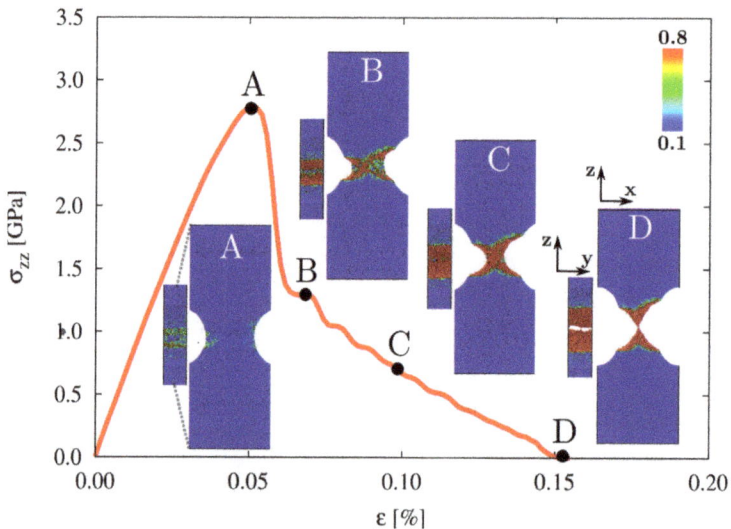

Figure 6. Deformation of a sample (in the z-direction) having pores of 18 nm in diameter. The front view (xz-plane) and side view (yz-plane) of each deformation pattern are displayed for different states of deformation. Note that the side view is scaled down, where the corresponding actual length is indicated by the dashed lines in the contour plots of Point A. Homogenous deformation (Point A), followed by two SBs intersecting each other (Point B) with a sudden decrease in strength. After this point, necking starts to prevail (Point C), and the failure occurs at a strain of 15% (Point D).

In order to investigate the deformation mechanism of nanoporous BMG, the most important four steps along the deformation are identified. Around 5% strain (Point A), when the maximum stress is achieved, two SBs start nucleating at the pore surface. This mechanism can be seen better by looking at the side of the plate (yz-plane). Increasing the strain to 7%, the two SB propagate along the x-direction of the BMG plate. Basically, the SBs occur perpendicular to the loading direction as found in the case of the BMG with micron-sized pores, except that a large number of SBs were observed at the micron level. The intersection of two SBs causes the serrated-type deformation as observed between Points B and C. Since the pores have a diameter of just 18 nm and the critical SB width is about 4–5 nm [31], only two SBs are mediating the plasticity. Increasing the strain forward to 10%, the two SB overlap and give rise to necking of the BMG plate; see Point C. Finally, the BMG plate fails at a strain of 15% normal to the loading direction, as can be seen at Point D. In general, thermally-activated processes are excluded in MD simulations, resulting in a higher ultimate yield strength. The trends for the strain and elastic modulus are comparable to the experimental results of the porous structures. The presented shear-driven-type deformation at the nanoscale also explains why the vein-like shear below 50 nm can be observed on the fracture surface of the actual samples (Figure 5b, inset).

3.3. Compositional Investigation of Pt-BMG

To reveal the compositional fluctuations during thermomechanical (*viz.* thermoplastic forming) and room temperature tensile testing, energy dispersive X-ray analysis (EDX attached to SEM) was carried out on the fracture surface. We have chosen Pt-BMG based on the fractographic analysis, which shows higher amounts of molten regions, radiating vein-like patterns and SBs between the neighboring pores compared to those of the Zr-BMG. The concentration of each element was determined from the smooth deformation region indicating the SB evolvement (Figure 7a), as well as the vein-like patterns (Figure 7b) and molten regions (Figure 7c) generated by fracture. In comparison, EDX analysis of the undeformed region of the thermoplastically formed sample (Figure 7d) and chemical analysis of the as-cast sample (Table 2) were likewise performed. The EDX results reveal 5 at.% higher P content observed at the undeformed regions of the specimen and in as-cast state at the expense of Ni and Cu compared to the P content of the molten regions with ridges. The phosphorus deficiency is pronounced for the core regions, which are, as mentioned before, counted as the onset of microfractures. On the other hand, an enrichment in the Pt content (mean value by $\approx$2.5 at.%) is attained as the deformation proceeds from the shear-type deformation toward the core of the radiating vein-like deformation patterns. When the as-cast state sample (chemical analysis given in Table 2) is taken as a basis, the results also suggest that the influence of thermoplastic forming on the compositional fluctuation (the mean of the undeformed region in Table 2, Figure 7a) is much less significant compared to the compositional fluctuation of the uniaxially deformed tensile samples.

Auger spectral analysis (JEOL JAMP 9500F Scanning Auger Microprobe) also validates the increase in content of phosphorus compared to the deformed regions (Figure 7e and Table 2). A Schottky field emission electron gun at 10 keV and 10 nA excited the metallic glass under 10^{-7} Pa vacuum, where a differentially-pumped ion gun with 1 keV was used to sputter argon ions onto the Pt-BMG at a sputtering rate of 8 nm/min for surface cleaning. Elastically-scattered electrons penetrated the material to a depth of 1 μm. Different from EDX, the Auger information comes only from the first nanometer thickness, where measurements are conducted after removing the surface contamination. Because of the use of a standard single element sensitivity factor, not validated for this special alloy, and possible concentration changes by preferential sputtering, the determined concentration are different from the EDX data. Moreover, the Ni peaks are overlaid by Cu and can only be detected by low efficiency peaks. For this reason, the comparison is made by the normalized n_P/n_{Pt} ratio. A decreasein this ratio (as in the case of EDX measurements) also confirms this P deficiency and Pt enrichment, particularly on the vein-like, core and molten regions.

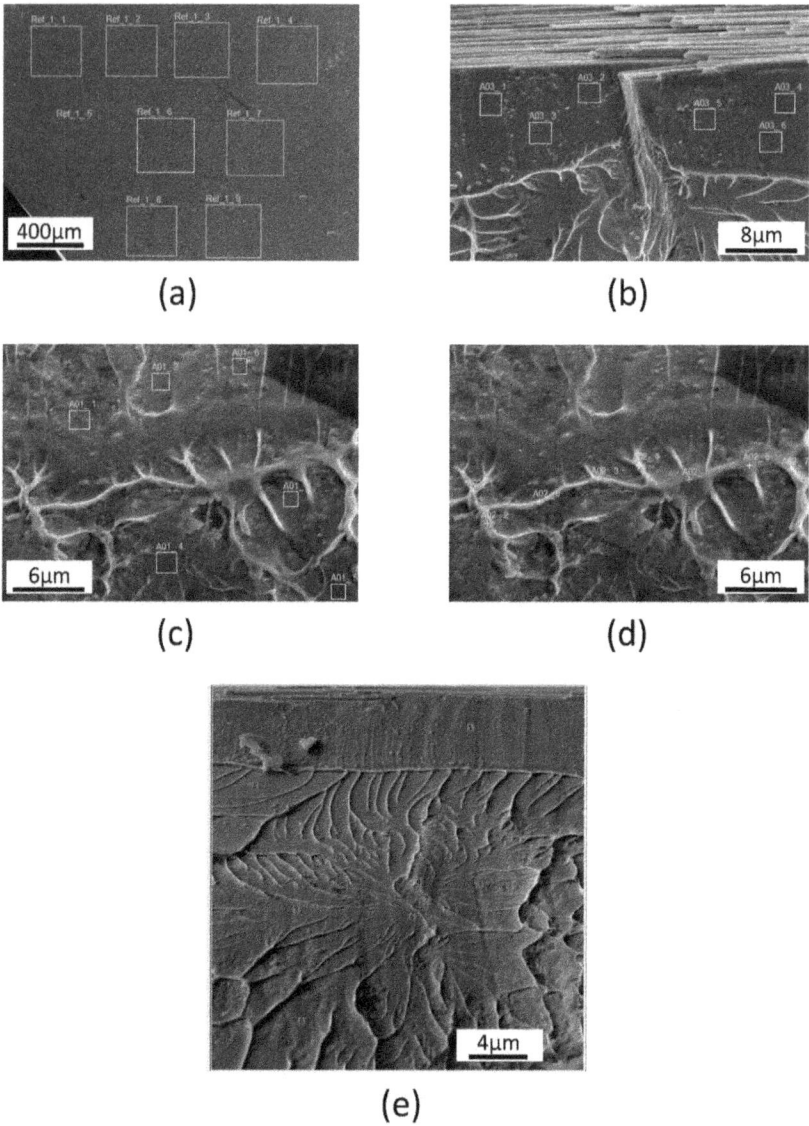

Figure 7. EDX analyses of the fracture surface of the porous Pt-BMG taken from the (a) undeformed surface, (b) smooth region, (c) vein-like region and (d) ridges of veins and molten regions. (e) Different fracture regions of interest obtained from Auger electron spectroscopy analysis. Point and region measurements were conducted depending on the specific area, where the numbers indicated are simply showing different measurements of interest within the same region.

Table 2. Concentration data of constituent elements of the Pt-BMG averaged from the deformed and undeformed regions. Chemical analysis reflects the as-cast state composition.

Element	Pt	Cu	Ni	P	n_P/n_{Pt}
Chemical analysis (at.%)	57.0 ± 0.4	14.4 ± 0.1	5.2 ± 0.1	23.4 ± 0.1	0.41
EDX Spectra (at.%) (Figure 7a–d)					
mean of undeformed region (Figure 7a)	56.0	14.6	5.5	23.9	0.43
mean of smooth region (Figure 7b)	55.5	15.4	6.8	22.3	0.40
mean of vein-like region (Figure 7c)	57.3	16.2	6.3	20.2	0.35
mean of ridges + molten regions(Figure 7d)	58.4	16.4	6.4	18.8	0.32
Auger spectra (at.%) (Figure 7e)					
mean of undeformed region	73.3	6.4	5.4	14.9	0.20
mean of smooth region	75.6	6.0	5.6	12.8	0.17
mean of vein-like region	76.9	6.1	5.3	11.7	0.15
mean of core + molten regions	76.4	6.4	5.2	12.0	0.16

4. Conclusions

This contribution highlighted the influence of pore design on the overall mechanical, fractographic and chemical properties. Replication of complicated patterns using a Si template provides a versatile toolbox to manipulate the properties of the metallic glasses in a controlled manner. The complex deformation state and the geometrical effect attributed to the pore lineup leads to multiple SB formations between neighboring pores perpendicular to loading, which increases the fracture strain by almost three times. Fracture surface analysis has identified three different deformation patterns, namely smooth region, vein-like pattern and core of the shear. Molten spots or regions were observed depending on the type of BMG analyzed. It has been confirmed by EDX and Auger spectra that the mechanically-induced plasticity causes variations among the constituent elements between deformed and undeformed regions up to 5 at%. Finally, MD simulations conducted with a similar d/s ratio sample depict SB formation followed by necking of the sample, where the SEM fracture surface observations have proven the shear-induced type of deformation below the 50 nm level.

Acknowledgments: The support for this work provided by the European Research Council under the Advanced Grant "INTELHYB–Next Generation of Complex Metallic Materials in Intelligent Hybrid Structures" (Grant ERC-2013-ADG-340025) is gratefully acknowledged. The authors thank Romy Keller for SEM imaging of BMGs, Andrea Voss for the chemical analysis of the as-cast Pt-BMG rod, Long Zhang for the XRD measurement of Pt-BMG sample and Junhee Han for the discussion and assistance regarding the EDX analysis.

Author Contributions: Baran Sarac wrote and edited the paper and contributed to all activities. Daniel Sopu performed the MD simulations. Eunmi Park and Julia Kristin Hufenbach contributed to the SEM and EDX analysis. Eunmi Park contributed to the XRD analysis. Steffen Oswald performed the Auger Spectroscopy analysis. Steffen Oswald, Daniel Sopu and Julia Kristin Hufenbach contributed to the discussions. Jürgen Eckert and Mihai Stoica contributed to the interpretation and discussion of the results.

Conflicts of Interest: The authors declare no conflict of interest.

References

1. Schwarz, M.; Karma, A.; Eckler, K.; Herlach, D.M. Physical-mechanism of grain-refinement in solidification of undercooled melts. *Phys. Rev. Lett.* **1994**, *73*, 1380–1383. [CrossRef] [PubMed]
2. Schroers, J.; HollandMoritz, D.; Herlach, D.M.; Grushko, B.; Urban, K. Undercooling and solidification behaviour of a metastable decagonal quasicrystalline phase and crystalline phases in Al-Co. *Mat. Sci. Eng. A* **1997**, *226*, 990–994. [CrossRef]
3. Hofmann, D.C.; Suh, J.Y.; Wiest, A.; Duan, G.; Lind, M.L.; Demetriou, M.D.; Johnson, W.L. Designing metallic glass matrix composites with high toughness and tensile ductility. *Nature* **2008**, *451*, 1085–1089. [CrossRef] [PubMed]

4. Hays, C.C.; Kim, C.P.; Johnson, W.L. Microstructure controlled shear band pattern formation and enhanced plasticity of bulk metallic glasses containing *in situ* formed ductile phase dendrite dispersions. *Phys. Rev. Lett.* **2000**, *84*, 2901–2904. [CrossRef] [PubMed]

5. Sarac, B.; Schroers, J. Designing tensile ductility in metallic glasses. *Nat. Commun.* **2013**, *4*, 1–7. [CrossRef] [PubMed]

6. Sarac, B.; Ketkaew, J.; Popnoe, D.O.; Schroers, J. Honeycomb structures of bulk metallic glasses. *Adv. Funct. Mater.* **2012**, *22*, 3161–3169. [CrossRef]

7. Sarac, B.; Schroers, J. From brittle to ductile: Density optimization for Zr-BMG cellular structures. *Scr. Mater.* **2013**, *68*, 921–924. [CrossRef]

8. Sarac, B. *Microstructure-Property Optimization in Metallic Glasses*; Springer Theses; Springer International Publishing: Cham, Switzerland, 2015; Volume XIII, p. 87.

9. Schroers, J. Bulk metallic glasses. *Phys. Today* **2013**, *66*, 32–37. [CrossRef]

10. Lewandowski, J.J.; Wang, W.H.; Greer, A.L. Intrinsic plasticity or brittleness of metallic glasses. *Philos. Mag. Lett.* **2005**, *85*, 77–87. [CrossRef]

11. Svendsen, B.; Bargmann, S. On the continuum thermodynamic rate variational formulation of models for extended crystal plasticity at large deformation. *J. Mech. Phys. Solids* **2010**, *58*, 1253–1271. [CrossRef]

12. Bargmann, S.; Ekh, M.; Runesson, K.; Svendsen, B. Modeling of polycrystals with gradient crystal plasticity: A comparison of strategies. *Philos. Mag.* **2010**, *90*, 1263–1288. [CrossRef]

13. Sarac, B.; Klusemann, B.; Xiao, T.; Bargmann, S. Materials by design: An experimental and computational investigation on the microanatomy arrangement of porous metallic glasses. *Acta Mater.* **2014**, *77*, 411–422. [CrossRef]

14. Sarac, B.; Wilmers, J.; Bargmann, S. Property optimization of porous metallic glasses via structural design. *Mater. Lett.* **2014**, *134*, 306–310. [CrossRef]

15. Dubach, A.; Raghavan, R.; Loffler, J.F.; Michler, J.; Ramamurty, U. Micropillar compression studies on a bulk metallic glass in different structural states. *Scr. Mater.* **2009**, *60*, 567–570. [CrossRef]

16. Kumar, G.; Desai, A.; Schroers, J. Bulk metallic glass: The smaller the better. *Adv. Mater.* **2011**, *23*, 461–476. [CrossRef] [PubMed]

17. Hasan, M.; Schroers, J.; Kumar, G. Functionalization of metallic glasses through hierarchical patterning. *Nano Lett.* **2015**, *15*, 963–968. [CrossRef] [PubMed]

18. Schroers, J. Processing of bulk metallic glass. *Adv. Mater.* **2010**, *22*, 1566–1597. [CrossRef] [PubMed]

19. Schroers, J. Ductile Bulk Metallic Glass. *Phys. Rev. Lett* **2004**, *93*, 255506. [CrossRef] [PubMed]

20. Schroers, J. On the formability of bulk metallic glass in its supercooled liquid state. *Acta Mater.* **2008**, *56*, 471–478. [CrossRef]

21. Legg, B.A.; Schroers, J.; Busch, R. Thermodynamics, kinetics, and crystallization of $Pt_{57.3}Cu_{14.6}Ni_{5.3}P_{22.8}$ bulk metallic glass. *Acta Mater.* **2007**, *55*, 1109–1116. [CrossRef]

22. Sarac, B.; Kumar, G.; Hodges, T.; Ding, S.Y.; Desai, A.; Schroers, J. Three-dimensional shell fabrication using blow molding of bulk metallic glass. *J. Microelectromech. S* **2011**, *20*, 28–36. [CrossRef]

23. Duan, G.; Wiest, A.; Lind, M.L.; Li, J.; Rhim, W.K.; Johnson, W.L. Bulk metallic glass with benchmark thermoplastic processability. *Adv. Mater.* **2007**, *19*, 4272–4275. [CrossRef]

24. Qu, R.T.; Stoica, M.; Eckert, J.; Zhang, Z.F. Tensile fracture morphologies of bulk metallic glass. *J. Appl. Phys.* **2010**, *108*, 063509. [CrossRef]

25. Zhang, Z.F.; Eckert, J.; Schultz, L. Difference in compressive and tensile fracture mechanisms of $Zr_{59}Cu_{20}Al_{10}Ni_8Ti_3$ bulk metallic glass. *Acta Mater.* **2003**, *51*, 1167–1179. [CrossRef]

26. Wright, W.J.; Saha, R.; Nix, W.D. Deformation mechanisms of the $Zr_{40}Ti_{14}Ni_{10}Cu_{12}Be_{24}$ bulk metallic glass. *Mater. Trans.* **2001**, *42*, 642–649. [CrossRef]

27. Liu, C.T.; Heatherly, L.; Easton, D.S.; Carmichael, C.A.; Schneibel, J.H.; Chen, C.H.; Wright, J.L.; Yoo, M.H.; Horton, J.A.; Inoue, A. Test environments and mechanical properties of Zr-base bulk amorphous alloys. *Metall. Mater. Trans. A* **1998**, *29*, 1811–1820. [CrossRef]

28. Mendelev, M.I.; Rehbein, D.K.; Ott, R.T.; Kramer, M.J.; Sordelet, D.J. Computer simulation and experimental study of elastic properties of amorphous Cu-Zr alloys. *J. Appl Phys.* **2007**, *102*, 093518. [CrossRef]

29. Sopu, D.; Karsten, A. Influence of grain size and composition, topology and excess free volume on the deformation behavior of Cu-Zr nanoglasses. *Beilstein J. Nanotechnol.* **2015**, *6*, 537–545. [CrossRef]

Metals **2015**, *5*, 920–933

30. Stukowski, A. Visualization and analysis of atomistic simulation data with OVITO—The open visualization tool. *Model. Simul. Mater. Sci.* **2010**, *18*, 015012. [CrossRef]
31. Shi, Y.F. Size-independent shear band formation in amorphous nanowires made from simulated casting. *Appl. Phys. Lett.* **2010**, *96*, 121909. [CrossRef]

![metals logo] *metals* MDPI

Article

Deformation-Induced Martensitic Transformation in Cu-Zr-Zn Bulk Metallic Glass Composites

Dianyu Wu [1], **Kaikai Song** [2,3,*], **Chongde Cao** [1,*], **Ran Li** [4], **Gang Wang** [5], **Yuan Wu** [6], **Feng Wan** [1], **Fuli Ding** [1], **Yue Shi** [1], **Xiaojun Bai** [1], **Ivan Kaban** [3,7] and **Jürgen Eckert** [8,9]

[1] Department of Physics, School of Science, Northwestern Polytechnical University, Youyi Xilu 127, 710072 Xi'an, China; 18740443890@163.com (D.W.); wanfeng1101@163.com (F.W.); dingfuli2008@163.com (F.D.); shiyue@hi123.com (Y.S.); xjbai@nwpu.edu.cn (X.B.)
[2] School of Mechanical, Electrical & Information Engineering, Shandong University (Weihai), Wenhua Xilu 180, 264209 Weihai, China
[3] IFW Dresden, Institute for Complex Materials, Helmholtzstraße 20, 01069 Dresden, Germany; i.kaban@ifw-dresden.com
[4] Key Laboratory of Aerospace Materials and Performance (Ministry of Education), School of Materials Science and Engineering, Beihang University, 100191 Beijing, China; liran@buaa.edu.cn
[5] Laboratory for Microstructures, Shanghai University, 200444 Shanghai, China; g.wang@shu.edu.cn
[6] State Key Laboratory for Advanced Metals and Materials, University of Science and Technology Beijing, 100191 Beijing, China; wuyuan@ustb.edu.cn
[7] TU Dresden, Institut für Werkstoffwissenschaft, 01062 Dresden, Germany
[8] Erich Schmid Institute of Materials Science, Austrian Academy of Sciences, Jahnstraße 12, A-8700 Leoben, Austria; juergen.eckert@unileoben.ac.at
[9] Department Materials Physics, Montanuniversität Leoben, Jahnstraße 12, A-8700 Leoben, Austria
* Authors to whom correspondence should be addressed; songkaikai8297@gmail.com (K.S.); caocd@nwpu.edu.cn (C.C.); Tel.: +86-029-8843-1657 (C.C.); Fax: +86-029-8843-1656 (C.C.).

Academic Editors: K.C. Chan and Jordi Sort Viñas
Received: 23 October 2015; Accepted: 11 November 2015; Published: 17 November 2015

Abstract: The microstructures and mechanical properties of $(Cu_{0.5}Zr_{0.5})_{100-x}Zn_x$ ($x = 0$, 1.5, 2.5, 4.5, 7, 10, and 14 at. %) bulk metallic glass (BMG) composites were studied. CuZr martensitic crystals together with minor B2 CuZr and amorphous phases dominate the microstructures of the as-quenched samples with low Zn additions ($x = 0$, 1.5, and 2.5 at. %), while B2 CuZr and amorphous phases being accompanied with minor martensitic crystals form at a higher Zn content ($x = 4.5$, 7, 10, and 14 at. %). The fabricated Cu-Zr-Zn BMG composites exhibit macroscopically appreciable compressive plastic strain and obvious work-hardening due to the formation of multiple shear bands and the deformation-induced martensitic transformation (MT) within B2 crystals. The present BMG composites could be a good candidate as high-performance structural materials.

Keywords: metallic glasses; composites; rapid solidification; martensitic transformation

1. Introduction

Although bulk metallic glasses (BMGs) exhibit attractive mechanical properties, such as high strength, high hardness, and a large elastic strain limit, they usually fail in an apparently brittle manner during deformation at room temperature [1–6]. A poor ductility, caused by shear localization and strain/thermal softening, severely restricts their practical applications as structural and functional materials [1–6]. In order to circumvent this shortcoming, BMG composites with ductile crystals precipitating in the glassy matrix have been developed in various alloy systems by *in situ* and *ex situ* fabrication methods [5–9]. It has been reported that improved toughness, and even tensile ductility, can be observed in such BMG composites by properly adjusting compositions, microstructures, and casting methods [5–18]. During deformation at room temperature, ductile crystals can effectively suppress the rapid propagation of the main shear bands and induce the formation of multiple shear bands in the glassy matrix [5–18]. However, these ductile crystals during deformation cannot provide sufficient work-hardening to eliminate the work-softening effect induced by the formation of shear bands in the glassy matrix, resulting in the microscopic strain-softening of BMG composites [10–18].

Recently, it has been found that, by introducing a shape memory ductile phase into the glassy matrix, the fabricated BMG composites show not only a pronounced tensile ductility but also a macroscopic work-hardening capability during deformation at room temperature [19–27]. Upon loading, besides the formation of multiple shear bands in the glassy matrix, ductile shape memory crystals experience the martensitic transformation (MT) from a cubic phase to monoclinic phases [19–27], which can provide the pronounced work hardening. Hence, the weakness of the work-softening effect during deformation of BMG composites can be effectively overcome.

Until now, CuZr- and TiNiCu-based alloy systems with different compositions have been fabricated into BMG composites [19–28]. For example, CuZr-based BMG composites with different sample sizes containing minor element additions such as Al, Ti, Ag, Ni, Co, Y, Er, V, W, or Ta have been successfully fabricated [20–25,29–37]. Two key requirements should be satisfied for the fabrication of CuZr-based BMG composites: (1) maintaining sufficiently high glass-forming ability (GFA) in order to fabricate large-sized BMG composites [20–25,29–37]; and (2) stabilization of metastable B2 CuZr phase in order to suppress its decomposition into equilibrium phases (EPs) [29–34]. Previous results [22,38] demonstrated that minor Co addition to CuZr alloys can dramatically enhance the thermal stability of the B2 CuZr phase but gravely deteriorate the GFA of CuZr-based alloys. Our recent work [39] has shown that minor Zn addition to binary CuZr alloys can not only improve their GFA but also effectively stabilize the B2 CuZr phase to low temperatures. Additionally, as a rule of thumb, Zn is much cheaper than Co, Al, Ag, Ti, Hf, or rare earth metals. Therefore, it is beneficial to fabricate Cu-Zr-Zn BMG composites and further investigate the correlation between microstructures and mechanical properties of Cu-Zr-Zn BMG composites. In this paper, the microstructures and mechanical properties of $(Cu_{0.5}Zr_{0.5})_{100-x}Zn_x$ ($x = 0, 2.5, 4.5, 7, 10$, and 14 at. %) BMG composites were investigated. Furthermore, the corresponding deformation mechanism at room temperature of Cu-Zr-Zn BMG composites was also discussed.

2. Experimental Section

$Cu_{50}Zr_{50}$ master alloys were prepared by arc-melting appropriate amounts of the constituting elements (at least purity 99.9%) under a Ti-gettered argon atmosphere. Each ingot was remelted at least three times in order to achieve chemical homogeneity. Afterwards, master alloys with nominal compositions of $(Cu_{0.5}Zr_{0.5})_{100-x}Zn_x$ ($x = 0, 1.5, 2.5, 4.5, 7, 10$, and 14 at. %) were melted using a high frequency furnace under an argon atmosphere. They were then cast into rods with a diameter of 2 mm using an injection casting machine. The microstructures of the as-cast samples were characterized using an optical microscope (OM, Olympus, Tokyo, Japan) and X-ray diffraction (XRD) in reflection geometry (Rigaku D/max-rB). The XRD measurements were performed on longitudinal sections of four polished rods with a final dimension of about 2 mm × 5 mm × 0.7 mm. Room-temperature compression tests were performed on specimens with a diameter of about Ø2 mm and a height of 4 mm using an electronic universal testing machine (New SANS, MTS System Corporation (China), Shenzhen, China) at an initial strain rate of $2.5 \times 10^{-4} \cdot s^{-1}$. The microstructures of the as-cast rods after deformation were checked using a scanning electron microscope (SEM) and XRD, respectively.

3. Results and Discussion

3.1. Phase Formation in the As-Cast Cu-Zr-Zn Rods

Figure 1 shows XRD patterns of the as-cast $(Cu_{0.5}Zr_{0.5})_{100-x}Zn_x$ ($x = 0, 1.5, 2.5, 4.5, 7, 10$, and 14 at. %) rods. For the samples with low Zn contents of 0, 1.5, and 2.5 at. %, the dominating phase is

Metals **2015**, *5*, 2134–2147

CuZr martensite, being similar to previous reports [19–27]. Additionally, a small quantity of B2 CuZr phase being accompanied with minor $Cu_{10}Zr_7$ crystals can be observed in the matrix. With Zn content increasing to 4.5 and 7 at. %, the volume fractions of CuZr martensitic crystals decrease quickly, while a large number of B2 CuZr crystals (P*m-3m*) precipitate [40–42]. However, it is difficult to conclude on the existence of amorphous phase due to much higher peak intensities of crystals than those of the amorphous phase for the as-cast $(Cu_{0.5}Zr_{0.5})_{100-x}Zn_x$ ($x = 0$, 1.5, and 2.5 at. %) samples. Meanwhile, for the Cu-Zr-Zn samples with 10 and 14 at.% Zn additions, a broad diffraction hump around $2\theta = 38.5°$ in the XRD patterns corresponding with amorphous phase is clearly seen together with crystalline reflexes of B2 CuZr and martensitic crystals in Figure 1. These observations reveal that minor Zn addition effectively improves the GFA of CuZr-based alloys. Recently, by using *in situ* high-accuracy X-ray diffraction analysis on $Cu_{50}Zr_{50}$ MG, Kalay *et al.* found that only the B2 CuZr phase is observed at 1045 K while at 1002 K a mixture of B2 CuZr, $Cu_{10}Zr_7$, and $CuZr_2$ crystals appear [43]. With decreasing temperature to 789 K, $Cu_{10}Zr_7$ and $CuZr_2$ crystals can be found, which further confirms previous *ex situ* experimental results [30,31,34]. Therefore, during quenching of Cu-Zr-Zn melts, the B2 CuZr phase should be kept to room temperature. However, MT would be induced within some B2 CuZr crystals due to the thermal stress originating from rapid quenching [29,41], resulting in the precipitation of monoclinic martensitic crystals. It has been shown that monoclinic martensitic phases consist of two kinds of structures, *i.e.*, basic structure (B19′, P2₁/m) and superstructures (B33, Cm) [40–42]. In our case, since Zn addition can also stabilize B2 CuZr phase to low temperatures [39], a larger number of B2 CuZr crystals could be obtained with Zn content increasing. On the other hand, the eutectoid decomposition of the B2 CuZr phase into other EPs can also be inhibited to some extent [39] since the thermal stability of the B2 CuZr phase is enhanced with the addition of Zn. That would result in the decrease of the volume fraction of $Cu_{10}Zr_7$ crystals in the matrix with Zn content increasing from 0 at. % to 14 at. %. As a result, martensitic crystals, B2 CuZr phase, amorphous phase, and/or a little $Cu_{10}Zr_7$ crystals can be obtained during solidification, ultimately resulting in the formation of Cu-Zr-Zn BMG composites.

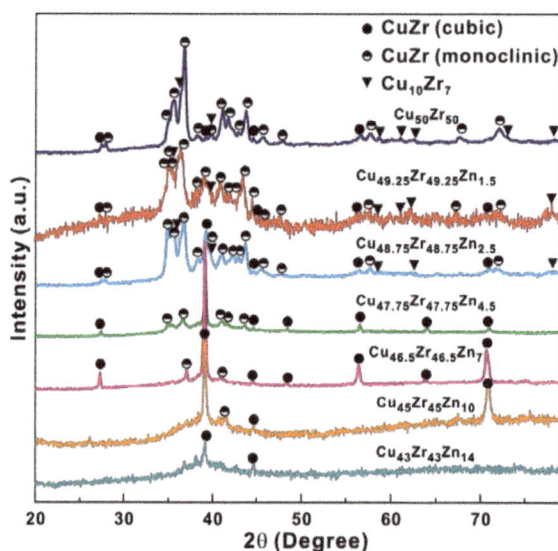

Figure 1. XRD patterns of the as-cast $(Cu_{0.5}Zr_{0.5})_{100-x}Zn_x$ ($x = 0$, 1.5, 2.5, 4.5, 7, 10, and 14 at. %) rods with a diameter of 2 mm.

In order to further illustrate the phase formation in the as-quenched CuZr-based alloys with Zn additions, the corresponding OM pictures are presented in Figure 2. In the $(Cu_{0.5}Zr_{0.5})_{100-x}Zn_x$ (x = 0, 1.5, and 2.5 at. %) samples, obvious martensitic plates were found in the matrix (marked by the red arrows in Figure 2a–c). Additionally, a little amorphous phase appears around the outer regions of the as-cast rods (marked by the blue dotted arrows in Figure 2a–c) due to a relatively higher cooling rate. For the as-cast $Cu_{47.25}Zr_{47.25}Zn_{4.5}$ sample, a large volume fraction of the B2 CuZr phase within the glassy matrix can be observed (see the red dot-dashed arrows in Figure 2d). Moreover, as it is seen in the insets in Figure 2d, some martensitic plates also appear within the B2 CuZr crystals. With increasing Zn content from 7 at. % to 10 at. %, the volume fraction of the B2 CuZr phase decreases while that of the amorphous phase gradually increases (Figure 2e,f). As it is shown in Figure 2f,g, the bottom and top parts of the $Cu_{45}Zr_{45}Zn_{10}$ rod consist of amorphous and crystalline phases. The corresponding volume fractions of the crystals at the bottom parts of the as-cast rods were less than those of the top parts due to somewhat different cooling rates. Furthermore, for the $Cu_{43}Zr_{43}Zn_{14}$ rod, the XRD pattern of the middle and top parts show a nature of BMG composites (Figure 1) while the bottom parts are fully amorphous from the OM images shown in Figure 2h. The differences for the volume fractions of amorphous phase between $Cu_{45}Zr_{45}Zn_{10}$ and $Cu_{43}Zr_{43}Zn_{14}$ rods should be contributed to the increase of GFA with Zn content increasing. In order to further ascertain heterogeneous distributions of the B2 CuZr crystals in the glassy matrix, OM measurements were performed on the longitudinal cross-sections of the as-cast $Cu_{48.75}Zr_{48.75}Zn_{2.5}$ rod. As shown in Figure 3a,b, almost fully crystalline microstructures are observed at the top parts of the investigated sample, whereas the bottom parts (Figure 3c,d) contain a larger fraction of the amorphous phase. Nevertheless, the changes of the volume fraction of the amorphous phase for the top part to bottom part in same rods reveal that the differences in the applied cooling rates may induce either vitrification or precipitation of crystals.

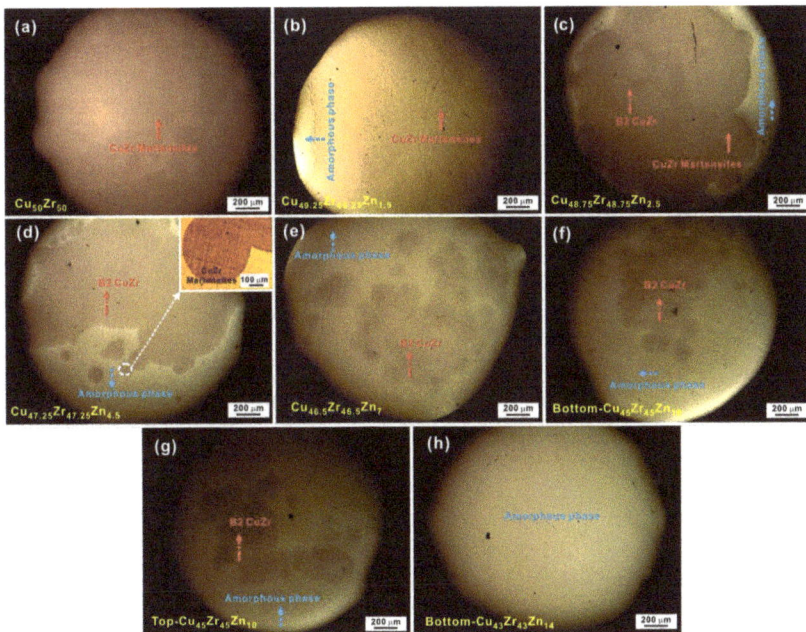

Figure 2. OM pictures of the as-cast $(Cu_{0.5}Zr_{0.5})_{100-x}Zn_x$ (x = (**a**) 0; (**b**) 1.5 at. %; (**c**) 2.5 at. %; (**d**) 4.5 at. %; (**e**) 7 at. %; (**f,g**) 10 at. %; and (**h**) 14 at. %) rods with a diameter of 2 mm [39]; Inset: martensitic plates within B2 CuZr crystals.

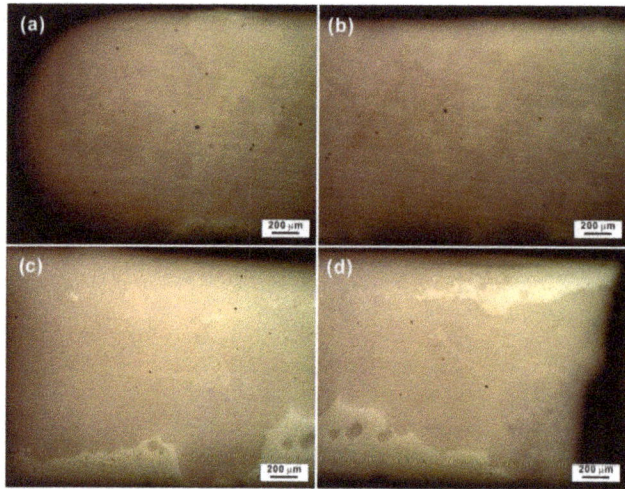

Figure 3. OM pictures of the longitudinal cross-sections cut from the top parts (**a,b**) to from the bottom parts (**c,d**) of the as-cast $Cu_{48.75}Zr_{48.75}Zn_{2.5}$ rod; The neighboring microstructures from the top parts and the bottom parts from a same rod were shown in Figure 3a,b and Figure 3c,d, respectively.

3.2. Mechanical Properties of the As-Cast Cu-Zr-Zn Samples

Figure 4 exhibits the strain-stress curves measured upon uniaxial compression at room temperature for the as-cast Cu-Zr-Zn rods. It has been reported that the yield stresses of the B2 CuZr phase and CuZr martensitic crystals are lower than 500 MPa while these of CuZr-based BMGs are higher than 1700 MPa [13,17,20–27]. Hence, with increasing volume fraction of the amorphous phase, as shown in Figure 4, the yield stress increases gradually. As shown in Figure 4a, the as-cast $(Cu_{0.5}Zr_{0.5})_{100-x}Zn_x$ (x = 0 and 2.5 at. %) rods show a similar compressive curves to CuZr martensites [21,24] since martensitic crystals dominate their microstructures. Their average yield stress is 819 ± 60 MPa and 839 ± 200 MPa, while their fracture strength is 1821 ± 75 MPa and 1772 ± 120 MPa, respectively (Figure 4a and Table 1). Furthermore, with Zn content increasing from 4.5 at. % to 7 at. %, the average yield stress and average fracture stress increase from 830 ± 220 MPa to 906 ± 115 MPa, 1713 ± 75 MPa, and 1884 ± 60 MPa, respectively. Meanwhile, an average plastic strain of $6.8\% \pm 0.8\%$ and $6.6\% \pm 0.8\%$ can be obtained for both of samples, respectively (Figure 4b and Table 1). As mentioned in Section 3.1, the as-cast $(Cu_{0.5}Zr_{0.5})_{100-x}Zn_x$ (x = 4.5 and 7 at. %) samples mainly consist of B2 CuZr crystals and amorphous phase. Hence, as listed in Table 1, the differences of mechanical properties for top and bottom parts for a same rod are not very obvious.

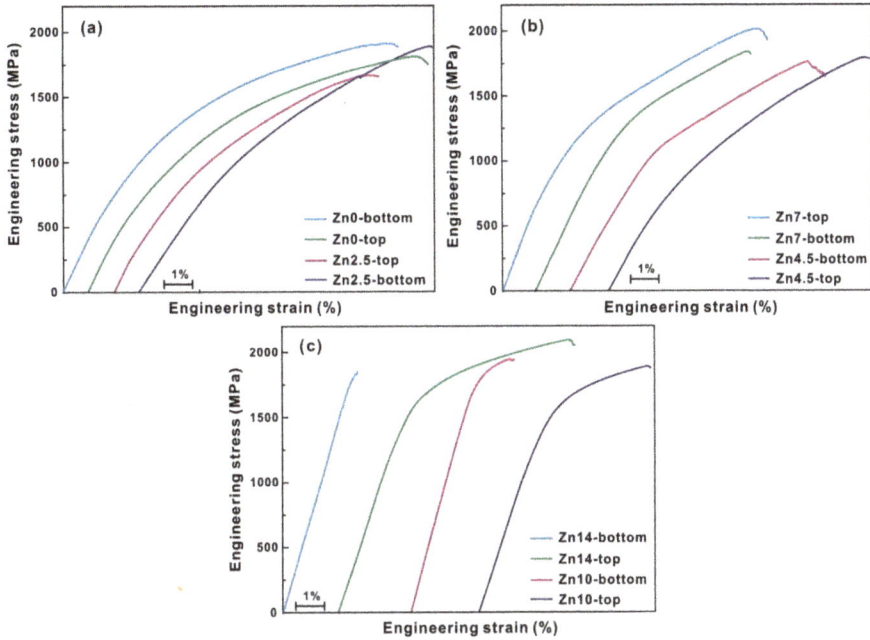

Figure 4. Compressive strain-stress curves of the as-cast $(Cu_{0.5}Zr_{0.5})_{100-x}Zn_x$ rods with a diameter of 2 mm: (**a**) $x = 0$ and 2.5 at. %, (**b**) $x = 4.5$ and 7 at. %, and (**c**) $x = 10$ and 14 at. %; Compressive curves of the top and bottom parts of the as-cast rods were blue and green curves, respectively.

Table 1. Yield strength, fracture strength, and plastic strain of different $(Cu_{0.5}Zr_{0.5})_{100-x}Zn_x$ ($x = 0$, 2.5, 4.5, 7, 10, and 14 at. %) BMGs and composites.

Samples	Yield Strength (MPa)	Fracture Strength (MPa)	Plastic Strain (%)	Average Yield Strength (MPa)	Average Fracture Strength (MPa)	Average Plastic Strain (%)
Zn0-Top	770 ± 10	1752 ± 10	10.0 ± 0.2	819 ± 60	1821 ± 75	9.7 ± 0.2
Zn0-Bottom	820 ± 10	1821 ± 10	9.7 ± 0.2			
Zn2.5-Top	834 ± 10	1646 ± 10	6.8 ± 0.2	839 ± 200	1772 ± 120	7.7 ± 0.5
Zn2.5-Bottom	843 ± 10	1892 ± 10	8.2 ± 0.2			
Zn4.5-Top	781 ± 10	1783 ± 10	6.9 ± 0.2	830 ± 220	1713 ± 75	6.8 ± 0.8
Zn4.5-Bottom	1058 ± 10	1643 ± 10	6.0 ± 0.5			
Zn7-Top	860 ± 10	1932 ± 10	7.6 ± 0.5	906 ± 115	1884 ± 60	6.6 ± 0.8
Zn7-Bottom	1086 ± 10	1819 ± 10	5.0 ± 0.5			
Zn10-Top	1420 ± 10	1881 ± 10	3.7 ± 0.5	1515 ± 200	1906 ± 100	2.8 ± 1.2
Zn10-Bottom	1773 ± 10	1940 ± 10	1.2 ± 0.5			
Zn14-Top	1464 ± 10	2051 ± 10	6.0 ± 0.5	1636 ± 200	1943 ± 100	3 ± 3
Zn14-Bottom	1815 ± 10	1840 ± 10	0.1 ± 0.02			

When the volume fraction of the amorphous phase is close to or larger than 50 vol. %, which was approximately estimated based on the area ratio of the amorphous and crystalline phases extracted from the OM images [21], the corresponding yield stresses of the as-cast $(Cu_{0.5}Zr_{0.5})_{100-x}Zn_x$ ($x = 10$ and 14 at. %) rods become relatively higher (Table 1). As shown in Figure 4c, an average plastic strain of

2.8% ± 1.2% can be achieved for the present $Cu_{45}Zr_{45}Zn_{10}$ rods. Moreover, due to the inhomogeneous distributions of B2 CuZr phase in the glassy matrix mentioned in the Section 3.1, the mechanical properties of Cu-Zr-Zn rods are extremely different, especially for the samples quenched at a cooling rate approaching the critical cooling rate for the fully amorphous phase. As the content of Zn is 10 at. %, the bottom parts show a high yield strength of 1773 ± 10 MPa together with a small plastic strain of 1.2% ± 0.5%, while the respective values for the top parts are approximately 1420 ± 10 MPa and 3.7% ± 0.5%, respectively. The corresponding facture strength changes from 1940 ± 10 MPa to 1881 ± 10 MPa. As shown in Figure 4c and listed in Table 1, the bottom parts of the as-cast $Cu_{43}Zr_{43}Zn_{14}$ rods show a higher yield strength of 1815 ± 10 MPa and a smaller plastic strain of 0.1% ± 0.02%, while the respective values for the top parts are equal to 1464 ± 10 MPa and 6% ± 0.5%, respectively. Meanwhile, the corresponding fracture strength increases from 1840 ± 10 MPa to 2051 ± 10 MPa. Nevertheless, all the present BMG composites exhibit macroscopically-detectable compressive plastic strains and obvious work-hardening.

In order to clarify the deformation mechanisms of Cu-Zr-Zn BMG composites at room temperature, the XRD measurement was conducted on the $Cu_{45}Zr_{45}Zn_{10}$ rods after deformation (Figure 5). It is obvious that more martensitic crystals appear after fracture compared with the as-cast rod (Figure 1), which indicates that B2 CuZr crystals undergo a MT upon loading. It was shown [20,23,25,44] that B2 CuZr crystals can provide a large plastic strain together with pronounced work hardening, being called the transformation-induced plasticity (TRIP) effect. MT within austenitic B2 CuZr crystals during deformation can remarkably release stress concentrations at the interface between the B2 CuZr phase and the glassy matrix [24,25,45–47]. This means that the applied stress is effectively transferred from the glassy matrix to a ductile B2 CuZr phase during deformation, resulting in the restriction of free volume accumulation and the partial release of the elastic energy stored in whole samples [45–47]. As a result, primary shear bands cannot rapidly traverse through an entire sample but need more energy to drive their continuous movement [24,25,45–47], which requires an external stress to activate shear bands. Previous results have shown that there exists an invariant critical stress for continuous shear banding [47]. In order to further investigate shear banding evolutions and MT behaviors during deformation, SEM measurements were also carried out on external surfaces and fractures of the deformed samples (Figure 6). All the samples exhibit a shear fracture and the compressive fracture angle, *i.e.*, the angle between the fracture surface and the loading direction, is approximately equal to 40°–45°. This implies that the compressive fractures of BMG and their composites do not occur along the plane of the maximum shear stress and, accordingly, does not follow the von Mises criterion [48]. As shown in Figure 6, no obvious shear bands can be observed for the deformed $Cu_{49.25}Zr_{49.25}Zn_{1.5}$ sample which is almost fully crystalline. At the same time, martensitic plates appeared within B2 crystals (see the red arrow in Figure 6a). For the deformed $Cu_{47.25}Zr_{47.25}Zn_{4.5}$ sample, not only martensitic plates can be observed within the B2 CuZr phase (see the red arrow and inset in Figure 6b), but also multiple shear bands appeared (see the yellow dotted arrow in Figure 6b). For the fully amorphous $Cu_{43}Zr_{43}Zn_{14}$ sample, only multiple shear bands can be observed (see the yellow dotted arrows in Figure 6c).

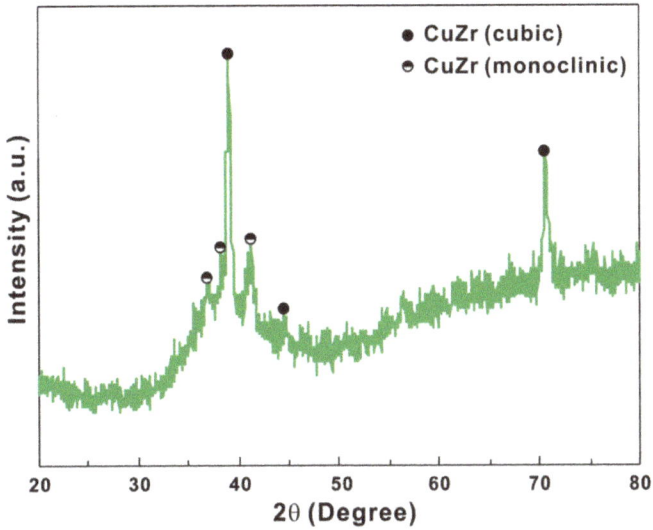

Figure 5. XRD patterns of the as-deformed $Cu_{45}Zr_{45}Zn_{10}$ rods.

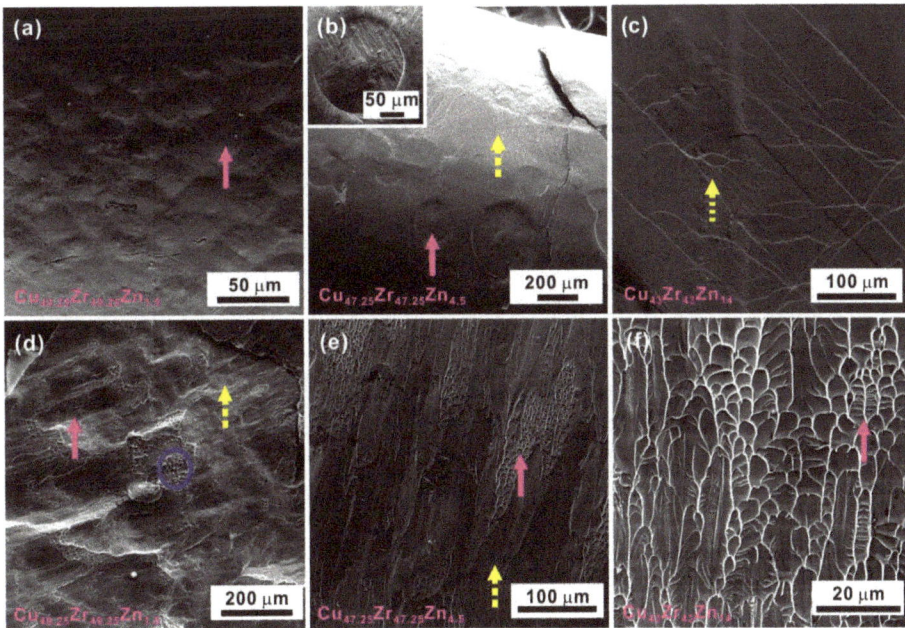

Figure 6. SEM pictures of the external (**a–c**) and fracture (**d–f**) surfaces of the deformed $(Cu_{0.5}Zr_{0.5})_{100-x}Zn_x$ samples: (**a**) $x = 1.5$ at. %; (**b**) $x = 4.5$ at. %; (**c**) $x = 10$ at. %; (**d**) $x = 1.5$ at. %; (**e**) $x = 4.5$ at. %; and (**f**) $x = 14$ at. %. Inset: martensitic plates within the B2 CuZr crystals.

The corresponding fracture images are shown in Figure 6d,f. For almost fully crystalline samples, some slurry-like structures (see the yellow dotted arrows) and dimples were observed (marked by the blue circle) beside a small vein-like structures (see the red arrow in Figure 6d). Cu-Zr-Zn BMG composites after deformation showed not only slurry-like structures but a number of vein- and river-like structures (see the red arrows in Figure 6e, respectively). For the fully amorphous sample, only vein- and river-like structures can be seen (Figure 6f). As we know, fracture surfaces of different BMGs are correlated with their ductility, toughness, deformation rate, and fracture mode [49–52]. Usually, vein-like patterns can be found on the whole fracture surface for BMGs, which is closely related with the significant softening or reduced viscosity in shear bands [49–53]. The origin responsible for the softening has been manifested to be shear-induced structural disordering or temperature rise [49–53]. In our case, by introducing ductile B2 CuZr crystals into the glassy matrix, the MT is quite pronounced during deformation before the formation of shear bands [24,25], resulting in the decrease of the elastic energy density of serration events related with the shear-banding process and then leading to the deceleration of shear-banding [24,25,45,54,55]. Hence, the serration events in the compressive curves become very tiny and almost invisible (Figure 4). Therefore, it is reasonably believed that the second crystals introduced into the glassy matrix could affect the shear-induced structural disordering or temperature rise, resulting in the slightly changes of vein-like patterns. Furthermore, it was shown [56] that typical facture features of ductile metallic alloys, *i.e.*, dimples, can be observed in ductile samples with pure B2 CuZr crystals. Therefore, for CuZr-based BMG composites, dimples, river- and vein-like patterns were observed. Nevertheless, during deformation, the formation of multiple shear bands, the deformation-induced MT within B2 CuZr crystals, and the interactions between the deformation of B2 CuZr crystals and shear-banding for Cu-Zr-Zn BMG composites should be responsible for their ductility and obvious work hardening.

4. Conclusions

In this paper, $(Cu_{0.5}Zr_{0.5})_{100-x}Zn_x$ (x = 0, 1.5, 2.5, 4.5, 7, 10, and 14 at. %) bulk metallic glass (BMG) composites were fabricated. Based on the XRD and OM measurements, it was found that for the CuZr-based alloys with 0, 1.5, and 2.5 at. % Zn, the dominant phase is CuZr martensite together with minor B2 CuZr crystals and amorphous phase. With increasing Zn content from 4.5 at. % to 14 at. %, B2 CuZr and amorphous phases accompanying minor martensitic crystals can be obtained during quenching. All the investigated samples showed a relatively large compressive plastic strain and obvious work-hardening, which should result from the deformation-induced martensitic transformation within B2 CuZr crystals, the multiplication of shear bands in the glassy matrix, and the interactions between the deformation of B2 CuZr crystals and shear-banding.

Acknowledgments: The authors are grateful to B. Bartusch, B. Opitz, L. B. Bruno, M. Frey, S. Donath, and U. Wilke for technical assistance and enlightening discussions. D.Y. Wu and K.K. Song also thanks the support from the Fundamental Research Funds for the Central Universities on Northwestern Polytechnical University (3102014JCQ01090 and 3102015ZY078), the National Natural Science Foundation of China (51501103, 51471135, 51171152 and 51301133) and the Fundamental Research Funds of Shandong University (1050501315006).

Author Contributions: K.S. and C.C. conceived and designed the experiments, wrote and edited the manuscript, and contributed in all activities. D.W. performed XRD, OM, and SEM experiment. R.L., G.W., Y.W, F.W., F.D., Y.S., X.B., I.K., and J.E. helped finishing casting experiments, analyzing the results and revising the manuscript.

Conflicts of Interest: The authors declare no conflict of interest.

References

1. Schuh, C.A.; Hufnagel, T.C.; Ramamurty, U. Mechanical behavior of amorphous alloys. *Acta Mater.* **2007**, *55*, 4067–4109. [CrossRef]
2. Greer, A.L.; Cheng, Y.Q.; Ma, E. Shear bands in metallic glasses. *Mater. Sci. Eng. R* **2013**, *74*, 71–132. [CrossRef]
3. Yavari, A.R.; Lewandowski, J.J.; Eckert, J. Mechanical properties of bulk metallic glasses. *MRS Bull.* **2007**, *32*, 635–638. [CrossRef]
4. Inoue, A.; Takeuchi, A. Recent Progress in Bulk Glassy Alloys. *Mater. Trans.* **2002**, *43*, 1892–1906. [CrossRef]
5. Egami, T.; Iwashita, T.; Dmowski, W. Mechanical Properties of Metallic Glasses. *Metals* **2013**, *3*, 77–113. [CrossRef]

6. Louzguine-Luzgin, D.V.; Louzguina-Luzgina, L.V.; Churyumo, A.Y. Mechanical properties and deformation behavior of bulk metallic glasses. *Metals* **2013**, *3*, 1–22. [CrossRef]

7. Mu, J.; Zhu, Z.W.; Su, R.; Wang, Y.D.; Zhang, H.F.; Ren, Y. *In situ* high-energy X-ray diffraction studies of deformation-induced phase transformation in Ti-based amorphous alloy composites containing ductile dendrites. *Acta Mater.* **2013**, *61*, 5008–5017. [CrossRef]

8. Pan, D.G.; Zhang, H.F.; Wang, A.M.; Hu, Z.Q. Enhanced plasticity in Mg-based bulk metallic glass composite reinforced with ductile Nb particles. *Appl. Phys. Lett.* **2006**. [CrossRef]

9. Fu, H.M.; Wang, H.; Zhang, H.F.; Hu, Z.Q. *In situ* TiB-reinforced Cu-based bulk metallic glass composites. *Scr. Mater.* **2006**, *54*, 1961–1966. [CrossRef]

10. Choi-Yim, H.; Conner, R.D.; Szuecs, F.; Johnson, W.L. Processing, microstructure and properties of ductile metal particulate reinforced $Zr_{57}Nb_5Al_{10}Cu_{15.4}Ni_{12.6}$ bulk metallic glass composites. *Acta Mater.* **2002**, *50*, 2737–2745. [CrossRef]

11. Oh, Y.S.; Kim, C.P.; Lee, S.; Kim, N.J. Microstructure and tensile properties of high-strength high-ductility Ti-based amorphous matrix composites containing ductile dendrites. *Acta Mater.* **2011**, *59*, 7277–7286. [CrossRef]

12. Qiao, J.W.; Sun, A.C.; Huang, E.W.; Zhang, Y.; Liaw, P.K.; Chuang, C.P. Tensile deformation micromechanisms for bulk metallic glass matrix composites: From work-hardening to softening. *Acta Mater.* **2011**, *59*, 4126–4137. [CrossRef]

13. Eckert, J.; Das, J.; Pauly, S.; Duhamel, C. Processing routes, microstructure and mechanical properties of metallic glasses and their composites. *Adv. Eng. Mater.* **2007**, *9*, 443–453. [CrossRef]

14. Choi-Yim, H.; Johnson, W.L. Bulk metallic glass matrix composites. *Appl. Phys. Lett.* **1997**, *71*, 3808–3810. [CrossRef]

15. Ma, H.; Xu, J.; Ma, E. Mg-based bulk metallic glass composites with plasticity and high strength. *Appl. Phys. Lett.* **2003**, *83*, 2793–2795. [CrossRef]

16. Hofmann, D.C.; Suh, J.Y.; Wiest, A.; Duan, G.; Lind, M.L.; Demetriou, M.D.; Johnson, W.L. Designing metallic glass matrix composites with high toughness and tensile ductility. *Nature* **2008**, *451*, 1085–1089. [CrossRef] [PubMed]

17. Wu, F.F.; Li, S.T.; Zhang, G.A.; Wu, X.F.; Lin, P. Plastic stability of metallic glass composites under tension. *Appl. Phys. Lett.* **2013**. [CrossRef]

18. Qiao, J.W.; Zhang, Y.; Liaw, P.K.; Chen, G.L. Micromechanisms of plastic deformation of a dendrite/Zr-based bulk-metallic-glass composite. *Scr. Mater.* **2009**, *61*, 1087–1090. [CrossRef]

19. Gargarella, P.; Pauly, S.; Song, K.K.; Hu, J.; Barekar, N.S.; Khoshkhoo, M.S.; Teresiak, A.; Wendrock, H.; Kühn, U.; Ruffing, C.; *et al.* Ti-Cu-Ni shape memory bulk metallic glass composites. *Acta Mater.* **2013**, *61*, 151–162. [CrossRef]

20. Pauly, S.; Liu, G.; Wang, G.; Kühn, U.; Mattern, N.; Eckert, J. Microstructural heterogeneities governing the deformation of $Cu_{47.5}Zr_{47.5}Al_5$ bulk metallic glass composites. *Acta Mater.* **2009**, *57*, 5445–5453. [CrossRef]

21. Song, K.K.; Pauly, S.; Zhang, Y.; Li, R.; Gorantla, S.; Narayanan, N.; Kühn, U.; Gemming, T.; Eckert, J. Triple yielding and deformation mechanisms in metastable $Cu_{47.5}Zr_{47.5}Al_5$ composites. *Acta Mater.* **2012**, *60*, 6000–6012. [CrossRef]

22. Wu, Y.; Wang, H.; Wu, H.H.; Zhang, Z.Y.; Hui, X.D.; Chen, G.L.; Ma, D.; Wang, X.L.; Lu, Z.P. Formation of Cu-Zr-Al bulk metallic glass composites with improved tensile properties. *Acta Mater.* **2011**, *59*, 2928–2936. [CrossRef]

23. Wu, Y.; Xiao, Y.H.; Chen, G.L.; Liu, C.T.; Lu, Z.P. Bulk Metallic Glass Composites with Transformation-Mediated Work-Hardening and Ductility. *Adv. Mater.* **2010**, *22*, 2770–2773. [CrossRef] [PubMed]

24. Song, K.K.; Pauly, S.; Sun, B.A.; Tan, J.; Stoica, M.; Kühn, U.; Eckert, J. Correlation between the microstructures and the deformation mechanisms of CuZr-based bulk metallic glass composites. *AIP Adv.* **2013**. [CrossRef]

25. Wu, F.F.; Chan, K.C.; Jiang, S.S.; Chen, S.H.; Wang, G. Bulk metallic glass composite with good tensile ductility, high strength and large elastic strain limit. *Sci. Rep.* **2014**. [CrossRef] [PubMed]

26. Liu, Z.Q.; Li, R.; Liu, G.; Su, W.H.; Wang, H.; Li, Y.; Shi, M.J.; Luo, X.K.; Wu, G.J.; Zhang, T. Microstructural tailoring and improvement of mechanical properties in CuZr-based bulk metallic glass composites. *Acta Mater.* **2012**, *60*, 3128–3139. [CrossRef]

27. Liu, Z.Q.; Liu, G.; Qu, R.T.; Zhang, Z.F.; Wu, S.J.; Zhang, T. Microstructural percolation assisted breakthrough of trade-off between strength and ductility in CuZr-based metallic glass composites. *Sci. Rep.* **2014**. [CrossRef] [PubMed]

28. Li, C.J.; Tan, J.; Wang, G.; Bednarčík, J.; Zhu, X.K.; Zhang, Y.; Stoica, M.; Kühn, U.; Eckert, J. Enhanced strength and transformation-induced plasticity in rapidly solidified Zr-Co-(Al) alloys. *Scr. Mater.* **2013**, *68*, 897–900. [CrossRef]

29. Pauly, S.; Das, J.; Bednarcik, J.; Mattern, N.; Kim, K.B.; Kim, D.H.; Eckert, J. Deformation-induced martensitic transformation in Cu-Zr-(Al, Ti) bulk metallic glass composites. *Scr. Mater.* **2009**, *60*, 431–434. [CrossRef]

30. Song, K.K.; Pauly, S.; Zhang, Y.; Sun, B.A.; He, J.; Ma, G.Z.; Kühn, U.; Eckert, J. Thermal stability and mechanical properties of $Cu_{46}Zr_{46}Ag_8$ bulk metallic glass and its composites. *Mater. Sci. Eng. A* **2013**, *559*, 711–718. [CrossRef]

31. Song, K.K.; Pauly, S.; Sun, B.A.; Zhang, Y.; Tan, J.; Kühn, U.; Stoica, M.; Eckert, J. Formation of Cu-Zr-Al-Er bulk metallic glass composites with enhanced deformability. *Intermetallics* **2012**, *30*, 132–138. [CrossRef]

32. Wu, Y.; Song, W.L.; Zhang, Z.Y.; Hui, X.D.; Ma, D.; Wang, X.L.; Shang, X.C.; Lu, Z.P. Relationship between composite structures and compressive properties in CuZr-based bulk metallic glass system. *Chin. Sci. Bull.* **2011**, *56*, 3960–3964. [CrossRef]

33. Sun, Y.F.; Wei, B.C.; Wang, Y.R.; Li, W.H.; Cheung, T.L.; Shek, C.H. Plasticity-improved Zr-Cu-Al bulk metallic glass matrix composites containing martensite phase. *Appl. Phys. Lett.* **2005**. [CrossRef]

34. Song, K.K.; Pauly, S.; Zhang, Y.; Gargarella, P.; Li, R.; Barekar, N.S.; Kühn, U.; Stoica, M.; Eckert, J. Strategy for pinpointing the formation of B2 CuZr in metastable CuZr-based shape memory alloys. *Acta Mater.* **2011**, *59*, 6620–6630. [CrossRef]

35. Kuo, C.N.; Huang, J.C.; Du, X.H.; Chen, Y.C.; Liu, X.J.; Nieh, T.G. Effects of V on phase formation and plasticity improvement in Cu-Zr-Al glassy alloys. *Mater. Sci. Eng. A* **2013**, *561*, 245–251. [CrossRef]

36. Xie, M.T.; Zhang, P.N.; Song, K.K. Thermal Stability and Transformation-mediated Deformability of Cu-Zr-Al-Ni Bulk Metallic Glass Composite. *J. Mater. Sci. Technol.* **2013**, *29*, 868–872. [CrossRef]

37. Song, K.K.; Pauly, S.; Wang, Z.; Kühn, U.; Eckert, J. Effect of TaW particles on the microstructure and mechanical properties of metastable $Cu_{47.5}Zr_{47.5}Al_5$ alloys. *Mater. Sci. Eng. A* **2013**, *587*, 372–380. [CrossRef]

38. Javid, F.A.; Mattern, N.; Khoshkhoo, M.S.; Stoica, M.; Pauly, S.; Eckert, J. Phase formation of $Cu_{50-x}Co_xZr_{50}$ (x = 0–20 at. %) alloys: Influence of cooling rate. *J. Alloys Compd.* **2014**, *590*, 428–434. [CrossRef]

39. Wu, D.Y.; Song, K.K.; Gargarella, P.; Cao, C.D.; Li, R.; Kaban, I.; Eckert, J. Glass-forming Ability, Thermal stability of B2 CuZr phase, and crystallization kinetics for rapidly solidified Cu-Zr-Zn alloys. *J. Alloys Compd.* submitted for publication. **2015**.

40. Seo, J.W.; Schryvers, D. TEM investigation of the microstructure and defects of CuZr martensite. Part I: Morphology and twin systems. *Acta Mater.* **1998**, *46*, 1165–1175. [CrossRef]

41. Schryvers, D.; Firstov, G.S.; Seo, J.W.; Humbeeck, J.V.; Koval, Y.N. Unit cell determination in CuZr martensite by electron microscopy and X-ray diffraction. *Scr. Mater.* **1997**, *36*, 1119–1125. [CrossRef]

42. Firstov, G.S.; Humbeeck, J.V.; Koval, Y.N. Peculiarities of the martensitic transformation in ZrCu intermetallic compound - potential high temperature SMA. *J. Phys. IV* **2001**, *11*, 481–486. [CrossRef]

43. Kalay, I.; Kramer, M.; Napolitano, R. Crystallization Kinetics and Phase Transformation Mechanisms in $Cu_{56}Zr_{44}$ Glassy Alloy. *Metall. Mater. Trans. A* **2015**, *46*, 3356–3364. [CrossRef]

44. Otsuka, K.; Wayman, C.M. *Shape Memory Materials*, 1st ed.; Cambridge University Press: Cambridge, UK, 1998; pp. 1–49.

45. Wang, G.; Pauly, S.; Gorantla, S.; Mattern, N.; Eckert, J. Plastic Flow of a $Cu_{50}Zr_{45}Ti_5$ Bulk Metallic Glass Composite. *J. Mater. Sci. Tech.* **2014**, *30*, 609–615. [CrossRef]

46. Pauly, S.; Gorantla, S.; Wang, G.; Kühn, U.; Eckert, J. Transformation-mediated ductility in CuZr-based bulk metallic glasses. *Nat. Mater.* **2010**, *9*, 473–477. [CrossRef] [PubMed]

47. Han, Z.; Yang, H.; Wu, W.F.; Li, Y. Invariant critical stress for shear banding in a bulk metallic glass. *Appl. Phys. Lett.* **2008**. [CrossRef]

48. Zhang, Z.F.; Eckert, J.; Schultz, L. Difference in compressive and tensile fracture mechanisms of $Zr_{59}Cu_{20}Al_{10}Ni_8Ti_3$ bulk metallic glass. *Acta Mater.* **2003**, *51*, 1167–1179. [CrossRef]

49. Sun, B.A.; Wang, W.H. The fracture of bulk metallic glasses. *Prog. Mater. Sci.* **2015**, *74*, 211–307. [CrossRef]

50. Argon, A.S.; Salama, M. The mechanism of fracture in glassy materials capable of some inelastic deformation. *Mater. Sci. Eng.* **1976**, *23*, 219–230. [CrossRef]

51. Sun, B.A.; Tan, J.; Pauly, S.; Kuhn, U.; Eckert, J. Stable fracture of a malleable Zr-based bulk metallic glass. *J. Appl. Phys.* **2012**. [CrossRef]
52. Deibler, L.A.; Lewandowski, J.J. Model experiments to mimic fracture surface features in metallic glasses. *Mater. Sci. Eng. A* **2010**, *527*, 2207–2213. [CrossRef]
53. Lewandoski, J.J.; Greer, A.L. Temperature rise at shear bands in metallic glasses. *Nat. Mater.* **2006**, *5*, 15–18. [CrossRef]
54. Wu, F.F.; Chan, K.C.; Li, S.T.; Wang, G. Stabilized shear banding of ZrCu-based metallic glass composites under tensile loading. *J. Mater. Sci.* **2014**, *49*, 2164–2170. [CrossRef]
55. Tong, X.; Wang, G.; Yi, J.; Ren, J.L.; Pauly, S.; Gao, Y.L.; Zhai, Q.J.; Mattern, N.; Dahmen, K.A.; Liaw, P.K.; *et al.* Shear avalanches in plastic deformation of a metallic glass composite. *Int. J. Plast.* **2016**, *77*, 141–155. [CrossRef]
56. Song, K.K. Synthesis, microstructure, and deformation mechanisms of CuZr-based bulk metallic glass composites. In *Doctoral Thesis Work*; TU Dresden: Dresden, Germany, 2013.

metals

MDPI

Article

Effect of Milling Time and the Consolidation Process on the Properties of Al Matrix Composites Reinforced with Fe-Based Glassy Particles

Özge Balcı [1,2], Konda Gokuldoss Prashanth [2,†,*], Sergio Scudino [2], Duygu Ağaoğulları [1], İsmail Duman [1], M. Lütfi Öveçoğlu [1], Volker Uhlenwinkel [3] and Jürgen Eckert [2,4]

[1] Particulate Materials Laboratories (PML), Department of Metallurgical and Materials Engineering, İstanbul Technical University, 34469 İstanbul, Turkey; balciozg@itu.edu.tr (O.B.); bozkurtdu@itu.edu.tr (D.A.); iduman@itu.edu.tr (I.D.); ovecoglu@itu.edu.tr (M.L.O.)

[2] Institute for Complex Materials, IFW Dresden, 270116 Dresden, Germany; s.scudino@ifw-dresden.de (S.S.); j.eckert@ifw-dresden.de (J.E.)

[3] Institut für Werkstofftechnik, Universität Bremen, D-28359 Bremen, Germany; uhl@iwt.uni-bremen.de

[4] TU Dresden, Institut für Werkstoffwissenschaft, D-01062 Dresden, Germany

* Author to whom correspondence should be addressed; kgprashanth@gmail.com; Tel.: +46-7277-91612; Fax: +46-2626-6125.

† Present Address: R&D Engineer, Additive manufacturing Center, Sandvik AB, 81181 Sandviken, Sweden; prashanth.konda_gokuldoss@sandvik.com.

Academic Editors: K. C. Chan and Jordi Sort Viñas

Received: 27 March 2015; Accepted: 22 April 2015; Published: 27 April 2015

Abstract: Al matrix composites reinforced with 40 vol% $Fe_{50.1}Co_{35.1}Nb_{7.7}B_{4.3}Si_{2.8}$ glassy particles have been produced by powder metallurgy, and their microstructure and mechanical properties have been investigated in detail. Different processing routes (hot pressing and hot extrusion) are used in order to consolidate the composite powders. The homogeneous distribution of the glassy reinforcement in the Al matrix and the decrease of the particle size are obtained through ball milling. This has a positive effect on the hardness and strength of the composites. Mechanical tests show that the hardness of the hot pressed samples increases from 51–155 HV, and the strength rises from 220–630 MPa by extending the milling time from 1–50 h. The use of hot extrusion after hot pressing reduces both the strength and hardness of the composites: however, it enhances the plastic deformation significantly.

Keywords: metallic glasses; composites; powder metallurgy; mechanical characterization

1. Introduction

Al-based metal matrix composites (MMCs) have attracted considerable interest due to their superior properties, including high strength and good fatigue and wear resistance [1–3]. They are advanced engineering materials able to meet the increasing demand for structural and thermal applications, particularly in the aerospace and automotive industries [4,5]. Al-based MMCs offer a unique combination of properties, including the ductility of the matrix and the strength of the reinforcement, that cannot be found in conventional unreinforced materials [4,6]. Various materials comprised of ceramics (e.g., SiO_2, Al_2O_3, SiC, TiC, TiB_2, ZrB_2 and AlN), metallic glasses and complex metallic alloys have been successfully used as reinforcements in Al-based MMCs in the form of fibers, flakes or particulates [7–15]. Amongst them, particulate-reinforced MMCs are particularly attractive due to their easier fabrication routes and lower costs compared to fibers or flakes [4,13].

Metallic glasses have been recently proposed as an effective type of reinforcement due to their exceptional mechanical properties, such as high strength, hardness and corrosion resistance [16–18].

Metallic glasses are believed to be more compatible with the metal matrix and may lead to improved interface strength between the matrix and reinforcement than their ceramic counterparts [16–18]. In addition, the sintering process conducted within supercooled liquid (SCL), where metallic glasses display a significant decrease of viscosity, can assist with the consolidation, resulting in bulk samples with reduced porosity [18,19]. Thus, in order to obtain highly-dense materials with increased mechanical properties, Al-, Zr-, Fe-, Ni- and Cu-based metallic glass particles or ribbons have been used as reinforcements in Al-based metal matrix composites [17–24]. Amongst the different types of metallic glasses, Fe-based glasses are of considerable interest, because of their ultrahigh strength, good corrosion resistance, low cost, excellent soft magnetic properties, good glass forming ability and large SCL region [25–29]. Aljerf *et al.* reported that the use of $[(Fe_{1/2}Co_{1/2})_{75}B_{20}Si_5]_{96}Nb_4$ glassy particles as reinforcement in the Al-6061 alloy leads to a remarkable combination of high strength and plasticity [19]. Similar results have been recently reported for Al-2024 matrix composites reinforced with $Fe_{73}Nb_5Ge_2P_{10}C_6B_4$ glassy particles [30].

Powder metallurgy is one of the methods successfully used for the preparation of MMCs [1,7,31]. The main advantage of powder metallurgy over other techniques is the low processing temperature, which may prevent unwanted interfacial reactions between the matrix and reinforcement and which permits the economic feasibility of large-scale production, thus allowing the commercial processing of MMCs [32,33]. Furthermore, it provides a homogeneous distribution of the reinforcements within the matrix, and it enables a high degree of control over the product microstructure (volume fraction, size, shape, *etc.*), which is comparatively limited in the casting or diffusion welding routes [7,24]. The current studies related to the metallic glass-reinforced Al-based MMCs produced via powder metallurgy are mainly focused on the effect of the glassy particle content on the properties of the composites [23–25]. On the other hand, the effect of microstructural modifications, such as particle shape and interparticle distance, induced by mechanical treatments, like ball milling, as well as the influence of the consolidation method on the properties of Al-based MMCs have received relatively little attention.

In this study, Al-based MMCs reinforced with Fe-based glassy particles have been produced via powder metallurgy and the effects of milling time (1, 10, 30 and 50 h) and the consolidation process (hot press or hot press followed by hot extrusion) on the microstructure and mechanical properties are investigated.

2. Experimental Section

2.1. Raw Materials

Glassy particles with a nominal composition of $Fe_{50.1}Co_{35.1}Nb_{7.7}B_{4.3}Si_{2.8}$ were produced by high-pressure N_2 gas atomization. The samples produced by this method are powders with sizes ranging from 38–112 μm. Al powders with a purity of 99.5% and an average particle size of 125 μm were used as the matrix material.

2.2. Mechanical Milling

Milling experiments on the powder mixtures comprised of pure Al and 40 vol% glassy particles were performed using a Retsch PM400 planetary ball mill (Retsch, Dusseldorf, Germany) equipped with hardened steel balls and vials and without any process control agents. The powders were milled at room temperature for 1, 10, 30 and 50 h using a ball-to-powder mass ratio (BPR) of 10:1 and at a milling speed of 150 rpm. Such a low milling speed was used only to have uniform dispersion of the reinforcement in the matrix, unlike conventional mechanical milling or alloying. Milling was carried out as a sequence of 15-min milling intervals interrupted by 15-min breaks to avoid a strong temperature rise during milling. All sample handling was carried out in a Braun MB 150B-G glove box under purified Ar atmosphere (less than 0.1 ppm O_2 and H_2O) in order to minimize atmospheric contamination.

2.3. Consolidation

Consolidation of the composite powders was done by uni-axial hot pressing (HP) or hot pressing followed by hot extrusion (HE) under Ar atmosphere at 673 K and 640 MPa. Hot pressing time and the hot extrusion ratio were 30 min and 4:1, respectively.

2.4. Characterization of the Powders and Consolidated Samples

Phase analysis of the powders and consolidated samples was performed by the X-ray diffraction technique (XRD) using a D3290 PANalytical X'pert PRO (PANalytical, Almelo, The Netherlands) with Co-Kα radiation (λ = 0.17889 nm) in the Bragg–Brentano configuration. After applying a series of metallographic treatments, microstructural characterization and elemental mapping of the consolidated samples were carried out by scanning electron microscopy (SEM, Zeiss, Oberkochen, Germany) using a Gemini 1530 microscope (operated at 15 kV) equipped with energy-dispersive X-ray detection (EDX). The matrix ligament size (λ = L/N) was calculated from the arithmetic mean of ten measurements by superposing random lines on the high magnification SEM micrographs of the composites. It is determined by the total length falling in the matrix (L) and by counting the number of matrix region intercepts per unit length of test line (N). The thermal behavior of the powders was investigated by differential scanning calorimetry (DSC) with a Perkin-Elmer DSC7 calorimeter (Perkin Elmer, Waltham, MA, USA) at a heating rate of 20 K/min under a continuous flow of purified Ar.

The experimental densities of the consolidated samples were evaluated by the Archimedes principle, and the relative densities were calculated as the percent value from the ratio of the experimental to the theoretical density of Al matrix composite reinforced with 40 vol% $Fe_{50.1}Co_{35.1}Nb_{7.7}B_{4.3}Si_{2.8}$ (4.43 g/cm^3). Microhardness measurements were conducted by a computer-controlled Struers Duramin 5 Vickers hardness tester using a load of 10 g and indenter dwell time of 10 s. The microhardness test result of each sample includes the arithmetic mean of twenty successive indentations and standard deviations. Optical micrographs of the representative hardness indentations performed on the glassy phase distributed Al matrix were also given for each sample. Since the hardness comparison of the composites was aimed at a constant load, measurements were not directly applied on the glassy phase, whose indentation requires a higher value of load than the one utilized. Five different cylindrical specimens 2 mm diameter and 4 mm length were prepared from each of the hot pressed and hot extruded samples and tested at room temperature using an Instron 8562 testing facility under quasistatic compressive loading (strain rate 8 × 10^{-4} s^{-1}). Both ends of these specimens were carefully polished to make them parallel to each other prior to the compression tests. The strain during compression was measured directly on the specimens using a Fiedler laser-extensometer.

3. Results and Discussion

3.1. Processing and Characterization of the Composite Powders

Figure 1a–c illustrates the XRD patterns of the starting and composite powders milled for different times. Figure 1a shows the XRD patterns of the gas-atomized $Fe_{50.1}Co_{35.1}Nb_{7.7}B_{4.3}Si_{2.8}$ and pure Al (The International Centre for Diffraction Data (ICDD) Card No: 01-072-3440) powders. The glassy powder is not completely amorphous and shows the typical broad maxima characteristic for glassy materials at angles between 2θ = 40° and 2θ = 60° together with small diffraction peaks at about 2θ = 41°, 45°, 50° and 53°. As previously reported for gas-atomized Al-based glassy powders, the cooling rate may not be sufficient to completely suppress the formation of crystalline phases during gas atomization [34]. Figure 1b shows the XRD patterns of the composite powders with 40 vol% $Fe_{50.1}Co_{35.1}Nb_{7.7}B_{4.3}Si_{2.8}$ milled for 1, 10, 30 and 50 h, revealing sharp Bragg peaks, which correspond to the Al matrix. No peaks belonging to additional phases can be observed even after milling for 50 h. Moreover, the Al peaks are broadened, and their intensities gradually decrease with increasing milling time from 1–50 h, indicating a gradual decrease of the crystallite size, as well as the increase in the lattice strain. The weak and diffuse peak at about 2θ = 50°–53° indicates the existence of the glassy phase in the Al matrix and proves the amorphous structure of the glassy phase after milling for 1 h. Further milling does not result in significant structural changes.

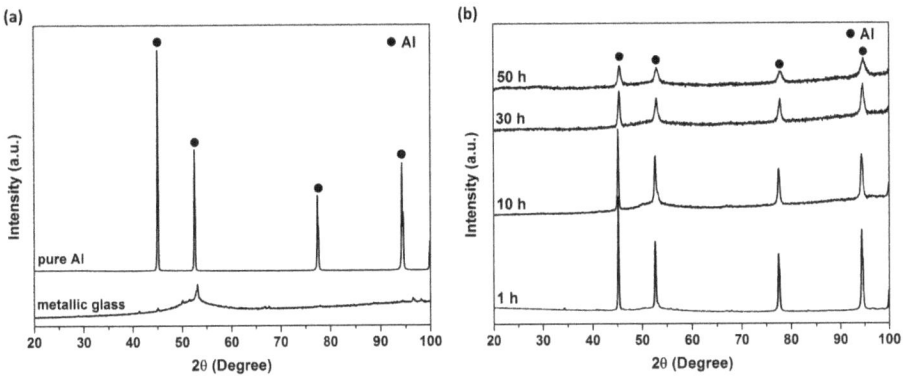

Figure 1. XRD patterns of the starting and composite powders milled for different times: (**a**) pure Al and metallic glass powders; and (**b**) composite powders milled for 1, 10, 30 and 50 h.

Figure 2a displays the DSC scan of the gas-atomized $Fe_{50.1}Co_{35.1}Nb_{7.7}B_{4.3}Si_{2.8}$ glassy powder. The scan displays a glass transition (T_g) at 838 K followed by a crystallization event with the onset (T_x) at 853 K. This is similar to what was observed for the $[(Fe_{1/2}Co_{1/2})_{75}B_{20}Si_5]_{96}Nb_4$ metallic glass, which has T_g and T_x values of 821 and 861 K, respectively [19]. DSC experiments were also carried out for the milled composite powders (Figure 2b) to analyze the effect of milling on T_g and T_x and to select the appropriate consolidation temperature. The DSC curves reveal a glass transition temperature (T_g) followed by the supercooled liquid (SCL) region (defined as $\Delta T_x = T_x - T_g$) before crystallization occurs at higher temperatures (T_x). The SCL regions lie below the melting temperature of the Al matrix, except for the powder milled for 1 h, which has T_x above 920 K. T_g values peaking with very weak endotherms are located at about 837, 800, 733 and 720 K, respectively, for the powders milled for 1, 10, 30 and 50 h. The prolonged milling time from 1–50 h results in a shift of T_g of about 117 K to lower temperatures. Continuous mechanical deformation of powder particles disturbs the bonds, creates dislocations, increases the fresh reactive surfaces of the particles and improves the chemical reactivity [32]. The shift of the glass transition temperature is larger (104 K) for the milling period between 1 and 30 h, whereas it is smaller for the powder milled between 30 and 50 h (13 K). Therefore, milling for 50 h is sufficient to get a desirable reduction in the T_g value of the composite powder, which correspondingly increases the temperature range of the SCL region. Figure 2b also displays broad exothermic events (corresponding to the crystallization of the glassy phase) with onset (T_x) at about 853, 840 and 837 K for the powders milled for 10, 30 and 50 h, therefore decreasing gradually with increasing the milling time. T_x determines the upper temperature limit for the sintering process, because at T_x, metallic glasses lose their liquid-like behavior, and the viscosity increases as crystallization starts [35].

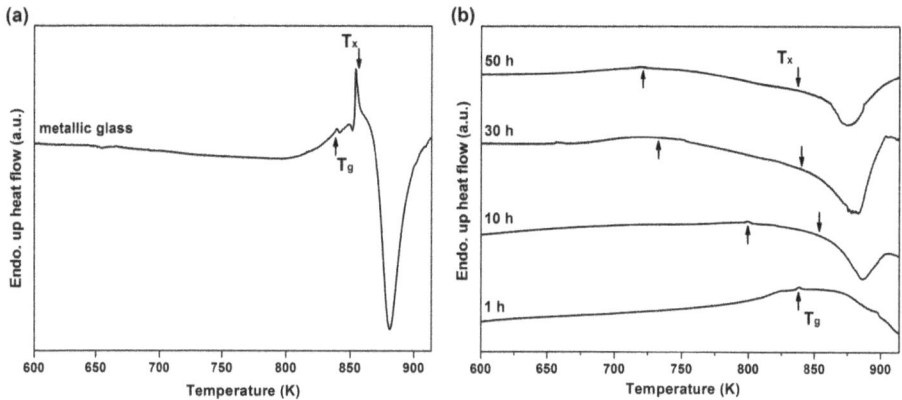

Figure 2. DSC scans (**a**) of the gas-atomized $Fe_{50.1}Co_{35.1}Nb_{7.7}B_{4.3}Si_{2.8}$ powder and (**b**) of the composite powders milled for 1, 10, 30 and 50 h.

The exothermic peaks in Figure 2b are rather broad, in contrast with what is generally observed for the crystallization of metallic glasses. This can be explained by the peak overlap for the peaks related to the crystallization of the glass and to the reaction of the Al matrix with the metallic glass induced by partial Al melting. When the crystallization temperature of the glassy phase in the composites is higher than that of Al melting, the corresponding DSC scan exhibits an endothermic event corresponding to the Al melting at about 835–897 K [21]. On the other hand, similar crystallization and Al melting temperatures were observed for Al-based composites reinforced with $Ni_{60}Nb_{40}$ metallic glass particles [18]. Similar to the present study, they showed broad exotherms in the DSC scans instead of the sharp peaks corresponding to the crystallization of the metallic glass [18]. The reason for this behavior was attributed to the amount of heat release originating from the $Ni_{60}Nb_{40}$ reinforcement, which was diluted in the Al matrix [18]. Furthermore, the reaction of the Al matrix with the metallic glass was observed, supported by the appearance of low-intensity diffraction peaks of aluminide intermetallics ($NiAl_3$ and $NbAl_3$) in the XRD patterns after annealing the composites at temperatures between 893 and 913 K [18].

3.2. Characterization of the Consolidated Samples

Figure 3a exhibits the XRD patterns of the hot pressed composites milled for different times. The patterns display the peaks of the Al matrix along with the diffuse amorphous halo at about $2\theta = 53°$. This indicates that, after consolidation, the glassy reinforcement remains amorphous, corroborating the DSC results showing the exothermic event due to crystallization of the glassy phase. Figure 3c shows the XRD patterns of the hot pressed and hot extruded composites milled for different times. The comparison of Figures 3a and 3c reveals the effect of hot extrusion on the microstructure: after extrusion, the characteristic Al peaks become narrower, and their intensities increase with respect to the hot pressed samples, indicating an increase in the crystallite size, probably due to the stress-induced grain growth during secondary consolidation [36]. As seen in Figure 3c, a small amount of Al_5Fe_2 (ICDD Card No: 00-029-0043) is formed in the extruded specimens. The amount of the Al_5Fe_2 phase increases with increasing the milling time, as expected from the contribution of deformation-induced crystallization.

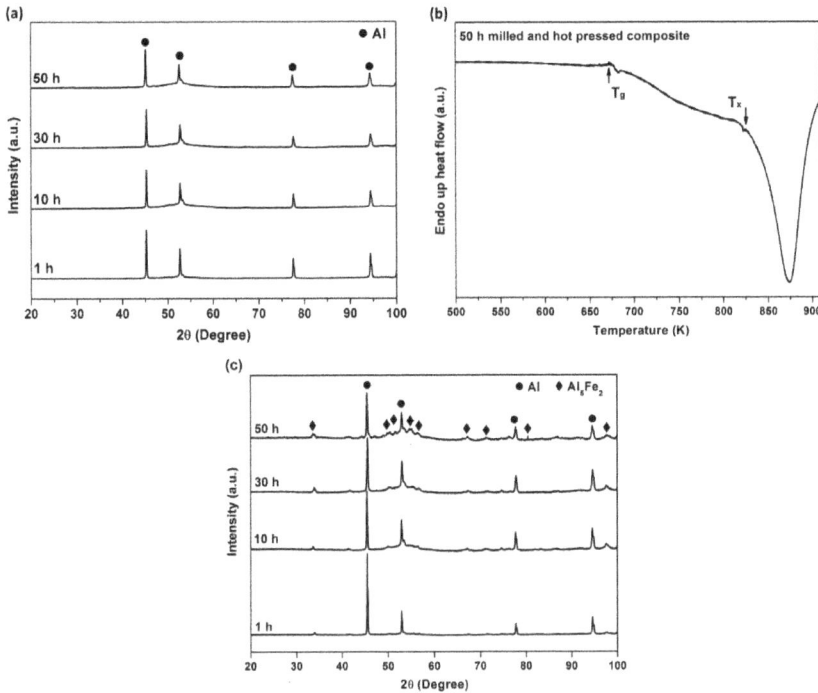

Figure 3. XRD patterns and DSC scan of the milled and consolidated samples: (**a**) XRD patterns of the hot pressed, (**b**) DSC scan of the 50-h milled and hot pressed and (**c**) XRD patterns of the hot pressed and hot extruded composites.

Figure 4a–d shows the SEM micrographs and the EDX elemental maps of the hot pressed composites. The SEM micrographs consist of dark and bright regions corresponding to the Al matrix and Fe-based glassy particles, as shown by the EDX analysis. The images also reveal the irregular shape and size of the Fe-based glassy particles embedded in the continuous Al matrix without observable cracks. The hot pressed composite produced from the powders milled for 1 h is characterized by a microstructure containing spherical glassy particles with an average size of 34 $\pm$ 8 μm and only a few pores (marked by arrows in Figure 4a). With increasing the milling time to 50 h (Figure 4d), repeated cold-welding, fracturing and re-welding [32] led to the formation of a heterogeneous microstructure consisting of large spheroidal particles along with small layered glassy particles. This gives rise to a smaller average size and to a broader particle size distribution (15 $\pm$ 12 μm). The decrease in the particle size of the Fe-based glassy particles resulting from the mechanically-induced fragmentation and the corresponding decrease of the inter-particle distance can be clearly seen in Figure 4a–d. The reduction of the distance between the particles can contribute considerably to the strength of the composites, because the matrix/particle interface can effectively inhibit dislocation movement [14].

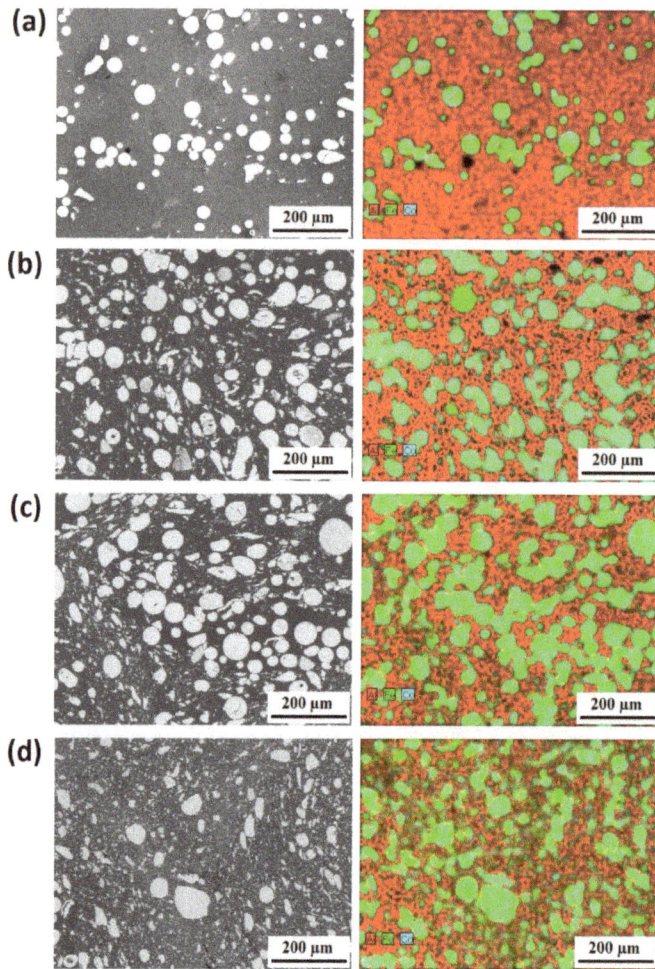

Figure 4. SEM micrographs and EDX maps of the hot pressed composites obtained from (**a**) 1-h, (**b**) 10-h, (**c**) 30-h and (**d**) 50-h milled powders.

Figure 5a–d illustrates the SEM micrographs of the hot extruded composites. The effects of milling on the microstructures of hot extruded samples are very similar to what was observed for the hot pressed samples. Compared to the hot pressed composites (Figure 4), hot extrusion leads to samples with enhanced density, as no pores are observed in the SEM micrographs (Figure 5). Although the XRD patterns in Figure 3c show the Al_5Fe_2 intermetallic product after hot extrusion, the matrix/particle interface of the hot extruded samples is clean, and no reaction zone is observed in Figure 5a–d. The Al_5Fe_2 phase cannot be detected by SEM, most likely because of the ultrafine dimension, as supported by the broad XRD peaks for this phase (Figure 3c). Furthermore, elongation of some large glassy particles occurs during hot extrusion with respect to the corresponding particles observed previously in the hot pressed composites (Figure 4a–d).

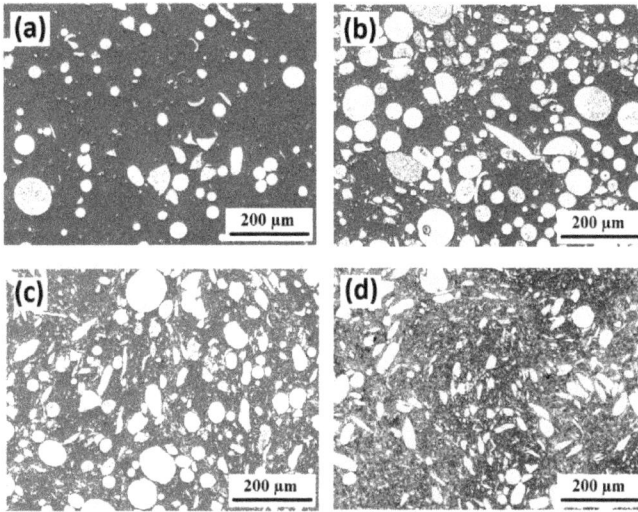

Figure 5. SEM micrographs of the hot pressed and hot extruded composites obtained from (**a**) 1-h, (**b**) 10-h, (**c**) 30-h and (**d**) 50-h milled powders.

The relative density (theoretical-experimental density) of the hot pressed and hot pressed and hot extruded composites are respectively ~98% and ~99%, as also supported by the SEM micrographs given in Figures 4 and 5, which show few or no pores. Therefore, hot pressing and hot extrusion provide nearly fully-dense specimens, since the combination of temperature and pressure stimulates the accelerated densification process and the elimination of residual porosity [37]. No significant effect of the milling time on the relative density of the composites is observed.

3.3. Mechanical Properties of the Consolidated Samples

Figure 6a,b displays the microhardness values of the composites as a function of milling time along with the optical micrographs of the indentations (under the same applied load). With increasing the milling time from 1–50 h, the microhardness increases from 51 ± 2.3–155 ± 6.5 HV for the hot pressed composites (Figure 6a) and from 45.1 ± 2.2–144 ± 10.5 HV for the hot pressed and hot extruded samples (Figure 6b). For a given milling time, the hardness of the hot pressed samples is higher than those of the hot pressed and hot extruded material. This can be explained by the smaller particle sizes (15 ± 12 μm) and more homogeneous distribution of the hard Fe-based glassy particles throughout the hot pressed microstructure than those consolidated by hot extrusion (22 ± 10 μm) [38]. A maximum average hardness value of 155 HV was measured for the 50-h milled and hot pressed sample, which is compatible with its microstructure, presenting a uniform distribution of fine glassy particles without significant clustering (Figure 4d). Moreover, the pyramidal indentations are not distorted, and no significant cracks are observed around the indentation, despite the difference in hardness between the soft matrix and the hard reinforcing phase (Figure 7). This indicates that the Fe-based glassy particles are strongly bonded with the Al matrix, which can play a significant role in enhancing the mechanical properties of the composites [30].

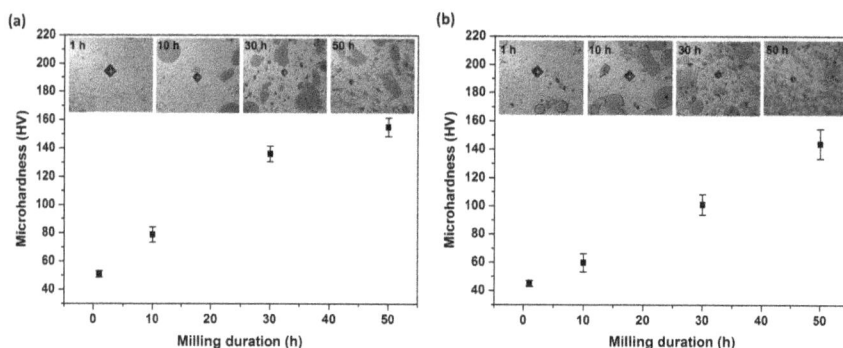

Figure 6. Microhardness values as a function of milling time and optical micrographs of the indentation marks obtained from: (**a**) hot pressed and (**b**) hot pressed and hot extruded composites.

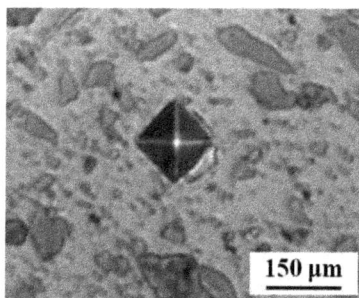

Figure 7. Optical microscopy image showing the indentation mark taken from the 50-h milled and hot pressed composite.

Room temperature compression true stress-true strain curves of the tests under quasistatic loading for the composite materials are shown in Figure 8. Data for the yield strength (0.2% offset), ultimate strength, strain at fracture and microhardness of the composites are summarized in Table 1. The yield and ultimate strength values of the composites are higher than those of the hot pressed and hot extruded pure Al (125 and 155 MPa, respectively), indicating the positive effect of Fe-based metallic glass reinforcements on the mechanical properties of Al [27,30]. These results also reveal the significant effect of milling and the consolidation process on the strength and plasticity of the composites. With increasing milling time from 1–50 h, the strength of the composites increases remarkably for both consolidation processes, showing a similar tendency as the hardness.

The ultimate strength of the hot pressed materials increases from 220 MPa for the composite milled for 1 h to 340, 440 and 630 MPa for the composites milled for 10, 30 and 50 h, respectively. In contrast, the plastic deformation of the hot pressed materials decreases from 35% for the composite milled for 1 h to 8.5%, 6.5% and 1.2% for the hot pressed composites produced from the powders milled for 10, 30 and 50 h. On the other hand, the ultimate strength of the hot extruded materials increases from 220 MPa for the composite milled for 1 h to 295 MPa for the composites milled for 10 h, while retaining appreciable plastic deformation, reaching an ultimate strain of 30% before fracture occurs. The 30 h milled, hot pressed and hot extruded composite gives a good combination of high strength and remarkable plasticity, exhibiting yield and ultimate strengths at 285 and 390 MPa and fracture strain at 21%.

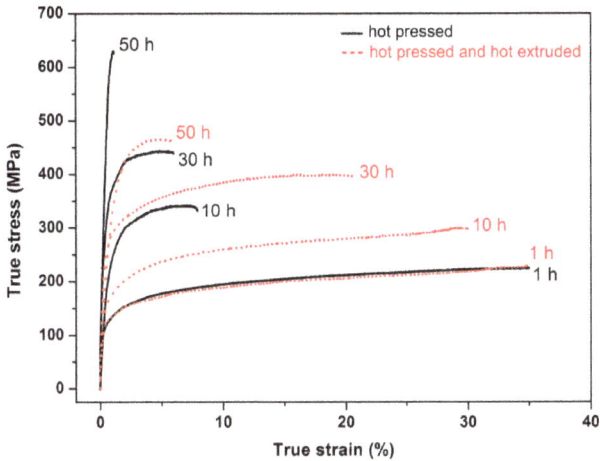

Figure 8. Room temperature compression true stress-true strain curves of the consolidated samples obtained from 1-, 10-, 30- and 50-h milled powders: (━) hot pressed and (━) hot pressed and hot extruded composites.

Table 1. Mechanical properties of Al metal matrix composites (MMCs) reinforced with 40 vol% Fe50.1Co35.1Nb7.7B4.3Si2.8 metallic glass particles produced by mechanical milling for different times, hot pressing (HP) and hot pressing followed by hot extrusion (HP + HE) at 673 K and 640 MPa, as derived from the uniaxial compression and Vickers microhardness tests.

Milling time (h)	Consolidation process	Yield Strength (MPa)	Ultimate Strength (MPa)	Strain at Fracture (%)	Vickers Microhardness (HV)
1	HP	125	220	35.0	51 ± 2
	HP + HE	125	220	35.0	45 ± 2
10	HP	275	340	8.5	79 ± 5
	HP + HE	180	295	30.5	60 ± 6
30	HP	350	440	6.5	136 ± 5
	HP + HE	285	390	21.0	101 ± 7
50	HP	595	630	1.2	155 ± 7
	HP + HE	340	470	6.5	144 ± 11

These results show that the use of hot extrusion after hot pressing slightly reduces both the strength and hardness of the composites, while raising the plastic deformation, which ranges between 35 and 6.5%. The values of strength for the 50-h milled and consolidated samples are higher than those reported for Al/Al$_2$O$_3$ composites [39] and Al MMCs reinforced with different types of metallic glasses [21,25]. This indicates that Fe-based glassy particles may be a valid alternative to the conventional ceramic or other metallic glass reinforcements.

As revealed by the XRD patterns and SEM micrographs of the powders and consolidated samples (Figures 1 and 3, Figures 4 and 5), milling leads to the reduction of the reinforcement particle size (Figure 9), which is most likely the reason for the increase of the yield and ultimate strengths with increasing milling time (Table 2). The reduction of the particle size leads to an improvement in the mechanical properties if a clean particle-matrix interface is obtained [4]. The average inter-particle distance of the composites decreases to 3 μm after milling for 50 h. This observation can be translated into an effective grain size of the Al matrix in the composites smaller than 3.5 µm [30]. Besides the reduced particle size, the improvement of the strength and hardness of the composites resulting from milling can also be ascribed to the corresponding reduced distance between the glassy particles, as observed in the SEM micrographs (Figures 4 and 5). This phenomenon was reported for the Al and brass matrix composites reinforced with different types of metallic glasses [18,35].

Table 2. The average inter-particle distance for the hot pressed and hot pressed and hot extruded composites as a function of milling time.

Milling Time (h)	Average inter-particle distance (µm)	
	Hot pressed composites	Hot pressed and hot extruded composites
1	21 ± 7	26 ± 7
10	8 ± 1	10 ± 2
30	6 ± 2	6 ± 1
50	3 ± 1	4 ± 1

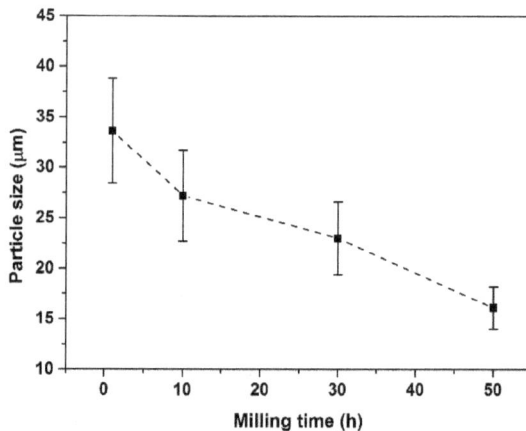

Figure 9. Average particle sizes of the hot pressed composites milled for different times.

3.4. Evaluation of the Mechanical Behavior by Theoretical Predictions

The prediction of the mechanical properties of a composite is an important prerequisite for material design and application. Recently, a model proposed by Gurland [40], which considers ($\sigma \propto V^{1/3} d^{-1/2}$) the combined effects of the reinforcement volume fraction (V) and the particle size (d), has been rearranged to consider the effect of the matrix ligament size λ, as [35]:

$$\sigma \propto V^{1/3} \lambda^{-1/2} \tag{1}$$

Figure 10a,b shows the yield strength as a function of $V^{1/3}\lambda^{-1/2}$ for the present consolidated samples. The matrix ligament size depends on the milling time of the composite powders, and for the current volume fraction of reinforcement (40 vol%), the relationship between strength and $V^{1/3}\lambda^{-1/2}$ is linear for both hot pressed and hot extruded samples. The reduction of λ can give a considerable contribution to the strength of the composites because the matrix/particle interface can effectively reduce the movement of dislocations [15]. This corroborates the validity of the model (Equation (1)) for the present composites and further indicates that the strength can be accurately modeled by considering the effect of the matrix ligament size. Furthermore, the reason for the differences in strength and plasticity of the hot pressed and hot extruded composites can be explained by the grain growth of the Al matrix and the formation of particle clusters during hot extrusion, as previously observed in Figures 3c and 5. The grain refinement provides an additional strengthening effect resulting from the dislocation piling-up at the grain boundaries [41–43].

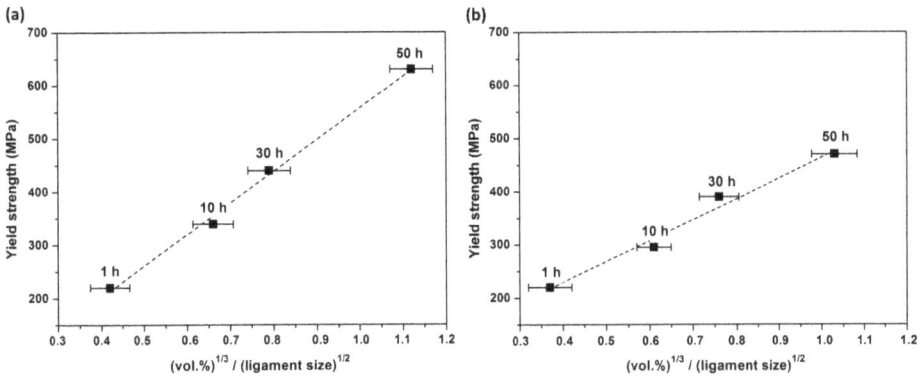

Figure 10. Yield strength as a function of volume fraction and matrix ligament size ($\sigma \propto V^{1/3} \lambda^{-1/2}$) for consolidated samples: (**a**) milled and hot pressed and (**b**) milled, hot pressed and hot extruded composites.

Eventually, composites fabricated by milling for different times, hot pressing and hot extrusion can be considered as potential candidates for applications demanding various mechanical properties. The hot pressed composites exhibit high strength and high hardness, which are of primary importance for structural applications [1]. On the other hand, hot extruded MMCs are potential materials for aerospace and automotive industries due to a superior balance of improved strength and plasticity [1].

4. Conclusions

Al matrix composites reinforced with 40 vol% $Fe_{50.1}Co_{35.1}Nb_{7.7}B_{4.3}Si_{2.8}$ glassy particles have been successfully produced via powder metallurgy routes using different milling times and consolidation processes. Based on the results of the present study, the following conclusions can be drawn:

1. Milled and hot pressed composites revealed no formation of intermetallic compounds even after milling for 50 h, within the sensitivity of the XRD measurements. Only a small amount of the Al_5Fe_2 intermetallic compound was observed in the XRD patterns of all milled, hot pressed and hot extruded samples. DSC scans revealed that milling time changes the overall crystallization behavior of the composite powders.
2. Ball milling resulted in the decrease of the grain sizes of the Al matrix, in the reduction of the particle size of the glassy reinforcements and in their homogeneous distribution in the Al matrix. This has a positive effect on the hardness and strength of the composites produced by both hot pressing and hot pressing followed by hot extrusion. With increasing the milling time from 1–50 h, the microhardness values of the hot pressed and hot extruded samples increase from 51 ± 2.26–155 ± 6.5 HV and from 45.1 ± 2.24–144 ± 10.5 HV, respectively. With increasing milling time from 1–50 h, the strength of the composites increases remarkably for both consolidation processes, showing a similar tendency as observed for the hardness.
3. The ultimate strength of the hot pressed materials increases from 220 MPa for the composite milled for 1 h to 340, 440 and 630 MPa for the composites milled for 10, 30 and 50 h, respectively. The 50-h milled and hot pressed composite exhibits small plastic deformation of 1.2% and a maximum strength of 630 MPa, which is in agreement with the highest Vickers microhardness of 155 ± 6.5 HV among all composites.

4.   The use of hot extrusion after hot pressing slightly reduces both the strength and hardness of the composites, while raising the plastic deformation ranging between 35 and 6.5%. The 30-h milled, hot pressed and hot extruded composite gives a combination of high strength (390 MPa) and remarkable plasticity (21%).

The strength of both the hot pressed or hot pressed and hot extruded composites can be accurately described by a simple model considering the effect of matrix ligament size on the strengthening of the composites.

Acknowledgments: Özge Balcı would like to express her appreciation to German Academic Exchange Service (DAAD) for the financial support during her stay at IFW Dresden.

Author Contributions: Özge Balcı performed all the experiments and characterization studies and created the initial draft. Konda Gokuldoss Prashanth aided in SEM analyses and mechanical tests and conceived the final manuscript. Sergio Scudino formulated the idea of this research, designed the experiments and supervised the discussion of the results. Duygu Ağaoğulları, İsmail Duman and M. Lütfi Öveçoğlu aided the review of the paper and provided critical comments. Volker Uhlenwinkel prepared the gas-atomized glassy powders used in the experiments. Jürgen Eckert contributed to the overall development of the main concepts of this study.

Conflicts of Interest: The authors declare no conflict of interest.

References

1.   Epple, M. *Biomaterialien und Biomineralisation—Eine Einführung für Naturwissenschaftler, Mediziner und Ingenieure*; Springer: Wiesbaden, Germany, 2003.
2.   Miracle, D.B. Metal matrix composites—From science to technological significance. *Compos. Sci. Technol.* **2005**, *65*, 2526–2540. [CrossRef]
3.   Miracle, D.B.; Donaldson, S.L. Composites. In *ASM Handbook*; ASM International: Materials Park, OH, USA, 2001.
4.   Davis, J.R. Aluminum and Aluminum alloys. In *ASM Specialty Handbook*; ASM International: Materials Park, OH, USA, 1993.
5.   Clyne, T.W.; Withers, P.J. *An Introduction to Metal Matrix Composites*; Cambridge University Press: New York, NY, USA, 1993.
6.   Kainer, K.U. *Metal Matrix Composites: Custom-Made Materials for Automotive and Aerospace Engineering*; WILEY-VCH: Weinheim, Germany, 2006.
7.   Christman, T.; Needleman, A.; Suresh, S. An experimental and numerical study of deformation in metal-ceramic composites. *Acta Metall.* **1998**, *37*, 3029–3050. [CrossRef]
8.   Slipenyuk, A.; Kuprin, V.; Milman, Y.; Goncharuk, V.; Eckert, J. Properties of P/M processed particle reinforced metal matrix composites specified by reinforcement concentration and matrix-to-reinforcement particle size ration. *Acta Mater.* **2006**, *54*, 157–166. [CrossRef]
9.   Song, M.S.; Zhang, M.X.; Zhang, S.G.; Huang, B.; Li, L.G. In situ fabrication of TiC particulates locally reinforced aluminum matrix composites by self-propagating reaction during casting. *Mater. Sci. Eng. A* **2008**, *473*, 166–171. [CrossRef]
10.   Wang, J.; Yi, D.; Su, X.; Yin, F.; Li, H. Properties of submicron AlN particulate reinforced aluminium matrix composite. *Mater. Des.* **2009**, *30*, 78–81. [CrossRef]
11.   Feng, C.F.; Froyen, L. In situ synthesis of Al_2O_3 and TiB_2 particulate mixture reinforced aluminium matrix composites. *Scr. Mater.* **1997**, *36*, 467–473. [CrossRef]
12.   Arsenault, R.J. The strengthening of aluminum 6061 by fiber and platelet silicon carbide. *Mater. Sci. Eng. A* **1984**, *64*, 171–181. [CrossRef]
13.   Balcı, Ö.; Ağaoğulları, D.; Gökçe, H.; Duman, İ.; Öveçoğlu, M.L. Influence of TiB_2 particle size on the microstructure and properties of Al matrix composites prepared via mechanical alloying and pressureless sintering. *J. Alloys Compd.* **2013**, *586*, S78–S84.
14.   Ibrahim, I.A.; Mohammed, F.A.; Lavernia, E.J. Particulate reinforce metal matrix composites: Review. *J. Mater. Sci.* **1991**, *26*, 1137–1156. [CrossRef]
15.   Scudino, S.; Liu, G.; Sakaliyska, M.; Surreddi, K.B.; Eckert, J. Powder metallurgy of Al-based metal matrix composites reinforced with β-Al_3Mg_2 intermetallic particles: Analysis and modeling of mechanical properties. *Acta Mater.* **2009**, *57*, 4529–4538. [CrossRef]

16. Inoue, A. Bulk amorphous and nanocrystalline alloys with high functional properties. *Mater. Sci. Eng. A* **2001**, *304–306*, 1–10.

17. Ashby, M.F.; Greer, A.L. Metallic glasses as structural materials. *Scr. Mater.* **2006**, *54*, 321–326. [CrossRef]

18. Yu, P.; Zhang, L.C.; Zhang, W.Y.; Das, J.; Kim, K.B.; Eckert, J. Interfacial reaction during the fabrication of $Ni_{60}Nb_{40}$ metallic glass particles-reinforced Al based MMCs. *Mater. Sci. Eng. A* **2007**, *444*, 206–213. [CrossRef]

19. Aljerf, M.; Georgarakis, K.; Louzguine-Luzgin, D.; le Moulec, A.; Inoue, A.; Yavari, A.R. Strong and light metal matrix composites with metallic glass particulate reinforcement. *Mater. Sci. Eng. A* **2012**, *532*, 325–330. [CrossRef]

20. Dudina, D.V.; Georgarakis, K.; Aljerf, M.; Li, Y.; Braccini, M.; Yavari, A.R.; Inoue, A. Cu-based metallic glass particle additions to significantly improve overall compressive properties of an Al alloy. *Compos. A* **2010**, *41*, 1551–1557. [CrossRef]

21. Lee, M.H.; Kim, J.H.; Park, J.S.; Kim, J.C.; Kim, W.T.; Kim, W.T.; Kim, D.H. Fabrication of Ni–Nb–Ta metallic glass reinforced Al-based alloy matrix composites by infiltration casting process. *Scr. Mater.* **2004**, *50*, 1367–1371. [CrossRef]

22. Yu, P.; Kim, K.B.; Das, J.; Baier, F.; Xu, W.; Eckert, J. Fabrication and mechanical properties of Ni–Nb metallic glass particle-reinforced Al-based metal matrix composite. *Scr. Mater.* **2006**, *54*, 1445–1450. [CrossRef]

23. Scudino, S.; Surreddi, K.B.; Sager, S.; Sakaliyska, M.; Kim, J.S.; Löser, W.; Eckert, J. Production and mechanical properties of metallic glass-reinforced Al-based metal matrix composites. *J. Mater. Sci.* **2008**, *43*, 4518–4526. [CrossRef]

24. Scudino, S.; Liu, G.; Prashanth, K.G.; Bartusch, B.; Surredi, K.B.; Murty, B.S.; Eckert, J. Mechanical properties of Al-based metal matrix composites reinforced with Zr-based glassy particles produced by powder metallurgy. *Acta Mater.* **2009**, *57*, 2029–2039. [CrossRef]

25. Prashanth, K.G.; Kumar, S.; Scudino, S.; Murty, B.S.; Eckert, J. Fabrication and response of $Al_{70}Y_{16}Ni_{10}Co_4$ glass reinforced metal matrix composites. *Mater. Manuf. Processes* **2011**, *26*, 1242–1247. [CrossRef]

26. Shen, T.D.; Schwarz, R.B. Bulk ferromagnetic glasses prepared by flux melting and water quenching. *Appl. Phys. Lett.* **1999**, *75*, 49–51. [CrossRef]

27. Fujii, H.; Sun, Y.; Inada, K.; Ji, Y.; Yokoyama, Y.; Kimura, H.; Inoue, A. Fabrication of Fe-based metallic glass particle reinforced Al-based composite materials by Friction Stir processing. *Mater. Trans.* **2011**, *52*, 1634–1640. [CrossRef]

28. Suryanarayana, C.; Inoue, A. Iron-based bulk metallic glasses. *Int. Mater. Rev.* **2013**, *58*, 131–166. [CrossRef]

29. Kaban, I.; Jovari, P.; Waske, A.; Stoica, M.; Bednarcik, J.; Beuneu, B.; Mattern, N.; Eckert, J. Atomic structure and magnetic properties of Fe–Nb–B metallic glasses. *J. Alloys Compd.* **2014**, *586*, S189–S193. [CrossRef]

30. Zheng, R.; Yang, H.; Liu, T.; Ameyama, K.; Ma, C. Microstructure and mechanical properties of aluminum alloy matrix composites reinforced with Fe-based metallic glass particles. *Mater. Des.* **2014**, *53*, 512–518. [CrossRef]

31. Murty, B.S.; Ranganathan, S. Novel materials synthesis by mechanical alloying/milling. *Int. Mater. Rev.* **1998**, *43*, 101–141. [CrossRef]

32. Suryanarayana, C. Mechanical alloying and milling. *Prog. Mater. Sci.* **2001**, *46*, 1–184. [CrossRef]

33. Harrigan, W.C., Jr. Commercial processing of metal matrix composites. *Mater. Sci. Eng. A* **1998**, *244*, 75–79. [CrossRef]

34. Surreddi, K.B.; Scudino, S.; Sakaliyska, M.; Prashanth, K.G.; Sordelet, D.J.; Eckert, J. Crystallization behavior and consolidation of gas-atomized $Al_{84}Gd_6Ni_7Co_3$ glassy powder. *J. Alloys Compd.* **2010**, *491*, 137–142. [CrossRef]

35. Kim, J.Y.; Scudino, S.; Kühn, U.; Kim, B.S.; Lee, M.H.; Eckert, J. Production and characterization of Brass-matrix composites reinforced with $Ni_{59}Zr_{20}Ti_{16}Si_2Sn_3$ glassy particles. *Metals* **2012**, *2*, 79–94. [CrossRef]

36. German, R.M. *Sintering Theory and Practice*; Wiley-Interscience: Weinheim, Germany, 1996.

37. Prashanth, K.G.; Murty, B.S. Production, kinetic study and properties of Fe-based glass and its composites. *Mat. Manuf. Processes* **2010**, *25*, 592–597. [CrossRef]

38. Keryvin, V.; Hoang, V.H.; Shen, J. Hardness, toughness, brittleness and cracking systems in an iron-based bulk metallic glass by indentation. *Intermetallics* **2009**, *17*, 211–217. [CrossRef]

39. San Marchi, C.; Cao, F.; Kouzeli, M.; Mortensen, A. Quasistatic and dynamic compression of aluminum-oxide particle reinforced pure aluminum. *Mater. Sci. Eng. A* **2002**, *337*, 202–211. [CrossRef]

40. Gurland, J. The fracture strength of sintered tungsten carbide-cobalt alloys in relation to composition and particle spacing. *Trans. Metall. Soc. AMIE* **1963**, *227*, 1146–1149.
41. Chawla, K.K. *Composite Materials: Science and Engineering*; Springer-Verlag: New York, NY, USA, 1987.
42. Kim, H.S. On the rule of mixtures for the hardness of particle reinforced composites. *Mater. Sci. Eng. A* **2000**, *289*, 30–33. [CrossRef]
43. Hirth, J.P. *Physical Metallurgy*; Cahn, R.W., Haasen, P., Eds.; North-Holland: Amstdam, The Netherlands, 1996.

metals

MDPI

Article

On the Stability of the Melt Jet Stream during Casting of Metallic Glass Wires

Ayo Olofinjana [1,*] and Nyuk Yoong Voo [2]

[1] School of Science and Engineering, University of the Sunshine Coast, Maroochydore DC, QLD 4558, Australia
[2] Faculty of Science, Universiti Brunei Darussalam, Jalan Tungku Link, BE1410, Brunei Darussalam; nyukyoong.voo@ubd.edu.bn
* Author to whom correspondence should be addressed; aolofinj@usc.edu.au; Tel.: +61-7-5456-5987; Fax: +61-7-5456-5453.

Academic Editors: K. C. Chan and Jordi Sort Viñas
Received: 19 May 2015; Accepted: 2 June 2015; Published: 8 June 2015

Abstract: The factors that affect the stability of the melt stream during the casting of wire directly from the melt have been investigated. It is shown that the criticality of process parameters centres mostly on the forces imposed on the melt stream at confluence with the cooling water. The analysis of these forces indicated that the shear component of the disturbance is dependent on the ratio of the velocity of the melt stream (v_m) to that of the cooling water (v_w) in accord with results obtained from previous experiments. The role of oxide-forming elements in widening the process parameters range is attributed to the increased stability of the melt stream due to the additional shear force resistance offered by the solid oxide layer. The roles of Cr and Si oxides in stabilising the melt stream are confirmed by X-ray photoelectron spectroscopy (XPS) of wire indicating the presence of these oxides on fresh as-cast wires. Melt superheat and nozzle clearance distance are not strictly stream stability factors, but rather their role in glass formation prescribes optimal limits for fully amorphous wire.

Keywords: metallic glass; amorphous alloy; wire casting; multi–strand casting; jet stability

1. Introduction

The ability to cast wire directly from melt provides such an obvious production process advantage that it has long attracted attention. One of the earliest methods proposed dates back to 1882 [1] and, more recently, many more processes have been proposed [2–9] to achieve this. Since molten metal unlike oxide melts are known to have very low viscosities, the success of the proposed methods relies heavily on maintaining continuous flow of mass on the transition from liquid to solid without stream breaking. Many of the early attempts, however, were fraught with difficulties in maintaining a stable melt jet at the interaction of the melt stream and the static quenching medium which inevitably has an impact due to the velocity resistance when the jet comes into contact with the quenching medium. Recent reports on the production of round sectioned wire from melt are therefore restricted to a few more modern processes [10–14] utilising a dynamic quenching medium. In these, the non-static nature of the quenching medium reduced the momentum difference and, thus, significantly reduced the impact on the stream and, to a large extent, determined the flow and shape of the final product.

Though the early attempts were not aimed at rapid solidification, the task of forming wires of small diameters directly from melt, apart from providing a technological shortcut in production route, also implies the inherent rapid solidification of the melt which could lead to the vitrification of a readily glass-forming alloy. The range of wire castable alloys [7,10] are mostly (except for a few) readily glass forming alloys. Magnetic Fe- and Co-based amorphous alloys are now routinely cast in diameters in the order of 100 μm. The potential applications of amorphous wire in magnetic devices are now well known [13–19] and improvements in production techniques are being pursued. Presently, only the Taylor-Ulitovsky and the rotating water bath processes are consistently used [14,20] to make directly cast amorphous wires for magnetic applications. The versatility of Taylor–Ulitovsky method is demonstrated in the wide range of compositions [11,21–24] from which metallic glass microwires can be produced, including Heusler-type alloys [25–29] that are being proposed for magnetic shape memory alloys (MSMA) owing to their giant magnetic field induced strain.

These new interests in magnetic applications (such as microwires for giant magneto impedance [30]) have led to continually increased interest in improving methods for casting metallic glass wires. A recent review [22] of metallic glass microwires as a multifunctional composite suggest melt extraction and the Taylor–Ulitovsky methods as the techniques of choice for making suitable microwires. In the Taylor–Ulitovsky process, the use of a glass coating removed the inherent problem of melt stream stability and, in principle, could be used for any alloy for which a matching glass could be found. It remains the method of choice for casting metallic glass microwires with diameters typically less than 60 μm. It is however limited in its applicability, especially regarding the production of metallic glass wires with larger diameters. To allow the imposition of an amorphous structure in the Taylor–Ulitovsky methods, critical cooling rates imposed by thermo-kinetic factors required the diameters of the cast wires to be less than 60 μm, translating to cooling rates greater than 10^5 Ks^{-1}. The chemical removal of the glass coating by hydrofluoric acid (HF) dissolution [30–32] is a necessary part of this process and can further complicate the development. Nonetheless, it has successfully been used [9,13,16,21–24,28,33,34] to prepare high quality metallic glass wires with diameters less than 60 μm. However, for larger diameter glassy wire, direct containerless solidification of the melt stream is required.

The rotating water bath process, however, has been used to produce wires of approximately 100 μm diameter in a variety of compositions [6,10,13]. The cooling rates for the diameter achieved is of the order of 10^5–10^6 Ks^{-1}, and is enough to vitrify most readily glass-forming alloys. We have explored the development of direct casting in a multi-stream mode [35,36] where it was demonstrated that the productivity of wire production could be increased through multi-streaming without adversely affecting the wire forming and vitrification processes. This process has potential for scale up, but inconsistency in products after several attempts required the clarification of the many varied factors that affect the melt stream stability and, consequently, the quality of the cast metallic glass wires. It represents the only method by which larger diameters of up to 150 μm glassy wires can be produced directly from melt. The main problem is to understand factors that affect the stability of melt jet stream in order to consistently produce high quality metallic glass wires. Here, discussion is focussed on how the critical process parameters might have affected the stability of the melt streams. It is hoped that an understanding of such problems would contribute to new developments in the process and alloy selection for multi-stream casting of metallic glass wires directly from melt. Such relatively thick metallic glass wires are known for their magnetic bi-stability and very high strength. It is expected that a high volume production would promote their use as structural reinforcement and possible magnetic applications.

2. Experimental Section

2.1. Wire Casting

Thirty-five grams ingots of $Fe_{100-x-y}Cr_xSi_yB_{15}$ alloys were produced by melting the pure constituent elements in a boron nitride crucible in an induction furnace. Each ingot was re-melted in a quartz crucible in the base of which were laser drilled nozzles as described previously [35]. Single and multi-orifice nozzles were used to study the fluid flow in order to establish the flow coefficients determined by the specific crucible nozzle geometry. The temperature of the melt was monitored via a boron nitride shielded Pt/PtRh (13%) thermocouple. The mV signal from the thermocouple was connected via a modified IC to the electronic control module of the rf generator, such that it was possible to program the generator through the thermocouple feedback. In all cases, the melt temperature could be maintained within ±10 K of specified temperature.

The flow of liquid metal through the nozzles was initiated by applying an argon gas pressure above the melt surface in the quartz crucible. The gas pressure was monitored and controlled by a pressure transducer. The rotating speed of the bath was monitored by an optical tachometer. The melt streams were projected into the rotating water bath. The success of filament cast depended on the stability of the melt stream which was affected by a number of factors which are discussed herein.

2.2. Determining Melt Stream Velocity

The accurate determination of the jet velocity is an important factor in optimising other process variables for successful wire casting. The effects of gravity and surface tension have opposing effects on the magnitude of the final velocity of the melt stream, and in any case, their combined effect is negligible compared to the applied pressure head. The stream velocity v_m as a result of gas pressure P was calculated from a modified Bernoulli equation [37]:

$$v_m = C_f \left(\frac{2P}{\rho} \right)^{1/2} \tag{1}$$

where ρ is the density of melt and C_f is a flow coefficient determined from flow curves by plotting the square of the flow rate against nozzle diameters in water experiments. Water was used as the experimental liquid to avoid the effects of high temperatures on thermal distortion of the geometrical factor that is being corrected for. Typical flow curves using single orifice laser drilled nozzles are shown in Figure 1. The linearity of the curves validated the use of flow coefficient C_f to determine the final velocity according to Equation (1). The derived relationship was used to calibrate the velocity of jets for a given pressure and was found to be adequate in the pressure range of 0.3–0.4 MPa. Typically, C_f varied from 0.90–0.97, indicative of how slight variations in crucible-nozzle geometry affected the stream flow even when the same pressure was applied.

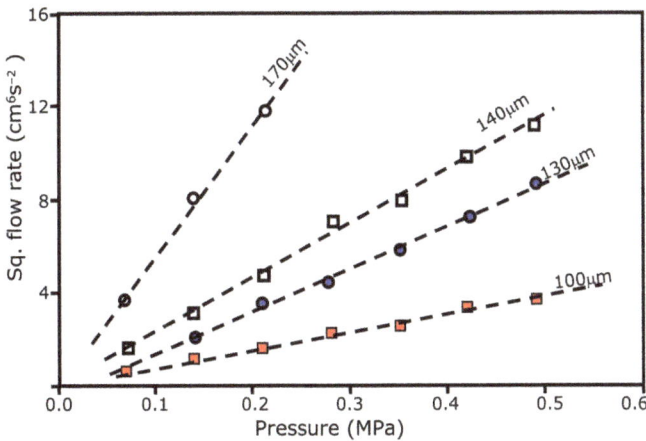

Figure 1. Flow curves showing relationships between square flow rate and applied pressure from water experiments for characterising flow coefficients.

3. Results and Discussion

3.1. Factors Affecting Stability of Melt Jet

In the rotating water bath process, it has long been recognised that the success of wire formation depends on (e.g., refs [6,7,35,36]):

- Nozzle clearance distance from cooling liquid
- Nozzle/melt jet diameter
- Ratio of melt jet velocity to that of the cooling liquid
- Melt super heat
- Alloy composition

All these are somehow related to the stability of the jet and are required to be optimised in order to cast wire continuously. The case of optimisation conditions in rotating water bath process is emphasised in the early reports for amorphous wire [6,13,35] and recently for crystalline Cu-Al-based alloys [38]. The casting conditions when using laser drilled multiple nozzles were summarised earlier [35]. These are concerned largely with stabilising the melt jets and maintaining their geometry in the transition from liquid melt to solid metallic glass wire. Here, each of these process parameters are considered as to how they affect the melt stream stability and how they determine the critical limits in the wire casting process.

3.2. Nozzle Clearance Distance

The distance between the nozzle tip and the water surface is important to wire formation. The optimum distance was found to be 2–4 mm. In the multi-stream configuration [35], it was important to use optically flat glass discs for drilling of orifices located in the nozzle base with the line of the orifice perpendicular to the direction of water flow, in order to maintain the same distance to the water surface by all of the melt streams. A slight curvature tended to develop on the flat disc as a result of fusion to the crucible base, but this generally had no perceptible effect on the ability to cast geometrically acceptable glassy wires. It was practically difficult to bring the nozzle tip closer than 2 mm to the surface of the water because of the risk of contact with the cooling water which results in cracking due to thermal shock. On the other hand, distances of greater than 4 mm usually led to stream break up and the formation of powder or short pieces of fibres.

It is important to consider the contribution of nozzle clearance distance to the stability of melt stream. Due to the effect of surface tension and viscosity, a liquid jet will tend to break into droplets. The theoretical and experimental studies of break up length of a free jet has attracted lots of attention [39,40] and the critical parameter to establish stability is conventionally expressed as the length to diameter ratio (L/d). The critical (L/d) ratio for a coherent isothermal liquid jet depends on the streaming conditions characterised by dimensionless numbers and can be expressed by the modified Weber equation [12,39]:

$$L/d = \left[\sqrt{We} + 3\frac{We}{Re}\right] \ln\left(\frac{d}{2\varepsilon_0}\right) \qquad (2)$$

where We is the Weber number given by $We = \left(\frac{\rho v_m^2 d}{\sigma}\right)$ and Re is the Reynolds number given by $Re = \frac{\rho v_m d}{\eta}$, σ, ρ, and η, are respectively the surface tension, density and dynamic viscosity of the melt.

The log term $\ln\left(\frac{d}{2\varepsilon_0}\right)$ is an experimental parameter relating to the perturbation at the emergence of the jet from the nozzle and is approximately 12 for a wide variety of liquids and streaming conditions [38]. Typical wire casting conditions, for which $v_m \approx 10$ ms^{-1}, and taking the values of melt physical properties given in [41], imply $We \approx 10$ and $Re \approx 10^4$, thus giving L/d ratio of approximately 40. A critical L/d of 40 is in accord with experimental observation and model prediction given in ref [40] for $Re = 10^3$–10^4. For a 100 μm diameter melt jet, this corresponds to a jet break-up length of 4 mm. This is ostensibly in fairly good accord with the present experimentally observed critical nozzle distance of less than 4 mm for casting continuous metallic glass wire. However, it has been shown that the break-up length for an oxide-forming melt is at least one order of magnitude greater than that predicted from Equation (2). Figure 2 shows a photograph of melt streams maintained at the typical wire casting condition, and observed unbroken for lengths well over 20 mm. It is seen that a coherent jet could still be maintained under casting conditions for lengths well above the critical value predicted from Equation (2). Nevertheless, the inability to produce continuous amorphous wire with clearance distances over 4 mm was an indication of instability of the flow setting-in irrespective of oxide protection. The fact that, even beyond this predicted limiting length of 4 mm, short brittle fibres could still be produced suggests that nozzle-water clearance distance in itself is not strictly a limiting factor on the stability of the stream up to the point of entry of the jet into the quenching water. However, the distance needs to be kept to less than 4 mm, partly to minimise velocity rise due to gravitational acceleration and also to minimise air cooling of the melt stream and, thus, maximising the quenching rate required for the vitrification of the melt. For longer nozzle-water clearance, the drop in temperature imposed from air-cooling would reduce the temperature differential between the melt stream and the quenching medium and may, thus, result in a cooling rate below the critical cooling rate for glassy wire production.

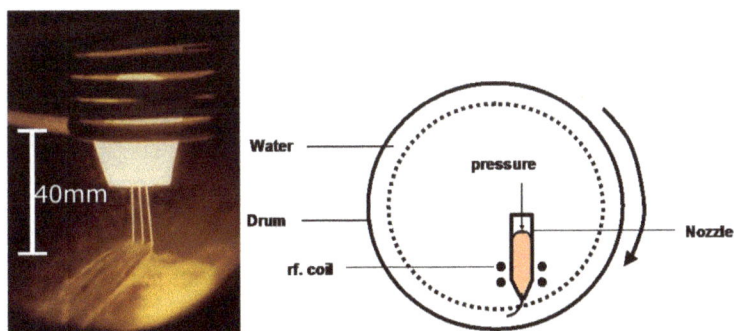

Figure 2. (a) Unbroken melt jet streams maintained for over 20 mm under streaming conditions. (b) Schematic diagram of the wire casting.

3.3. Nozzle Diameter and Jet Contraction

For most compositions investigated, continuous wire production was limited to diameters between 80 and 150 μm. Outside these ranges, powder or short pieces of fibres were produced, indicating an unstable melt stream during casting attempts. The lower limit of 80 μm was imposed by the shorter distance at which instability sets in and also by the difficulty of initiating flow of the viscous melt through the nozzle. In order to increase the L/d ratio (as predicted in Equation (2)), it would have been necessary to increase cast pressure which again was limited by the ability of the crucible material to withstand higher pressures at the casting temperatures. Additionally, at small diameters, blockage of nozzles was a problem that constantly disturbed and caused instability in the melt stream.

At larger nozzle diameters ($d > 150$ μm), the cooling rate becomes a problem. Although it was easier to maintain a stable stream (as predicted by Equation (2)), from heat transfer considerations, the average cooling rate of the melt was reduced for large diameters. This had implications for the solidification rate. The success of forming continuous wire is ultimately dependent on rapid solidification of the molten jet stream. The longer the solidification times, experienced with larger diameters, the greater the effect of forces related to momentum change as the melt stream changed course in the stream on confluence. Incomplete solidification would therefore cause the break-up of the stream leading to powders or short fibres. Additionally, the reduced cooling rates may be below the critical rate for glass formation in that particular composition being cast, thus resulting in a partially or fully crystalline brittle wire.

Generally, it was observed that the wire diameter is about 10% smaller than the nozzle diameter, thus signifying a jet contraction in flight during casting. Such contraction can be explained if we consider the relaxation of the velocity profile that exists across the jet stream. A simplified approach to predicting such contraction from velocity relaxation would be to assume a fully developed laminar flow across the nozzle at the point of exit. The velocity profile as depicted in Figure 3 can then be expressed as that given for a laminar non-compressible fluid over a flow length L as [37]:

$$v = \frac{\Delta P R^2}{4\eta L}\left[1 - \left(\frac{r}{R}\right)^2\right]$$

(3)

where R is the radius of the nozzle and r represents the radial distance from the centreline.

It could be shown that the mass flow rate ($\dot{m}$) is [37]:

$$\dot{m} = \int_{r=0}^{r=R} 2\pi r \rho \frac{\Delta P R^2}{4\eta L}\left[1 - \left(\frac{r}{R}\right)^2\right]dr = \frac{\pi \rho \Delta P R^4}{8\eta L}$$

(4)

On emergence from the nozzle, the melt stream condition is equivalent to that in which the fixed boundary of the nozzle is suddenly removed. The streamlines near the boundary, having initially zero velocity, will accelerate. The most simplified scenario is to assume acceleration to the centre line velocity which is the maximum of the velocity profile equation (Equation (3)) and is given by [37]:

$$v_{\max} = \frac{\Delta P R^2}{4\eta L}$$

(5)

The mass flow rate $\dot{m}'$ of a contracted free jet with reduced radius R' is given by:

$$\dot{m}' = \rho v_{\max} \times Area = \frac{\rho \Delta P R^2}{4\eta L} \times \pi (R')^2$$

(6)

For mass conservation, we expect that flow rate remains the same before and after jet contraction and this implies that:

$$\frac{\rho \Delta P R^2}{4\eta L} \times \pi (R')^2 = \frac{\rho \Delta P R^4}{8\eta L} \text{ or } R' = \frac{1}{\sqrt{2}}R$$

(7)

This analysis predicts a jet contraction of $1 - 1/\sqrt{2}$ (~30%) which clearly is an overestimation when compared with the experimental results of about 10%. A drawback to the theoretical prediction of jet contraction in short bore orifices as in nozzles for wire casting can be the incomplete description of hydrodynamic flow within the jet. A schematic model of flow development is shown in Figure 4. A semi-empirical fluid mechanics approach suggests [37] that, to fully establish (99%) laminar flow, an L/d ratio of 116 is required. This translates to bore lengths of greater than 12 mm for a typical nozzle orifice diameter of 100 μm, normally used for wire casting. In practice, nozzles are converging and with lengths of the order of a few millimeters. The assumption of a fully developed laminar flow in this analysis, therefore, cannot truly represent the streaming conditions of the nozzle in the wire casting process. Nevertheless, the analysis here qualitatively explains the observed contraction from a possible relaxation of a streamline velocity profile across the jet in free flight.

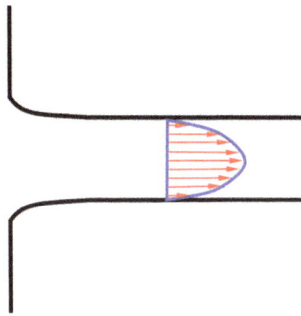

Figure 3. Schematic representation of velocity profile in a fully developed laminar flow.

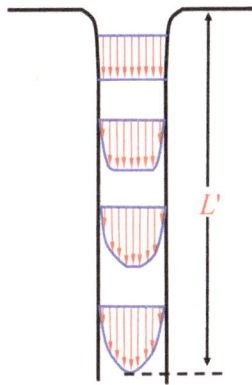

Figure 4. Schematic representation of flow development showing critical length for fully developed laminar flow.

3.4. Melt Superheat

The optimum superheat for continuous wire formation was found [35] to be 100–150 K. Lower superheats cause fluidity problems resulting in premature nozzle blockage that prevented the formation of a coherent jet. Excessive superheat could lead to nozzle distortion that affects the dynamics of fluid flow through the nozzle. Moreover, large superheats imply a longer solidification time which would be detrimental to maintaining a continuous flow of the melt stream before initial solidification occurs. It has been shown [35] that large superheat resulted in partially crystalline wire as indicated by the sharp crystalline peaks of XRD pattern for wire samples with superheat greater than 100 K. For an alloy composition of $Fe_{77.5}Si_{7.5}B_{15}$, the role of excessive superheat on melt in the cast wire is shown on the DSC crystallisation thermograms in Figure 5. Here, it is seen that crystallisation peaks for metallic glass cast at high melt temperature (1450 °C) are much lower smaller compared to those cast at lower temperatures (*i.e.*, with less superheat). The indication then is that for excessive superheat, complete vitrification is not achieved. The longer cooling intervals associated with a large superheat resulted in incomplete vitrification, and it is particularly more evident for larger diameters, for which the margin for glass formation is narrower for most compositions.

Ordinarily, for heat transfer considerations, we would expect a large superheat to translate to a higher average cooling rate of the melt if complete Newtonian cooling was responsible for the heat transfer. However, limitations exist in heat transfer coefficients when a solid is being cooled in a liquid medium from a temperature well above the boiling point of the cooling liquid. In this case, expected boiling (at 100 °C) would mean the wire would be encased in a steam jacket initially and the increased temperature potential from additional superheat would not lead to increased average cooling rate as would be expected from a fully Newtonian cooling; rather, the longer solidification time would prove the kinetic lag for diffusive atomic rearrangement into crystalline structure. Although superheat is not critical to melt jet stability, its role in reducing glass-forming can lead to the formation of partially crystalline and subsequently brittle wire that tends to break into short fibres or fragment into powders.

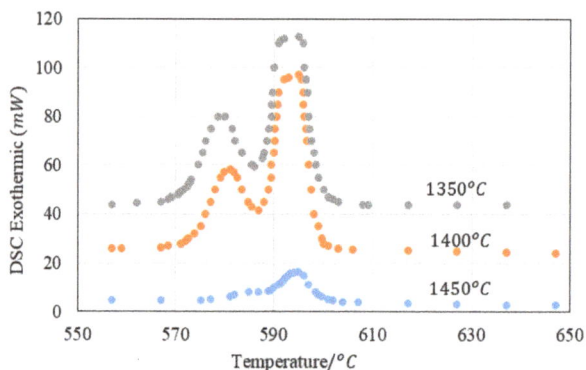

Figure 5. Effect excessive superheat on the degree of vitrification on as-cast wire.

3.5. Effect of Water and Jet Velocity

Velocity mismatch between melt stream and cooling stream has long been recognised [5] as one of the possible contributors to break up forces that cause instability in the melt stream. In the water bath process, the ratio of the velocities of the melt stream and the water bath (v_m/v_w) has been found to be a critical factor for continuous wire formation. Depending on alloy composition, we have found [35] that the critical values of v_m/v_w range from 1.1–1.21, still within the wider range of 1–2 proposed originally by Masumoto *et al.* [6]. However, these studies were directed at casting both crystalline and amorphous alloys wires. Some effects of the magnitude of v_m/v_w on the morphology of wire for the glass-forming alloy FeCrSiB are shown in Figure 6. These illustrate the effects of the various deformation forces on the liquid jet prior to solidifying.

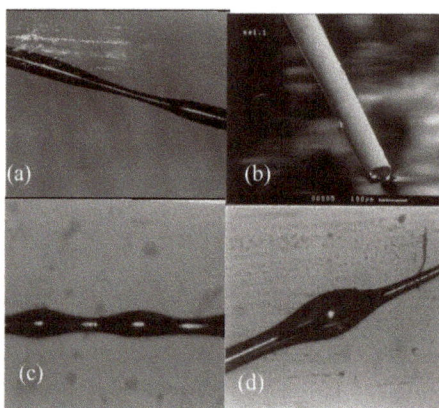

Figure 6. Some of the effects of v_m/v_w ratios on wire morphology (optical for (**a**) (**c**) and (**d**); and SEM for (**b**)) wire morphology (**a**) $v_m/v_w = 1.3$ (**b**) $v_m/v_w = 1.1$ (**c**) $v_m/v_w = 0.9$ (**d**) $v_m/v_w = 1.5$.

The significance of the velocity ratio v_m/v_w for this casting process lies in the forces imposed on the melt jet as it changes course from an initial stream velocity v_m to be in confluence with the quenching water bath having a velocity v_w. Previous studies suggest matching the two stream velocities would maintain continuity in the transition of the molten stream to solid filament. However, the change in direction of jet on entry into the water imposes a substantial force on it. The velocity change of the jet and the force exerted on it on entry into the water bath are depicted in Figure 7. Resolving the velocities, and applying Newton's second law to the jet trajectory, the force $\vec{F}$ imposed from the momentum change is given in vector form by the equation:

$$\vec{F} = \dot{m}\{(v_w \sin \phi)i + (v_m - v_w \cos \phi)j\} \tag{8}$$

where i and j are unit vectors respectively along the horizontal and vertical axes.

The direction and magnitude of $\vec{F}$ is most critical to the stability of the jet. The direction depicted by angle θ in Figure 7, is given by:

$$\theta = \tan^{-1}\left\{ \left(\frac{v_m}{v_w} - \cos \phi \right) \frac{1}{\sin \phi} \right\} \tag{9}$$

We can consider two limiting conditions for the stability of the jet; one for which magnitude of the force $|F|$ is minimum but completely shearing and the other where the direction indicates an equal distribution between the shear and tensile or compression; i.e. we choose $\theta = 45°$. In the case of the former, $v_m/v_w = \cos \phi$. For the limiting condition of $\theta = 45°$, $v_m/v_w = \cos \phi + \sin \phi$. In practice, wire casting successfully relies on minimising magnitude of the interaction force and optimising its distribution into shearing and direct. It is therefore expected that an optimum casting condition would be between these two limiting conditions as:

$$\cos \phi < v_m/v_w < (\cos \phi + \sin \phi) \tag{10}$$

It is clear that stability of jet due to momentum change is dependent on both the velocity ratio v_m/v_w and the jet entrant angle ϕ. Since angle ϕ is normally kept constant, (though in our experiments it was kept at 30°) we can redefine the range for which the critical v_m/v_w value must lie in the range: $0.87 < v_m/v_w < 1.37$. This is consistent with experimental observations for which critical v_m/v_w for FeCrSiB melt was found to be 1.1–1.21. Outside the optimal range, the direction θ of the imposed force would upset the force component allowing either of the shearing or compressive forces to predominate leading respectively to powder or deformed wire. The criticality of the ratio v_m/v_w is confined to a narrow range so that it gives a minimum shear component of force imposed by the momentum change.

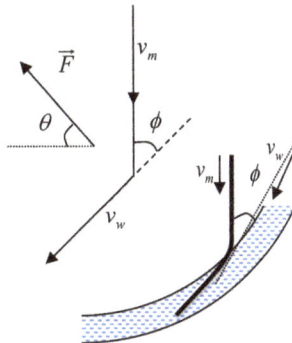

Figure 7. Velocity change and force exerted at the confluence of melt and water streams.

3.6. *Effect of Alloy Composition*

It was found [35] that for the compositional series $Fe_{85-x}Si_xB_{15}$, a minimum silicon content of 5% was required for continuous wire formation even with very strict process control. Replacement of Fe with Cr was also found to increase the wire forming ability, within a broader window of values of v_m/v_w for continuous wire formation.

The influence of oxide formers on wire forming ability has long been recognized [10]. The natural tendency of a melt stream to disintegrate into droplet decreases with the presence of a coherent oxide skin. It is consistent with the earlier analysis in that we expect a larger than predicted L/d ratio and thus a longer break-up length of free jet for a strong oxide-forming melts. The presence of an oxide skin on a melt would lead to greater shear resistance of the free jet. This would stabilize the jet against breakage notably when the momentum changed as the melt stream altered course as it travelled through confluence and finally achieved the speed of the quenching water.

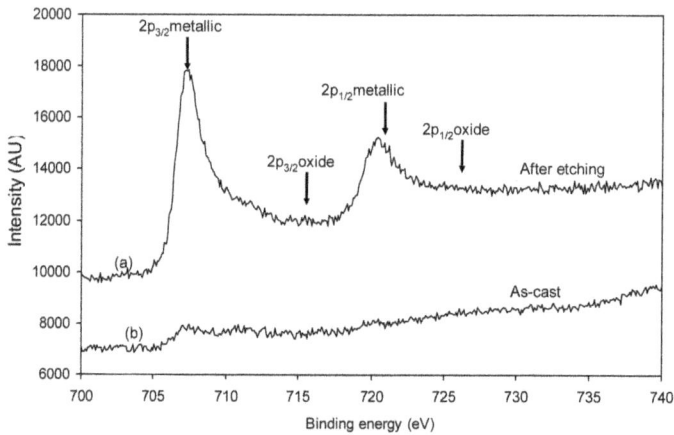

Figure 8. XPS Fe 2p in as-cast and after Ar$^+$ sputtering.

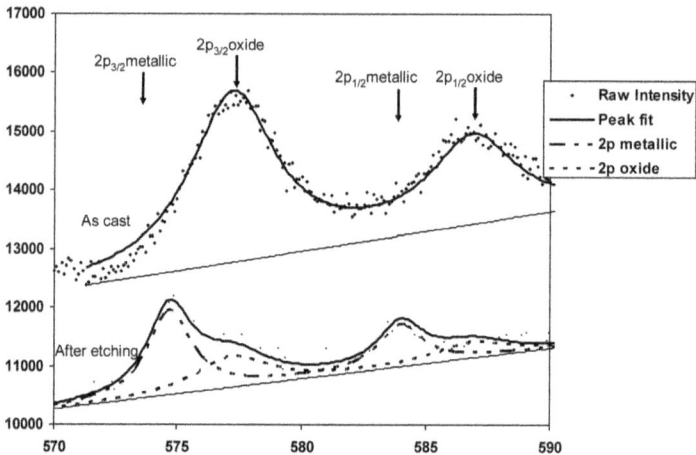

Figure 9. XPS Cr 2p peak fitting for standard Cr and Cr_2O_3 for as-cast wire and after ion sputtering.

The role of Si and Cr in forming oxides was confirmed through surface analysis of cast wires of composition $Fe_{69.5}Cr_8Si_{7.5}B_{15}$ with X-ray photo electron spectroscopy (XPS). Figure 8 shows the Fe 2p characteristic spectrum for as-cast and Ar^+ sputtered wire samples. It is clear that the surface has very weak Fe signal on the as-cast sample but the Fe 2p signal becomes more pronounced after prolonged etching that removed the surface oxide. The characteristic Fe $2p_{3/2}$ and $2p_{1/2}$ are clearly indicative of metallic, thus confirming that the surface composition is essentially different from the bulk. Similar Cr-2p and Si-2p lines are respectively shown in Figures 9 and 10. It is shown in these figures that both the Cr 2p and Si 2p XPS characteristic peaks for the as-cast wire fit more closely with standard oxide spectra. These characteristic peaks move closer to metallic after Ar^+ sputtering. The comparison of the XPS spectra (Figures 8–10) for as-cast and sputtered samples confirms the compositional difference of the surface and the core of the wire samples. The present XPS evidence suggests that the surface consisted essentially Si and Cr oxides. Semi-quantitative analyses of these peaks based on integrated areas and applying sensitivity factors, suggest that Cr/Fe and Si/Fe atomic ratios were respectively 1.2 and 3.9. These values are much higher than the nominal Cr/Fe and Si/Fe atomic ratios of $\approx$0.1 that are expected from the nominal composition of the bulk alloy.

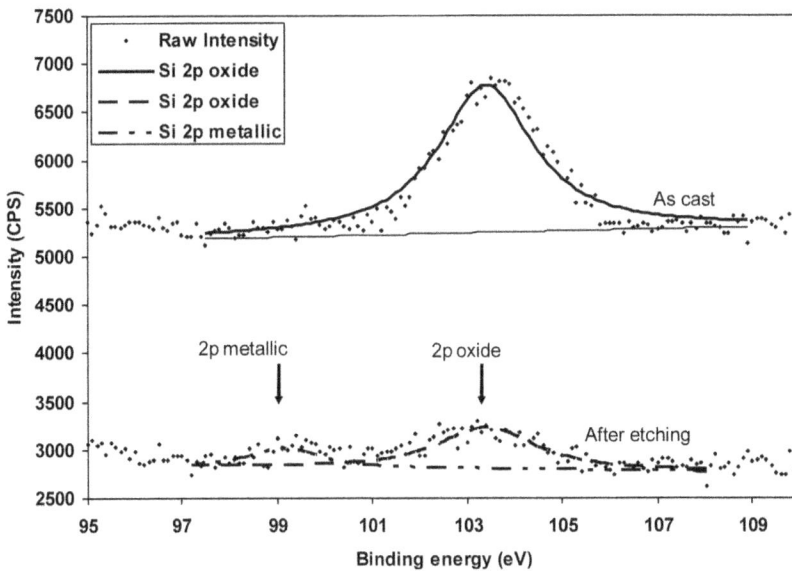

Figure 10. XPS Si 2p peak fitting for standard SiO_2 for as-cast wire and after ion sputtering.

It is somewhat surprising that the Fe 2p for the un-sputtered samples indicate almost no presence of Fe. The result of the Fe 2p due to light etching confirmed this is only limited to the surface and must be related to the oxide-forming tendencies of the constituents Si and Cr elements. The relative influence of elements on oxide formation according to thermodynamic drive [42] derived from free energies for oxide formation would be of the order Si > Cr > Fe and this is consistent with the present XPS observations. Since both Si and Cr at high temperatures have much higher negative free energies for oxide formation, even their presence in relatively small concentration (7–8 at%) would give a thermodynamic and kinetic preference for their oxide formation over the main (Fe) alloy constituent. This XPS evidence confirms the important roles of Si and Cr in forming a coherent oxide skin which stabilizes the free flight of the melt jet and, thus, allows more flexibility with the process variables in forming wires.

4. Conclusions

Successful wire formation by the rotating water bath melt spinning technique depends on maintaining good stability of the melt stream and on minimizing any disturbances of this stream as it freezes. The criticality of some of the process variables is determined by the narrow process window required for stabilizing and minimizing the disturbances to the melt stream. While nozzle diameters and melt superheat within the limits applicable are not strictly jet stability criteria, they play a critical role in complete glass formation and, thus, continuity of the final wire product. The most essential criteria for melt jet stability is the velocity ratio of melt to cooling water v_m/v_w as it relates to the direction of the force imposed on the melt as it comes into confluence with the cooling stream of water. The role of the oxide-forming elements Si and Cr is crucial to provide a strong Si/Cr oxide skin on the stream surface which resists the shearing force imparted on the melt stream as it is subjected to acceleration/deceleration when it enters the water bath at an oblique angle.

Acknowledgments: The wire casting experiments reported in this work were done at the laboratories of The University of Sheffield, UK, under the supervision of H.A. Davies.

Author Contributions: A.O. carried out the wire casting experiments and prepared the original draft. N.Y.V processed the XPS data, carried out the thermal analysis and contributed to literature review that form the basis for this manuscript.

Conflicts of Interest: The authors declare no conflict of interest.

References

1. Small, E. Aparatus for Making Wire Solder. U.S. Patent 262625, 1882.
2. Otstot, R.S.; Motern, J.W. Method and Apparatus for Improved Extrusion of Essentially Inviscid Jets. U.S. Patent 3,645,657, 1972.
3. Privott, W.J.; Cunningham, R.E. Low Viscousity Melt Spinning Process. U.S. Patent 3,715,419, 1973.
4. Kavesh, S. Apparatus for liquid quenching of free jet spun metal. U.S. Patent 3,845,805, 1976.
5. Adler, R.P.I. Melt Spinning Process and Machine. U.S. Patent 4,020,891, 1977.
6. Masumoto, T.; Hagiwara, M. Process for the Production of Fine Amorphous Metallic Wires. U.S. Patent 4,495,691, 1985.
7. Masumoto, T.; Hamashima, T.; Hagiwara, M. Method of Manufacturing Thin Metal Wire. U.S. Patent 4,614,221, 1986.
8. Hagiwara, M.; Menjiu, A.; Kohachi, N.; Masaru, K.; Yoshianao, Y.; Miyuri, S. Fine Amorphous Metal Wire. U.S. Patent 4,806,179, 1989.
9. Larin, V.S.; Torcunov, A.V.; Zhukov, A.; Gonzalez, J.; Vazquez, M.; Panina, L. Preparation and properties of glass-coated microwires. *J. Magn. Magn. Mater.* **2002**, *249*, 39–45. [CrossRef]
10. Inoue, A.; Krause, J.T.; Masumoto, T.; Hagiwara, M. Young's modulus of Fe-, Co-, Pd- and Pt-based amorphous wires produced by the in-rotating-water spinning method. *J. Mater. Sci.* **1983**, *18*, 2743–2751. [CrossRef]
11. Shalyginaa, E.E.; Umnova, N.V.; Umnov, P.P.; Molokanov, V.V.; Samsonova, V.V.; Shalygin, A.N.; Rozhnovskaya, A.A. Specific Features of Magnetic Properties of "Thick" Microwires Produced by the Ulitovsky–Taylor Method. *Phys. Solid State* **2012**, *54*, 287–292. [CrossRef]
12. Frommeyer, G.; Frech, W. Continuous casting and rapid solidification of wires produced by a newly developed shape flow casting technique. *Mater. Sci. Eng. A* **1997**, *226*, 1019–1024. [CrossRef]
13. Sarkar, P.; Roy, R.K.; Panda, A.K.; Mitra, A. Optimization of process parameters for developing FeCoSiB amorphous microwires through in-rotating-water quenching technique. *Appl. Phys. A* **2013**, *111*, 575–580. [CrossRef]
14. Vazquez, M. Soft magnetic wires. *Phys. B* **2001**, *299*, 302–313. [CrossRef]
15. Gavrilyuk, A.V.; Gavrilyuk, A.A.; Kovaleva, N.P.; Mokhovikov, A.Y.; Semenov, A.L.; Gavrilyuk, B.V. Magnetic properties of $Fe_{75}Si_{10}B_{15}$ amorphous metallic wires. *Phys. Metals Metallogr.* **2006**, *101*, 434–439. [CrossRef]

16. Sarkar, P.; Roy, R.K.; Mitra, A.; Panda, A.K.; Churyukanov, M.; Kaloshkin, S. Effect of Nb and Cr incorporation on the structural and magnetic properties of rapidly quenched FeCoSiB microwires. *J. Magn. Magn. Mater.* **2012**, *324*, 2543–2546. [CrossRef]

17. Zhang, D.; Chen, K.; Jia, X.; Wang, D.; Wang, S.; Luo, Y.; Ge, S. Bending fatigue behaviour of bearing ropes working around pulleys of different materials. *Eng. Fail. Anal.* **2013**, *33*, 37–47. [CrossRef]

18. Vazquez, M.; Marin, P.; Olofinjana, A.O.; Davies, H.A. The magnetic properties of FeSiBCuNb wires during the first stages to the nanocrystallization process. *Mater. Sci. Forum* **1995**, *179–181*, 521–526. [CrossRef]

19. Vazquez, M. Giant magneto-impedance in soft magnetic "wires". *J. Magn. Magn. Mater.* **2001**, *226*, 693–699. [CrossRef]

20. Mokhirev, I.I.; Chueva, T.R.; Zabolotnyi, V.T.; Umnov, P.P.; Umnova, N.V.; Molokanov, V.V. Strength and Plastic Properties of Amorphous Cobalt Alloy Wires Produced by Various Melt Quenching Methods. *Russ. Metall. (Metally)* **2010**, *2011*, 345–349. [CrossRef]

21. Luo, Y.; Peng, H.X.; Qin, F.X.; Ipatov, M.; Zhukova, V.; Zhukov, A.; Gonzalez, J. Fe-based ferromagnetic microwires enabled meta-composites. *Appl. Phys. Lett.* **2013**, *103*, 251092. [CrossRef]

22. Qin, F.; Peng, H.-X. Ferromagnetic microwires enabled multifunctional composite materials. *Prog. Mater. Sci.* **2013**, *58*, 183–259. [CrossRef]

23. Zhao, Y.Y.; Li, H.; Hao, H.Y.; Li, M.; Zhang, Y.; Liaw, P.K. Microwires fabricated by glass-coated melt spinning. *Rev. Sci. Instrum.* **2013**, *84*, 075102. [CrossRef] [PubMed]

24. Zhukov, A.; Chichay, K.; Talaat, A.; Rodionova, V.; Blanco, J.M.; Ipatov, M.; Zhukova, V. Manipulation of magnetic properties of glass-coated microwires by annealing. *J. Magn. Magn. Mater.* **2015**, *383*, 232–236. [CrossRef]

25. Varga, R.; Ryba, T.; Vargova, Z.; Saksl, K.; Zhukova, V.; Zhukov, A. Magnetic and structural properties of Ni-Mn-Ga Heusler-type microwires. *Scr. Mater.* **2011**, *65*, 703–706. [CrossRef]

26. Zhukov, A.; Rodionova, V.; Ilyn, M.; Aliev, A.M.; Varga, R.; Michalik, S.; Aronin, A.; Abrosimova, G.; Kiselev, A.; Ipatov, M.; Zhukova, V. Magnetic properties and magnetocaloric effect in Heusler-type glass-coated NiMnGa microwires. *J. Alloys Compd.* **2013**, *575*, 73–79. [CrossRef]

27. Zhukova, V.; Aliev, A.M.; Varga, R.; Aronin, A.; Abrosimova, G.; Kiselev, A.; Zhukov, A. Magnetic Properties and MCE in Heusler-Type Glass-Coated Microwires. *J. Supercond. Novel Magn.* **2013**, *26*, 1415–1419. [CrossRef]

28. Zhukova, V.; Ipatov, M.; Granovsky, A.; Zhukov, A. Magnetic properties of Ni-Mn-In-Co Heusler-type glass-coated microwires. *J. Appl. Phys.* **2014**, *115*, 17A939. [CrossRef]

29. Zhukova, V.; Rodionova, V.; Fetisov, L.; Grunin, A.; Goikhman, A.; Torcunov, A.; Aronin, A.; Abrosimova, G.; Kiselev, A.; Perov, N.; *et al.* Magnetic Properties of Heusler-Type Microwires and Thin Films. *IEEE Trans. Magn.* **2014**, *50*. [CrossRef]

30. Zhukov, A.; Gonzalez, J.; Blanco, J.M.; Vazquez, M.; Larin, V. Microwires coated by glass: A new family of soft and hard magnetic materials. *J. Mater. Res.* **2000**, *15*, 2107–2113. [CrossRef]

31. Chizhik, A.; Zhukov, A.; Gonzalez, J. Magnetic properties of sub-micrometric Fe-rich wires. *Thin Solid Films* **2013**, *543*, 130–132. [CrossRef]

32. Varga, R.; Zhukov, A.; Ipatov, M.; Blanco, J.M.; Gonzalez, J.; Zhukova, V.; Vojtanik, P. The influence of glass coating on the single domain wall potential in amorphous glass-coated Fe-based microwires. *J. Magn. Magn. Mater.* **2006**, *304*, E519–E521. [CrossRef]

33. Qin, F.; Peng, H.X.; Tang, J.; Qin, L.C. Ferromagnetic microwires enabled polymer composites for sensing applications. *Composites Part A* **2010**, *41*, 1823–1828. [CrossRef]

34. Zhukova, V.; Cobeno, A.F.; Zhukov, A.; Blanco, J.M.; Puerta, S.; Gonzalez, J.; Vazquez, M. Tailoring of magnetic properties of glass-coated microwires by current annealing. *J. Non-Cryst. Solids* **2001**, *287*, 31–36. [CrossRef]

35. Olofinjana, A.O.; Kern, J.H.; Daves, H.A. Effects of process variables on the multi-strand casting of high strength sub-millimetre metallic glass wire. *J. Mater. Process. Technol.* **2004**, *155*, 1344–1349. [CrossRef]

36. Olofinjana, A.O.; Kern, J.H.; Davies, H.A. Multistrand casting of amorphous alloy wire. *Mater. Lett.* **1995**, *23*, 55–57. [CrossRef]

37. Finnmore, E.; Franzini, J. *Fluid Mechanics with Engineering Applications*; McGraw-Hill Education: New York, NY, USA, 2001.

Metals **2015**, *5*, 1029–1044

38. Zeller, S.; Gnauk, J. Shape memory behaviour of Cu-Al wires produced by horizontal in-rotating-liquid-spinning. *Mater. Sci. Eng. A* **2008**, *481*, 562–566. [CrossRef]

39. Castrejon-Pita, A.A.; Castrejon-Pita, J.R.; Hutchings, I.M. Breakup of Liquid Filaments. *Phys. Rev. Lett.* **2012**, *108*, 074506. [CrossRef] [PubMed]

40. Richards, J.R.; Lenhoff, A.M.; Beris, A.N. Dynamic breakup of liquid jets. *Phys. Fluids* **1994**, *6*, 2640–2655. [CrossRef]

41. Liu, J.; Arnberg, N.; Backstrom, S.; Savage, S. Fundamental Parameters in the direct wire casting process. *Mater. Sci. Eng. A* **1988**, *98*, 21–24. [CrossRef]

42. Darken, L.S. *Physical Chemistry of Metals*; McGraw-Hill: New York, NY, USA, 1953.

metals

MDPI

Review

Toughness of Bulk Metallic Glasses

Shantanu V. Madge

CSIR-National Metallurgical Laboratory, Jamshedpur 831007, India; s.madge.99@cantab.net;
Tel.: +91-0657-234-5299; Fax: +91-0657-234-5213

Academic Editors: K. C. Chan and Jordi Sort Viñas
Received: 1 June 2015; Accepted: 10 July 2015; Published: 17 July 2015

Abstract: Bulk metallic glasses (BMGs) have desirable properties like high strength and low modulus, but their toughness can show much variation, depending on the kind of test as well as alloy chemistry. This article reviews the type of toughness tests commonly performed and the factors influencing the data obtained. It appears that even the less-tough metallic glasses are tougher than oxide glasses. The current theories describing the links between toughness and material parameters, including elastic constants and alloy chemistry (ordering in the glass), are discussed. Based on the current literature, a few important issues for further work are identified.

Keywords: bulk metallic glasses; toughness; elastic properties; shear transformation zones; plasticity; indentation

1. Introduction

The past 25 years have seen the emergence of newer alloys with a lower critical cooling rate for vitrification (0.1–1000 K/s), which permits the use of conventional casting techniques to obtain metallic glasses in bulk form, *i.e.*, >1 mm thickness [1,2]. The availability of these bulk metallic glasses (BMGs) has triggered intense research activity on various topics like their mechanical properties [3,4], studies on diffusion [5,6], and transformations like phase separation and crystallization [7–9]. BMGs exhibit certain attractive properties like high strength, hardness and wear resistance [10,11] and in some cases, good corrosion resistance [12]. Unlike crystalline alloys, BMGs do not work-harden and deformation in these materials tend to be localized into narrow regions called shear bands. Although the percent strain within a shear band is enormous, it contributes little to the overall plastic strain [13]. However, this is not to suggest that BMGs have low resistance to fracture initiation; indeed some compositions can show fracture toughness values comparable to engineering materials like Ti-6Al-4V, or maraging steels [14]. Other BMG compositions, like Mg-, or Fe-based show lower toughness and earlier reports suggested that their fracture energy approaches that of ideal brittle materials [15]. What really controls the toughness of BMGs has been a subject of intense research and debate. Furthermore, since BMGs are finding niche applications as in micro-gears, sensors or coatings [16–19], toughness is clearly of practical relevance. Excellent reviews exist on broader topics related to mechanical behavior of BMGs, including BMG-based composites [13,20] as well as a focused review on toughness [21]. However, the theories on intrinsic toughness of BMGs are still evolving and in light of the rapid progress made in recent years, it is worth summarizing the developments. The present short review is focused on toughness of monolithic BMGs—to put matters into context, it aims to first discuss the kind of toughness tests performed, including some current issues. Also included are some recent findings on the size-dependence of fracture toughness, particularly toughness and fracture in glassy thin films, since these are of fundamental significance as well as have implications for nano- and micro-scale applications of metallic glasses. A more detailed treatment of size-dependent mechanical properties in general will not be covered here since it has been extensively reviewed elsewhere [22]. Secondly, the evolving theories and open questions on correlations between toughness and material constants will be discussed, since such correlations are of importance in alloy design for enhanced toughness.

2. Techniques of Measuring Toughness

2.1. K_{Ic}/Notch Toughness Tests

The availability of amorphous alloys in bulk form has enabled the measurement of fracture toughness using standard techniques. The most investigated systems are those based on zirconium. Essentially, toughness has been measured either on fatigue pre-cracked specimens to give K_{Ic} values, or on notched samples (without pre-cracking) to yield notch toughness data. Conner *et al.* [14] first reported the notch toughness for Vitreloy 1 ($Zr_{41.25}Ti_{13.75}Cu_{12.5}Ni_{10}Be_{22.5}$) as 55–59 MPa.m$^{1/2}$, followed by tests on fatigue pre-cracked specimens of this alloy which showed K_{Ic}~30–68 MPa.m$^{1/2}$ [23]. Lowphaphandu and Lewandowski [24] reported that the toughness depends significantly on the notch root radius. In their work, fatigue pre-cracked samples showed a K_{Ic} ~18.4 MPa.m$^{1/2}$, whereas for the notched samples, it increased to 101–131 MPa.m$^{1/2}$, depending on the notch root radius, which varied from 65–250 µm. Kim *et al.* [25] reported toughness data for a range of newer Zr-based BMGs; the $Zr_{44}Ti_{11}Ni_{10.2}Cu_{9.3}Be_{25}Fe_{0.5}$ shows a K_{Ic} of ~27 MPa.m$^{1/2}$, whereas the notch toughness of $Zr_{33.5}Ti_{24}Cu_{15}Be_{27.5}$ is reported to be 80 MPa.m$^{1/2}$. It appears that usually notch toughness is higher than true K_{Ic} toughness because multiple shear bands form at the notch root and also fracture path does not remain planar, but instead shows bifurcation, processes that increase the energy absorbed during fracture [24].

A variety of other factors can also affect toughness. Compressive residual stresses in the surface of a cast sample can increase the K_{Ic}, e.g., from 34 MPa.m$^{1/2}$ to 51 MPa.m$^{1/2}$ for $Zr_{44}Ti_{11}Ni_{10}Cu_{10}Be_{25}$ [26]. Also, the state of relaxation, *i.e.*, free volume content can markedly affect BMG toughness, as amply demonstrated by Launey *et al.* [26], where the toughness can drop from 34 MPa.m$^{1/2}$ to 3 MPa.m$^{1/2}$ for relaxed samples. Another major factor is the oxygen content in the alloy, as demonstrated by Keryvin *et al.* [27] for a $Zr_{55}Cu_{30}Al_{10}Ni_5$ glass—at an oxygen level of 1000 appm, K_{Ic} is 37 MPa.m$^{1/2}$, whereas samples with 300 appm oxygen are much tougher leading to difficulties in fatigue pre-cracking and obtaining a valid K_{Ic}.

The stress state also plays a role. Flores and Dauskardt [28,29] were the first to report the mode II fracture toughness (K_{IIc}) of Vitreloy 1, which is 75 MPa.m$^{1/2}$, 4–5 times higher than K_{Ic} (16 MPa.m$^{1/2}$); suggesting that flow and fracture are affected by the stress normal to the failure plane. The hypothesis is that an applied tensile stress causes a local increase in free volume, which decreases the shear stress needed to cause flow.

It should be noted that for certain BMGs, toughness is almost independent of whether they are notched or fatigue pre-cracked. For instance, in $Ti_{40}Zr_{25}Cu_{12}Ni_3Be_{20}$, extensive shear banding blunts a pre-crack, effectively turning it into a notch, leading to high toughness [30]. Other recently developed alloys show similar traits. The $Pd_{79}Ag_{3.5}P_6Si_{9.5}Ge_2$ glass shows an apparent K_{Ic} of 150 MPa.m$^{1/2}$ and is one of the most damage-tolerant BMGs known [31]. Similarly, the $Zr_{61}Ti_2Cu_{25}Al_{12}$ glass has toughness of 130 ± 20 MPa.m$^{1/2}$ and a rising R-curve [32] which arises due to extensive shear banding at the crack tip that leads to crack deflection and a change in the local loading (at the crack tip) from pure mode I to mixed mode I/II. Mixed mode loading is known to enormously increase the toughness values [33]. Although the above toughness data are not strictly K_{Ic} values, it is impressive that certain compositions are able to show such profuse shear banding so as to change the loading mode. The key question is, why do only some compositions (and not all BMGs) show such behavior? Some insight has recently been provided by Xu *et al.* [34]. They postulated that the copious shear banding seen in glasses like $Zr_{61}Ti_2Cu_{25}Al_{12}$ is related to the presence of geometrically unfavored motifs (GUMs) in the glassy structure. Essentially, these are clusters of atoms that are more flexible and amenable to rearrangement upon application of stress. For example, the GUMs may include clusters with a higher or lower coordination number than expected from the alloy composition and atomic radii.

The hypothesis was supported by molecular dynamics simulations of binary $Cu_{64}Zr_{36}$ and $Cu_{20}Zr_{80}$ glasses. The former alloy had a higher fraction of the geometrically favored full icosahedra and these resist deformation. The $Cu_{20}Zr_{80}$ glass, however, possesses a greater variety of local motifs, which are amenable to change upon experiencing stress, thereby leading to a greater proliferation of shear bands in the alloy during deformation. A further discussion on compositional effects will be presented in Section 3.1.

An issue of particular interest is the sample size dependence of K_{Ic} values. Gludovatz *et al.* [35] investigated the $Zr_{52.5}Cu_{17.9}Ni_{14.6}Al_{10}Ti_5$ (Vitreloy 105) BMG and stated that K_{Ic} for compact tension (CT) specimens is 25.3 MPa.m$^{1/2}$, whereas it increases to 35.7 MPa.m$^{1/2}$ for single-edge (SE) notch bend specimens—a difference just below the threshold for statistical significance. Furthermore, their data suggest that toughness may increase with decreasing ligament size, even if the samples meet the size requirements specified by the ASTM E399 for K_{Ic} testing. This is quite unlike crystalline metals and alloys, and could be possibly related to the size-dependent ductility well known for glassy metals, e.g., their bending ductility increases with decreasing plate thickness, as shown by Conner *et al.* [36]. Also pertinent may be the fact that BMGs show strain softening behavior, unlike the work hardening seen in crystalline alloys. The authors recommend cautiously accepting ASTM E399 as providing specimen size-independent K_{Ic} data for BMGs.

2.2. Toughness from Fracture Surfaces

It is also possible to estimate toughness from the length scale of features on BMG fracture surfaces. Metallic glasses basically show two types of fracture, shear (ductile) or brittle (quasi-cleavage). Figure 1a shows an example of shear fracture in $Cu_{49}Hf_{42}Al_9$, characterized by shear band vein patterns. The mechanism of their formation will now be briefly considered. Shear bands form ahead of a crack tip and the material within these bands has a lowered viscosity due to structural changes and heating. Thus, the glass inside a shear band behaves like a viscous fluid, with the crack tip acting as a fluid meniscus that can advance under the action of a stress gradient, $d\sigma/dx$ ahead of the crack tip. Argon and Salama [37] showed that a perturbation in the fluid meniscus with wavelength λ will grow unstably, via fingering into the viscous material in the shear band, if $d\sigma/dx$ overcomes the surface tension χ. Specifically, the relation is as follows:

$$\lambda \geq \lambda_c = 2\pi\sqrt{\frac{\chi}{d\sigma/dx}} \tag{1}$$

where λ_c is a critical wavelength. In other words, only a perturbation with initial wavelength (λ) greater than λ_c will be able to grow. Once this condition is met, the crack tip (meniscus) breaks down into a series of parallel protrusions (fingers) that advance into the viscous material inside the shear band. These fingers grow and eventually, the ligaments connecting them rupture, thus causing the crack to advance. The corresponding fracture surface then shows the vein patterns typified by Figure 1a. These patterns resemble those found on separating two glass plates with a viscous medium in between and this mechanism fracture is called the fluid meniscus instability (FMI) mechanism [13,37,38]. A key parameter here is also the curvature radius of the crack tip, R. As shown by Jiang *et al.* [39], the FMI mechanism can operate only if the curvature radius, R is higher than λ_c. The curvature radius in turn depends on factors like crack speed, cracking mode and the intrinsic material toughness [39]. If $R < \lambda_c$, the glass will fail not through shear, but brittle (quasi-cleavage) fracture.

Figure 1b shows a typical example of brittle failure, as in a $Fe_{48}Cr_{15}Mo_{14}Er_2C_{15}B_6$ glass, with mirror, mist and hackle zones. Figure 1c,d show higher magnification views of the hackle zone—nanoscale corrugations (NCs) are visible. Brittle fracture features are very similar for all BMGs, ranging from the less tough Mg- or La-based alloys to the tougher Zr-based glasses. Their exact formation mechanism has been intensely debated. Earlier work [40] explained the features in terms of an FMI mechanism, similar to the vein patterns discussed above. The material at the crack tip is at a temperature close to the glass transition (T_g) and the crack tip acts as a fluid meniscus showing perturbations. Under the action of a stress gradient, perturbations with a critical wavelength will grow into the material ahead of the crack tip, in the process creating a new crack tip with a viscous zone ahead of it. The process repeats, leading to the formation of nanoscale corrugations (NCs) [38,40]. Others, however, hold different views and it has been asserted that FMI is unlikely to play a role in NC formation [38,41]. Narasimhan *et al.* [38], for example, have argued that the FMI mechanism predicts the formation of fingers that should run perpendicular to the crack front, whereas the NCs (as in Figure 1d) run parallel to the crack front. Molecular dynamics simulations by Murali *et al.* [42] on a ductile glass (CuZr) and a brittle glass (FeP) have shed considerable light on the possible mechanisms. They could show that extensive shear banding tends to dominate in the CuZr glass, causing ductile fracture. Brittle fracture in the FeP glass is preceded by cavitation in the zone ahead of the crack tip and it is correlated with nanoscale density fluctuations in the glass, where cavitation occurs preferentially in areas having a lower local density/strength.

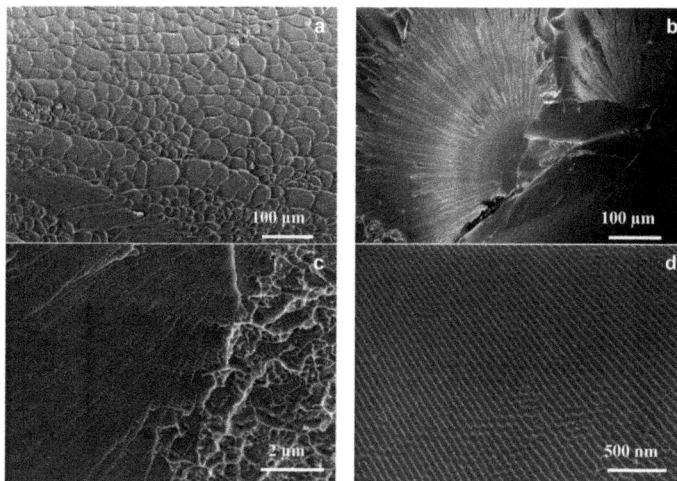

Figure 1. (**a**) An example of shear fracture in a $Cu_{49}Hf_{42}Al_9$ based glass tested in compression, showing vein patterns. (**b**) Brittle fracture in the $Fe_{48}Cr_{15}Mo_{14}Er_2C_{15}B_6$ glass, showing mirror, mist and hackle zones. (**c,d**) Higher magnification views showing nanoscale dimples and corrugations on the fracture surface of $Fe_{48}Cr_{15}Mo_{14}Er_2C_{15}B_6$.

The formation of NCs can then be explained as follows: Cavitation leads to the formation of nanovoids ahead of a crack tip, which grow under the action of a stress, and eventually, the ligaments between the crack tip and the voids break and the crack extends. The process continues, leading to a fracture surface as in Figure 1d, where broken ligaments are visible as nano-corrugations, and the crack has propagated perpendicular to these NCs. The periodicity of the NCs reflects the average wavelength of the strength fluctuations in the metallic glass.

As is apparent from this discussion, the scale of vein patterns (Figure 1a) should reflect the scale of the fracture process zone size. Hence, there has been interest in correlating fracture toughness with the scale of shear band vein patterns, as seen in early work by Kimura and Masumoto [43]. Recently, Xi *et al.* [44] utilized 3-point bending of single-edge notched specimens and showed that the scale of shear band vein patterns (w) represents the process zone size in a BMG and the fracture toughness (K_c) can be simply calculated from the yield strength (σ_y) using the following relationship

$$w = 0.025 \left(\frac{K_C}{\sigma_y} \right)^2 \tag{2}$$

It seems that the K_c is a mixed mode I/II toughness, and thus significantly larger than true K_{Ic}. For example, Xi *et al.* report K_c for Vitreloy 1 to be 86 MPa.m$^{1/2}$, compared to K_{IIc} of 75 MPa.m$^{1/2}$ and K_{Ic} of ~16 MPa.m$^{1/2}$ [28,29]. Although this relation was originally developed for bending fracture of notched samples, it has also been used for reliably estimating K_{IIc} toughness from compressive fracture surfaces *i.e.*, from the shear band vein patterns as in Figure 1a; but using the size of nanoscale features (Figure 1c) does not yield reliable estimates of fracture toughness [45].

The above discussion applies to glasses in bulk form. However, the toughness and fracture behavior of thin film metallic glasses can be very different, as reported by Ghidelli *et al.* [46,47]. They investigated fracture (in bending) of a range of Ni–Zr glassy films with extremely low thickness, varying between 200 to 900 nm. Although the glassy films are intrinsically (*i.e.*, structurally) similar to glasses in bulk form, a much lower toughness was noted, attributed to the very low film thickness, which restricts the development of a fracture process zone. It was also observed [46,47] that because of such low thickness and the resultant geometrical confinement (an extrinsic factor), the meniscus instability mechanism does not operate, leading to an absence of vein patterns on the fracture surface. Instead, nanoscale corrugations are seen for films with thickness down to 500 nm, below which even the corrugations disappear, leaving a featureless fracture surface. Additionally, toughness also depends on the film composition. These issues will be important in micro- or nano-scale applications of metallic glasses.

2.3. Compression Testing

Uniaxial compression testing has been one of the most popular techniques for assessing plasticity in BMGs, with compressive plastic strain being often used to evaluate BMGs. Interestingly, a larger plastic strain does not necessarily indicate higher fracture toughness, as seen in the work of Gu, *et al.* [30] on a Ti-based BMG. They studied samples with different sizes, with cross-sectional area ranging from 5×5 mm^2 to 1×1 mm^2. The latter were machined from the 5×5 mm^2 samples. Although the smaller samples showed higher compressive plasticity than the 5×5 mm^2 material (which showed no plasticity), toughness tests on fatigue pre-cracked specimens showed that the 5×5 mm^2 samples in fact had a high fracture toughness of 110 MPa.m$^{1/2}$. The findings emphasize that compressive plasticity may not always indicate a higher toughness. The larger plasticity for smaller samples seems related to a size effect, as also noted by Conner *et al.* [36], where thinner BMG plates show larger plastic strain in bending, arising from a greater number of operating shear bands.

The size-dependence of plasticity was more thoroughly investigated by Han *et al.* [48], who showed that in addition to sample size, stiffness of the testing machine plays a major role. Han *et al.* defined a fundamental parameter called the shear band instability index (S), as follows:

$$S = \frac{\pi E_Y d}{4 \rho k_M} \tag{3}$$

where E_Y is Young's modulus of the sample, d its diameter, ρ is the aspect ratio (height-to-diameter ratio) and k_M is the machine stiffness. They neatly showed that for S below a critical value (S_{cr}), deformation occurs through multiple shear banding leading to large plastic strain; but once $S > S_{cr}$, a single shear band dominates, leading to catastrophic failure. The S_{cr} cannot be determined from first principles and it has to be obtained experimentally. The key idea from their work is that S should be low for a given test, *i.e.*, plastic strain will be larger for smaller samples and higher machine stiffness. The question that remains is, what is the link between plasticity and fracture toughness? Intuitively, one can expect them to be directly related. As stated in [48], the parameter S_{cr} is a measure of the intrinsic toughness of a metallic glass—the higher the S_{cr}, the greater the toughness, as with Pd-, Pt-, and Zr-based BMGs and conversely, the lower the S_{cr}, the lower the toughness, as for Mg-based glasses.

The above discussion suggests that plasticity/toughness will increase with reducing sample size, but will this continue indefinitely, to extremely small sample sizes? This aspect was investigated by

Ghidelli *et al.* [46,47] on Ni–Zr thin films with sub-micron thickness. Interestingly, (as also stated in Section 2.2), for sizes below 1 μm, different mechanisms dominate. Unlike bulk samples, the fracture process zone cannot fully develop and the meniscus instability mechanism leading to shear band vein patterns does not operate anymore. Instead, brittle fracture occurs, with the formation of nanoscale corrugations and the material exhibits low fracture toughness.

Some glassy alloys display a very large compressive strain to failure, thereby suggesting high fracture toughness e.g., $Zr_{59}Cu_{18}Ni_8Ta_5Al_{10}$ shows about 20% plastic strain [49]. As shown by the authors, this can be misleading, because, in fact, the samples already crack at a lower strain (~10%), manifest as an inflection in the stress–strain curve, and the sample is held intact by a pattern of interlocking shear bands and cracks. The interlocking cracks are evident in scanning electron microscopy (SEM) observations of the tested samples. Thus, unusually high plasticity accompanied by an inflection in the stress–strain curves of monolithic BMGs should be treated with caution and microscopy should be used to ascertain that the large plastic strain really arises from multiple shear banding.

Recent work by Madge *et al.* [45] has shown that the mode II fracture toughness (K_{IIc}) can be reliably estimated from the compressive fracture surfaces, which show shear band vein patterns, such as those in Figure 1a for a $Cu_{49}Hf_{42}Al_9$ BMG. As mentioned in Section 2.2, the relevant parameters may be put into Equation (2) to estimate K_{IIc} fracture toughness [45].

As one would expect, another way of judging the toughness of a material is the area under the stress-strain curve. However, as mentioned earlier, data for a Ti-based BMG [30] reveal that higher compressive plasticity may not always indicate higher fracture toughness. In this regard, tensile tests could be more reliable, particularly for probing brittleness, which should be evident in low fracture strength (hence area under the curve). A fine example can be found in the work of Li *et al.* [50] on a Zr-based BMG. They conducted tensile and compressive tests on samples structurally relaxed by annealing below T_g for various times. The severely embrittled samples showed drastically reduced tensile strength (<600 MPa) instead of the 1660 MPa for as-cast samples. In contrast, the compressive strength remains unchanged and embrittlement is manifest only through fractography, with a change in fracture mode from shear (ductile) to brittle fracture. Embrittlement can also occur because of oxygen contamination, as demonstrated for the $Cu_{49}Hf_{42}Al_9$ BMG [51]. In this case too, the compressive fracture strength stays unchanged, and the embrittlement is only evident from the change in fracture mode to quasi-cleavage for samples with higher oxygen. Perhaps the tensile strength may be drastically lowered, although it was not tested in their work.

2.4. Indentation Fracture Toughness

For brittle materials, indentation tests can provide a rapid method of estimating toughness without the need for extensive fracture mechanics testing. Indentation fracture toughness (K_r) is related to the length of cracks emanating from the corners of a Vickers indent and can be estimated using Equation (4) for half-penny shaped cracks [52].

$$K_r = 0.016\sqrt{\frac{E}{H}} \cdot \frac{P}{c^{3/2}} \qquad (4)$$

where E is Young's modulus, H is hardness, P is the indentation load and c is the half crack length on the surface. If the cracks are radial, instead of half-penny, modified equations are used as discussed in [52].

The first data on indentation toughness were reported by Hess *et al.* [52] for an amorphous steel, $Fe_{48}Cr_{15}Mo_{14}Er_2C_{15}B_6$, which has a critical load for cracking between 31.4 and 41.2 N and an indentation toughness between 3.2–3.8 $MPa.m^{1/2}$, depending on the sample thickness. Previously, Gilbert *et al.* [24] had also evaluated indentation toughness of partially crystallized Vitreloy 1 specimens. Keryvin *et al.* [53] reported a lower critical load (between 5–10 N) for cracking and a toughness of 2.94 $MPa.m^{1/2}$ for a Fe–Co–Cr–Mo–C–B–Y bulk glass. It is important to note, however, that indentation toughness does not directly yield K_{Ic} values and its use has been criticized [54]. At best, these data can be used as a semi-quantitative estimate of toughness. One might expect the other supposedly brittle BMGs, like those based on Mg-, La- or Ce to also be suitable for indentation tests. Surprisingly, not all of these materials actually have such low toughness as to permit indentation tests. Figure 2 shows images of indents made in such BMGs—despite very high loads, no cracks are seen in the Ce-, La- and Fe-based BMGs; the measurements were repeated after re-polishing the indented specimens.

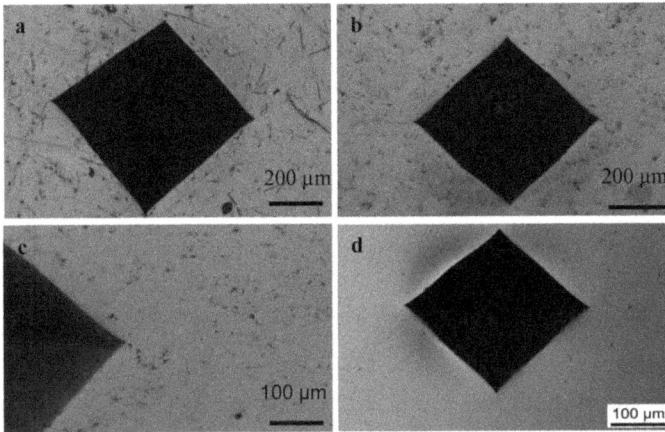

Figure 2. Vickers indents made in a variety of less-tough glasses, showing a surprising lack of crack initiation: (**a**) $Ce_{60}Al_{20}Cu_{10}Ni_{10}$ glass, at 589 N load; (**b,c**) $La_{55}Co_5Cu_{10}Ni_{10}Al_{20}$ at 491 N; and (**d**) $Fe_{64}Mo_{14}C_{15}B_6Er_1$ at 589 N. The present alloys appear to be tougher than other glasses with similar chemistries.

Table 1 summarizes the currently available indentation toughness data for a range of BMGs as well as their elastic properties, *i.e.*, Poisson's ratio, shear modulus and shear transformation zone (STZ) barrier energy densities. These parameters and their link with toughness will be discussed in Sections 3.1 and 3.3.

Table 1. Comparison of indentation toughness data.

Alloy Composition (at. %)	Poisson's Ratio (ν)	Yield Strength (σ_y) GPa	Shear Modulus (μ) GPa	STZ Barrier Energy Density. ρ (GJ/m³)	Indentation Toughness (MPa.m$^{1/2}$)	Reference
$Ce_{60}Al_{20}Cu_{10}Ni_{10}$ *	0.317 *	0.8	15 *	0.0259	Tough	This work, [55]
$La_{55}Co_5Cu_{10}Ni_{10}Al_{20}$	0.34	0.85	15.6	0.02815	Tough	This work, [55]
$Mg_{58}Cu_{31}Y_{11}$	0.318	0.986	20.4	0.02897	2.91	[55,56]
$Fe_{64}Mo_{14}C_{15}B_6Er_1$	0.316	3.9	75.4 **	~0.122	Tough	This work, [55,57]
$Fe_{48}Cr_{15}Mo_{14}Er_2C_{15}B_6$	0.318	3.75	80.8	0.1059	3.8 ± 0.3	[52,57]
$Fe_{41}Co_7Cr_{15}Mo_{14}C_{15}B_6Y_2$	0.334	3.5	84.1	0.0886	2.26 ± 0.4	[53,58]

* ν and μ are available in literature for the $Ce_{70}Al_{10}Cu_{10}Ni_{10}$ glass. For the present $Ce_{60}Al_{20}Cu_{10}Ni_{10}$, they were estimated using the approach given by Zhang and Greer [59], from the measured alloy density (6.479 g/cc). ** μ is taken as an average of two neighboring compositions $Fe_{65}Mo_{14}C_{15}B_6$ (73 GPa) and $Fe_{63}Mo_{14}C_{15}B_6Er_2$ (77.8 GPa) from [55].

2.5. Impact Toughness

This is yet another technique of measuring toughness of BMGs, at moderately high strain rates and is relatively simple to perform. Nagendra *et al.* [60] correlated the decrease in impact energy with crystallization in a La-based glass and showed that the formation of brittle intermetallics strongly reduces the impact toughness of the material. Degradation of toughness upon crystallization of a Zr–Ti–Cu–Ni–Be–Al glass was also noted by Raghavan *et al.* [61], as expected, because of the formation of brittle phases that aid crack nucleation. Surprising results were reported by Yokoyama *et al.* [62,63] for Zr–Cu–Al glasses—the hypo-eutectic $Zr_{60}Cu_{30}Al_{10}$ alloy shows an increase in impact toughness upon structural relaxation, in stark contrast to the usual embrittlement seen upon annealing. On the other hand, the eutectic $Zr_{50}Cu_{40}Al_{10}$ shows the usual reduced impact toughness upon annealing. They attributed this anomaly to structural changes, *i.e.*, short-range ordering in the glass.

2.6. Wear Resistance (An Indirect Indication of Toughness)

The wear resistance of any material is a property that derives from a combination of hardness and toughness. Abrasive wear resistance of BMGs has now been well characterized and representative data are shown in Figure 3, including data for conventional materials [10,64]. The plot shows data for pure metals, alloys and ceramics—within each class of material, wear resistance increases linearly with hardness, unless the material is brittle, in which case brittle fracture becomes a wear mechanism.

BMGs tend to obey Archard's wear law,

$$V_w = K\frac{SN}{H} \tag{5}$$

where V_w is wear volume, S is sliding distance, N is the normal load, H is hardness and K is the dimensionless wear coefficient, a fundamental measure of the wear severity [10]. If a material is less wear-resistant, say due to brittleness, it has higher K. Thus, the parameter K is important in detecting brittleness of a material. Upon embrittlement, metallic glasses show a lower wear resistance and a higher K, as discussed with examples in the comprehensive review by Greer *et al.* [10]. Table 2 summarizes the K values for 3-body abrasive wear, extracted from Figure 3. All BMGs, including the Mg- and La-based compositions, have rather similar wear coefficients and are in the category of hardened alloys. On the other hand, typically brittle materials like Si, have much larger K values, consistent with their brittleness. This shows that at least for the testing conditions used in these wear tests, all BMGs behave similarly, *i.e.*, are almost equally tough. Based on these data, past work raised the possibility that La- and Mg-based BMGs are probably much tougher than typically brittle materials like oxide glasses [45].

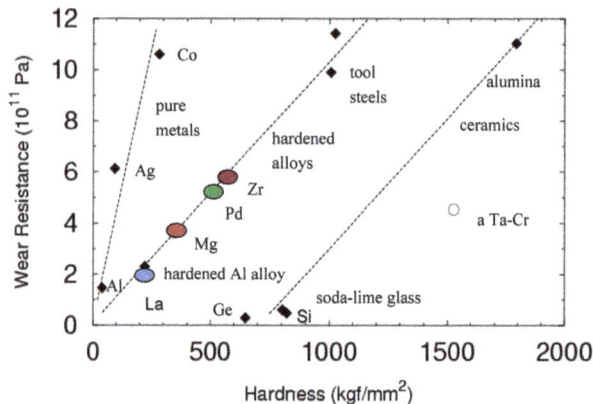

Figure 3. Correlation of the abrasive wear resistance with hardness for several different material classes [10,64]. Colored oval symbols show data for the bulk glasses with the alloy system being represented by the main element, e.g., Zr for Zr-based BMGs, *etc.* Within each class of material, the wear resistance scales linearly with hardness. All BMGs lie in the category of hardened alloys instead of ceramics.

Table 2. Dimensionless Wear Coefficients for selected materials shown in Figure 3.

Material	Wear Coefficient, K (3-body Abrasive Wear)
Zr–Cu–N–Al	1.03×10^{-2}
Pd-BMG	0.96×10^{-2}
Mg-BMG	0.9×10^{-2}
La-BMG	1.1×10^{-2}
Tool Steel	0.98×10^{-2}
Hardened Al alloy	0.93×10^{-2}
Pure Co	0.2×10^{-2}
Pure Si	0.15

3. What Controls Toughness of Bulk Metallic Glasses?

Bulk glasses can exhibit a range of toughness values, with alloys based on Zr, Cu, or Pd being typically tough, whereas the La-, Mg- and Fe-based systems are less tough [13]. What controls the toughness of glassy alloys has been an issue of much interest. Earlier work has shown that the free volume in a glass plays a major role. For instance, Wu and Spaepen [65] investigated embrittlement in a Fe-based glass upon structural relaxation and reported that, upon heating, the relaxed specimens undergo a brittle-to-ductile transition and the transition temperature (T_{DB}) is a measure of the degree of embrittlement. Furthermore, a neat correlation between T_{DB} and fractional free volume could be established; lower the free volume, greater the T_{DB}. However, it has also been noted that free volume alone may not always explain experimental findings, e.g., Raghavan *et al.* [66] reported that for relaxation-induced embrittlement in a Zr-based BMG, for the same free volume fraction, T_{DB} can vary widely. This has prompted detailed studies on the toughness of BMGs taking into account possible additional factors that may play a role.

The past decade has seen much activity in correlating toughness with physical properties like elastic moduli—which are not independent of free volume—but offer the advantage of being easily measurable. Moreover, this helps in developing tougher BMGs, since the moduli of most BMGs (apart from the metal-metalloid compositions) can be predicted reasonably well, based on the constituent elements [59]. In this section, we shall examine the various, often inter-related theories, on the toughness of BMGs.

3.1. Toughness–Poisson's Ratio Correlation

Chen *et al.* [67] first realized that the Poisson's ratio (ν) of a metallic glass influences toughness, also later proposed to explain the high notch toughness of a Pt-based BMG [68]. The brittleness or plasticity of a wide range of BMGs was correlated with ν (or equivalently, the ratio of the shear modulus, μ, to bulk modulus, B) of the alloy [69,70]. The idea is that μ represents the resistance to shear flow and B, the resistance to volume dilatation involved in cracking; ultimately, toughness will be controlled by whether the material undergoes shear flow, or shows cracking and the ratio μ/B (or ν) should play a role in deciding toughness. Figure 4 replots the data (taken from [69]) for various as-cast BMGs, which shows that fracture energy (G) reduces with decreasing ν and there appears to be a critical ν of 0.31–0.32, below which toughness apparently plummets, almost to the level of oxide glasses. The physical origin of the low toughness (and Poisson's ratio) of oxide glasses in comparison to most BMGs lies in their rigid covalent bonding, because of which cracking, rather than shearing (plastic flow) becomes the preferred deformation mechanism. In fact, recent work [71] has quantified the energy needed for the competing processes of fracture and shear in BMGs. Whether a glass undergoes shearing or cracking depends on two factors: (1) the ratio of the deformation strain energy density (U_D) to volume strain energy density (U_V) and (2) the ratio of the resistance of a glass to shearing (W_D) to cracking (W_V). These quantities are as under:

$$U_D/U_V = 2(1+\nu)/(1-2\nu) \tag{6}$$

$$W_D/W_V = 25.31(1 - 2v)/(1 + v) \tag{7}$$

If (U_D/U_V) exceeds (W_D/W_V), the glass undergoes shear flow (is plastic), else it shows cracking (is brittle). The transition is predicted to occur at a $v = 0.31$–0.32. The approach draws upon earlier work by Kelly *et al.* [72] on the ductility/brittleness of crystalline materials, who showed that the ratio of the largest tensile stress to the largest shear stress ahead of a crack tip controls ductile/brittle behavior. If this ratio exceeds the ratio of ideal cleavage stress to ideal shear stress of the material, brittle fracture would occur; for the converse case, some plasticity can be expected. However, for crystalline materials, the transition from ductile to brittle is not as sharp as the apparent transition that occurs for BMGs in Figure 4.

There have been mixed reports in the literature about the existence of a critical v for BMGs. The notch toughness of Fe-based BMGs and compressive plasticity could be improved through alloying to increase the Poisson's ratio to above 0.32 [57,73]. Conversely, the fracture toughness of Zr-based glasses, all with similar Poisson's ratios ($\sim$0.36) varies widely from 27.3 to 96.8 MPa.m$^{1/2}$ [25]. Likewise, the notch toughness of Cu-based BMGs [74] and annealing-induced embrittlement in two Zr-based BMGs was not found to bear any correlation with a critical v [66,75].

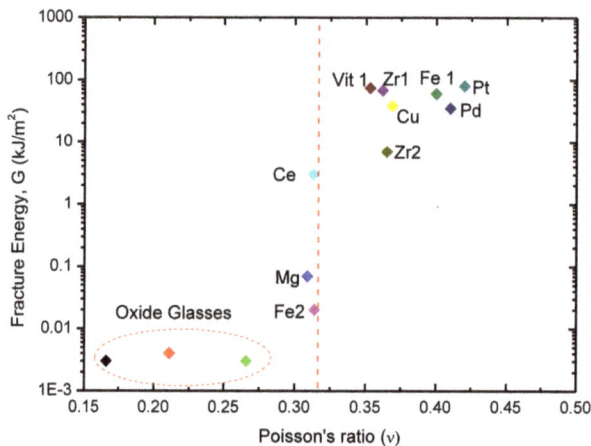

Figure 4. A plot of fracture energy *versus* Poisson's ratio for a range of bulk metallic glasses and oxide glasses. Data are taken from Lewandowski *et al.* [69]. There seems to be an apparent tough-to-brittle transition at a critical Poisson's ratio of 0.32, marked by the red dotted line. The symbols represent the following BMGs—**Pt**: $Pt_{57.5}Cu_{14.7}Ni_{5.3}P_{22.5}$; **Vit 1**: $Zr_{41}Ti_{14}Cu_{12.5}Ni_{10}Be_{22.5}$; **Zr 1**: $Zr_{57}Ti_{5}Cu_{20}Ni_{8}Al_{10}$; **Fe1**: $Fe_{80}P_{13}C_{7}$; **Cu**: $Cu_{60}Zr_{20}Hf_{10}Ti_{10}$; **Pd**: $Pd_{77.5}Cu_{6}Si_{16.5}$; **Zr2**: $Zr_{57}Nb_{5}Cu_{15.4}Ni_{12.6}Al_{10}$; **Ce**: $Ce_{70}Al_{10}Ni_{10}Cu_{10}$; and **Mg**: $Mg_{65}Cu_{25}Tb_{10}$; **Fe2**: $Fe_{50}Mn_{10}Mo_{14}Cr_{4}C_{16}B_{6}$.

Recent work [45] raised the issue of extrinsic effects on toughness data. In Figure 4, it has been assumed that G is an intrinsic material property. But it was noted that alloys like Mg- or rare earth-based compositions are very reactive and, unlike Zr- or Ti-based BMGs, have very low solubility for oxygen, inevitably leading to oxide inclusions. Figure 5a shows an example of such inclusions dispersed in a $Mg_{65}Cu_{25}Tb_{10}$ glass, containing $\sim$1000 appm oxygen. It was argued that the material is inherently capable of shear flow, but inclusions initiate cleavage fracture. Figure 5b–f shows fractographs for the Mg-based glass—clearly, the material can show micron scale shear band vein patterns, indicative of shear flow.

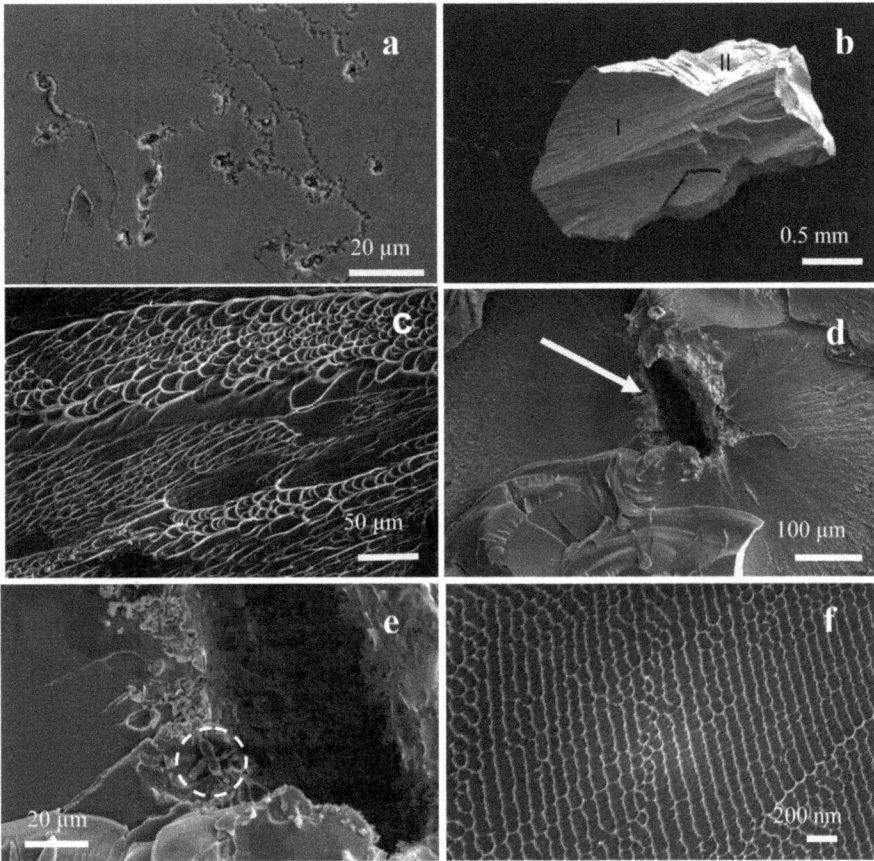

Figure 5. (a) Mixed oxides of Mg and Tb in a $Mg_{65}Cu_{25}Tb_{10}$ glassy matrix. (b) A fractured piece of the alloy after compression testing. Two regions, I and II are seen, that correspond to shear failure and brittle fracture respectively. (c) Region I shows vein patterns with a size of ~10 μm. (d) Region II consists of a flat, mirror-like fracture surface. Clear crack initiation sites are visible, as pointed by the arrow. (e) A closer view of a crack initiation site, which is a cluster of oxide particles. The dotted circle highlights an individual oxide particle. (f) Typical nano-scale fractographic features are seen in region II. Reprinted from [45], with permission from Elsevier.

The findings, *i.e.*, oxides nucleating cracks, were also found to hold good for the La-based glasses. It could be argued that the Mg- and La-based glasses are sensitive to oxides precisely because they lie close to the critical ν and hence have a low toughness. In other words, glasses having a higher ν would be immune to such embrittling effects. This aspect was investigated [45] and it was found that the tougher glasses like $Cu_{49}Hf_{42}Al_9$ (ν = 0.351) also embrittle when oxygen is intentionally added to the alloy, at levels (~1700 appm) sufficient to form oxygen-rich phases. In this Cu-based glass, the fracture mode is normally shear, which changes to quasi-cleavage upon introducing the oxygen-rich phases. Similarly, Zr-based glasses undergo embrittlement due to brittle phases, either oxygen-containing [76,77] or other intermetallics [78]. Even for Pd-based BMGs with high ν, Granata *et al.* [79] found plastic strain to decrease with oxygen content. It was thus argued in [45] that most of those BMGs in Figure 4 are capable of shear flow (at least under compression) but it is the presence of oxides, which induces quasi-cleavage fracture with nanoscale corrugations. The fundamental question is why do BMGs that usually form shear bands, should undergo quasi-cleavage fracture, when the glass is dispersed with just 1–3 volume percent brittle oxide phases? The author speculates that the oxides may be effectively behaving like cracks with a curvature radius (R) lower than the critical wavelength (λ_c) for meniscus instability essential for shear fracture. As stated in Section 2.2, shear fracture will be suppressed in such a scenario. However, this needs further investigation. Madge *et al.* [45] used the length scale of shear fracture features, *i.e.*, vein patterns, to estimate the mode II toughness of all BMGs via Equation (2). Figure 6 reproduces their data showing the re-calculated mode II fracture energies as a function of Poisson's ratio, for a range of BMGs. The trend now seen is quite different—the fracture energy spans two orders of magnitude, instead of four orders, and the sharpness of the ductile-brittle transition is lost; instead fracture energy gradually decreases with decreasing ν.

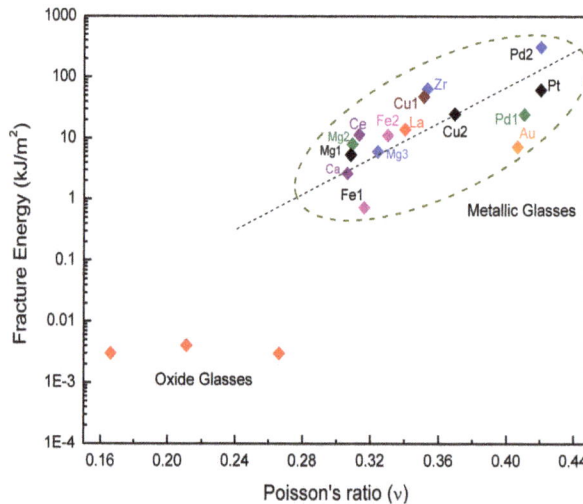

Figure 6. The correlation between mode II fracture energy (based on shear band vein patterns) and Poisson's ratio. The BMGs represented are, **Fe1**: $Fe_{48}Cr_{15}Mo_{14}Er_2C_{15}B_6$; **Fe2**: $Fe_{66}Cr_3Mo_{14}C_{15}B_6$; **Ca**: $Ca_{65}Mg_{15}Zn_{20}$; **Mg1**: $Mg_{65}Cu_{25}Gd_{10}$; **Mg2**: $Mg_{65}Cu_{25}Tb_{10}$; **Mg3**: $Mg_{64}Ni_{21}Nd_{15}$; **Ce**: $Ce_{60}Al_{20}Cu_{10}Ni_{10}$; **La**: $La_{55}Al_{25}Co_5Cu_{10}Ni_5$; **Cu1**: $Cu_{49}Hf_{42}Al_9$; **Cu2**: $Cu_{60}Zr_{20}Ti_{10}Hf_{10}$; **Pd1**: $Pd_{77.5}Cu_6Si_{16.5}$; **Pd2**: $Pd_{79}Ag_{3.5}P_6Si_{9.5}Ge_2$; **Pt**: $Pt_{57.5}Cu_{14.7}Ni_{5.3}P_{22.5}$; **Au**: $Au_{49}Ag_{5.5}Pd_{2.3}Cu_{26.9}Si_{16.3}$; and **Zr**: $Zr_{41.2}Ti_{13.8}Cu_{10}Ni_{12.5}Be_{22.5}$. All BMGs are much tougher than oxide glasses and their toughness gradually increases with Poisson's ratio instead of an abrupt increase at a critical value of the Poisson's ratio. Reprinted from [45], with permission from Elsevier.

It is noteworthy that glasses with similar ν show widely varying fracture energy (about an order of magnitude), suggesting that the alloy chemistry also plays a major role. Indeed, the indentation toughness data shown earlier (Figure 2 and Table 1) show much variation for alloys with similar Poisson's ratio. Another point is that BMGs are tougher than oxide glasses, also reflected in the wear data in Figure 3 and Table 2.

That Mg- or La-based glasses are capable of forming shear bands is also evident through the large compressive plasticity seen in composites based on these glasses. For example, $Mg_{65}Cu_{25}Tb_{10}$ reinforced with Ti shows extensive shear banding and large plastic strain under compression [80]. La-based BMGs also show similarly high plasticity when reinforced with Ti [81] or Ta [82] particles. Figure 7 shows an example of a La-based glass reinforced with Ta particles—under compression, multiple shear bands form that lead to large plastic strains of up to 40%. A profoundly brittle material, even when reinforced with 10–20 vol. % ductile particles, is unexpected to show legitimate shear flow.

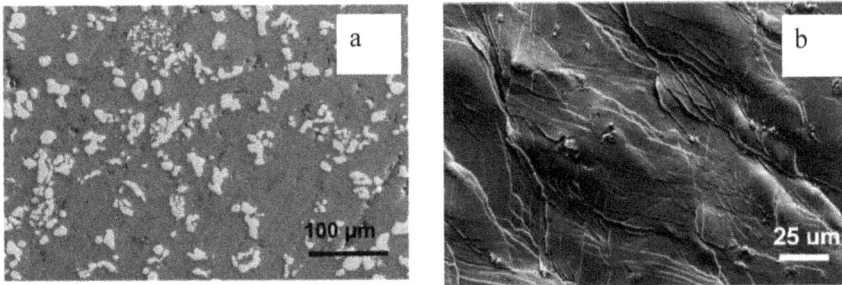

Figure 7. (a) A scanning electron micrograph showing Ta particles dispersed in a $La_{55}Al_{25}Cu_{10}Ni_{10}$ glassy matrix. (b) Multiple shear band offsets can be seen on samples tested under compression, which reach strains up to 40% (reproduced from [82]).

A characteristic of very brittle materials is a large difference in their tensile and compressive fracture strength, because any flaws in the material cause pre-mature fracture under tension. For BMGs, this can be seen in the work of Li *et al.* [50] who studied the effect of annealing on tensile/compressive behavior of a Zr-based glass (Vit 105). The embrittled Vit 105 shows a tensile strength <600 MPa compared to compressive strength of 1660 MPa. For some samples, tensile tests were not even possible due to extreme brittleness, although the compressive strength remained high, at 1876 MPa. So, a truly brittle BMG might be expected to possess a large asymmetry in the tensile and compressive fracture strength. Yet, Lee *et al.* [83] have reported identical tensile and compressive strengths of 550 MPa for a $La_{62}Al_{14}Cu_{12}Ni_{12}$ BMG. The compressive fracture strength for a $Mg_{65}Cu_{25}Tb_{10}$ glass is ~900 MPa [45], whereas the yield (fracture) strength for this glass under 3-point bending (which entails mixed mode I/II loading) is reported to be 660 MPa [44], which is probably not a huge difference, especially considering that the samples were prepared in different laboratories and may contain varying amount of oxide inclusions. These facts, though preliminary, suggest that La- or Mg-based BMGs may have much greater fracture energy under tension than oxide glasses. However, more data are necessary to draw any firm conclusions.

Also, the stress state is expected to significantly affect toughness [45,84]. Figure 6 relates only to mode II failure, but the trend may be very different for other states of loading. A good approach would be performing valid K_{Ic} tests on fatigue pre-cracked samples and comparing toughness data for all BMGs. However, this would mean first designing better casting/filtration techniques to ensure that samples are free from oxide inclusions, which would otherwise affect all toughness data.

So, the question is how does one reconcile the experimental facts in Figure 6 with the theory by Liu *et al.* [71], which predicts a critical Poisson's ratio? The answer may lie in local elastic moduli. It has been suggested that apart from Poisson's ratio, the local structure, *i.e.*, short-range order in the glass can affect toughness [84,85]. Poon *et al.* [85] have clearly shown that for a given global Poisson's ratio, local fluctuations in shear modulus at the STZ scale, arising from topological and chemical ordering in the glass, also influence the tendency to form shear bands and thus the toughness of the alloy. Their findings are reproduced in Figure 8; a glass can move between the ductile and brittle regimes depending on Poisson's ratio and local shear modulus. Modeling work on a variety of amorphous systems has indicated that a change in the local coordination number can affect toughness [86]. This seems the most plausible explanation for the scatter in fracture energy for a given Poisson's ratio in Figure 6, e.g., in spite of high ν, the Au-based glass possesses much lower toughness/fracture energy than Pt- or Pd-based systems.

Thus, it is reasonable to infer that toughness of BMGs depends significantly on elastic properties like Poisson's ratio. However, other factors like alloy chemistry, ordering and stress state also play a role such that there is significant variation in toughness/fracture energy at any given Poisson's ratio and hence the transition between tough and brittle glasses is probably less sharp than once thought.

In a sense, it may be similar to the ductile-brittle transition for crystalline materials [72], with the boundary between ductile and brittle materials not being very sharp.

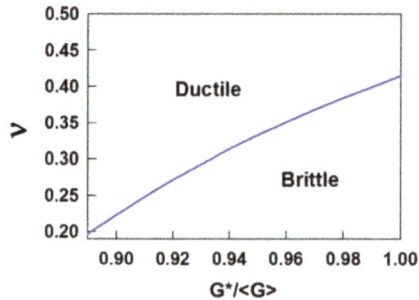

Figure 8. Plasticity or brittleness of BMGs depends on Poisson's ratio as well as local fluctuations in shear modulus. G* is the local shear modulus and <G> is the global shear modulus of a glassy alloy. Reproduced from [85] with permission from Applied Physics Letters. Copyright (2008), AIP Publishing LLC (Melville, NY, USA).

3.2. Toughness and Shear Transformation Zones (STZs)

The fundamental carriers of plasticity in metallic glasses are shear transformation zones (STZs), which are clusters of atoms in the glassy structure that, upon application of stress, undergo cooperative rearrangement from one low energy configuration to another (in the potential energy landscape), in the process surmounting a barrier corresponding to an activated state with greater energy [13,87,88]. In a metallic glass subjected to a critical shear stress, an STZ first forms at a site of greater free volume; secondary STZs may form around the primary STZ due to local strain fields and any free volume generated by the primary STZ. The process can repeat, leading to the formation of a shear band nucleus, which consists of a series of STZs. Above a critical shear strain, the nucleus propagates as a shear band, seen as macroscopic yielding in the metallic glass [13,20,88].

STZ size is now known to be important to BMG toughness and it has been experimentally estimated through nano-indentation at varying loading rates [88]. Pan et al. [88] estimated the activation volume for shear transformations through the strain rate sensitivity of hardness and the STZ volume was further calculated from the activation volume, using the cooperative shear model proposed earlier [89]. It was shown that the tougher BMGs have a larger STZ volume as well as Poisson's ratio (Figure 9). Physically, a larger STZ, in contrast to a smaller STZ, requires the activation of a lower number of STZs for the nucleation of a shear band. With smaller STZ sizes, a greater number of them need to cooperatively shear to generate a shear band, and instead, the competing process of local tensile failure may be favored [90]. The STZ size in BMGs appears to play a similar role as the width of dislocation cores in crystalline materials, where a wider core confers better ductility.

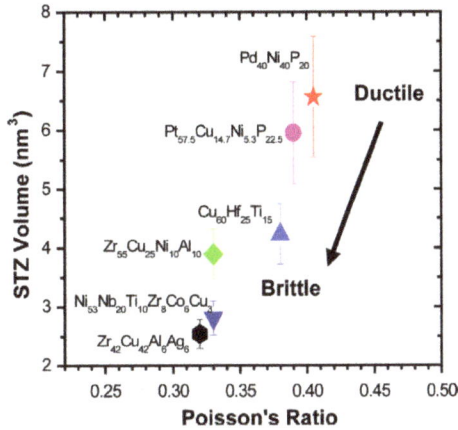

Figure 9. Correlation of shear transformation zone (STZ) volume with Poisson's ratio. The tougher glasses tend to have larger STZ volume and Poisson's ratio. Reprinted from [88], with permission from Proceedings of the National Academy of Sciences PNAS. Copyright (2008) National Academy of Sciences (Washington, DC, USA).

The ductile-brittle transition (DBT) in a Zr-based BMG could not be explained in terms of a critical ν nor free volume content, and it was suggested that STZ size is the parameter controlling DBT [66]. In crystalline materials, toughness is well known to depend upon the testing temperature and strain rate. Recent work has aimed to capture effects of such testing parameters on DBT in BMGs and correlating them with STZ size [50,90,91]. Li *et al.* [50] related the DBT, caused by structural relaxation, in the Vit 105 BMG to the STZ size and it was demonstrated that STZ size depends on three factors: (i) test temperature; (ii) strain rate; and (iii) free volume fraction in the glass, through Equation (8).

$$\Omega = f(T)\left(\ln\frac{\omega_0}{C\dot{\gamma}} - \frac{\Delta V^*}{V_f} \right) \tag{8}$$

where Ω is the STZ volume, $f(T)$ is a temperature-dependent function, ω_0 is an attempt frequency, C a constant, $\dot{\gamma}$ is the strain rate, $V_f/\Delta V^*$ is fractional free volume. The tough-brittle transition corresponds to a reduction in STZ volume ~0.17 nm^3, below which STZs do not operate, and fracture occurs not through shear but through the quasi-cleavage mechanism involving local tensile failure. As elegantly shown in their work, a higher strain rate, lower temperature or a decrease in free volume, all lead to a reduction in STZ volume, and ultimately to a DBT below a critical STZ size. Likewise, STZ size increases with temperature, lower strain rate and a larger free volume fraction (higher Poisson's ratio), leading to higher toughness. The theory takes the view that whether a BMG is brittle or tough depends on where it resides in the 3D space of temperature, strain rate and fractional free volume. An example of high strain rates causing brittle fracture in the otherwise tough Vitreloy-1 BMG is seen in the experiments of Jiang *et al.* [39]. The idea of STZ volume thus seems to offer a more complete picture of ductile-brittle behavior, compared to studies on say the effect of testing temperature [65], which effectively analyzed a 2D section, *i.e.*, free volume and temperature. For metallic glasses, this work is probably the first unified approach that illustrates the complexity of BMG toughness.

It would be interesting to use this approach in investigating the change in toughness of the less-tough glasses with low Poisson's ratio, *i.e.*, Mg- or La-based alloys. The Au-based BMG (Figure 6) presents a peculiar case—it has a T_g of ~110 °C, so room temperature represents a high homologous temperature ($T/T_g = 0.76$), and according to the present model, it should have a large STZ volume and toughness. Yet, it has strangely low fracture energy despite its high Poisson's ratio. Does this mean, for some reason, a low STZ volume? Or do other effects like local changes to the elastic moduli, arising out of ordering play a more dominant role? These questions may be of interest for future work.

3.3. Toughness–Shear Modulus

The shear modulus (μ) is another parameter reported to affect fracture toughness. Based on the cooperative shear model, Johnson and Samwer [89] proposed that the energy barrier to shear flow (W) is related to μ and the molar volume (V_m) for a glass configuration frozen at T_g through the following relation:

$$W(T_g) = \mu(T_g)V_m(T_g) \tag{9}$$

The idea is that a glass with low μ and low T_g should have a lower energy barrier to the operation of shear transformation zones (the precursors of shear bands) and hence higher toughness. Demetriou *et al.* [92] developed newer Fe-based BMGs starting from $Fe_{80}P_{12.5}C_{7.5}$ and found a greater notch toughness for alloys with lower μ.

The ideas presented in [89] were adapted by Liu *et al.* [55] to arrive at the barrier energy density for STZ activation (ρ), defined as the barrier energy that must be overcome per unit volume of STZs, for their operation. Alloys with lower values of ρ are expected to be tougher.

$$\rho = (6/\pi^2)(\sigma_y^2/\mu) \tag{10}$$

Their findings are shown schematically in Figure 10. In general, a higher Poisson's ratio correlates with a lower ρ [Figure 10a] and fracture energy (G) increases with $1/\rho$, as seen in Figure 10b. The Mg-, Ce- and La-based glasses are outliers in both graphs, attributed by the authors to their low T_g, meaning that deformation at room temperature (a significant fraction of T_g) may be in the transition region from shear bands to homogeneous flow. For any given ρ, there is significant scatter in fracture energy (about an order of magnitude), suggesting that some other factors are at work too. Likewise, indentation toughness varies considerably for BMGs with a similar ρ (Table 1). This is consistent with the scatter in toughness–Poisson's ratio correlation in Figure 6 and the "other" influencing factors may possibly be short-range ordering in the BMG, leading to fluctuations in the local shear modulus as depicted in Figure 8. Chemistry effects on toughness are clearly seen for a series of Zr–TM–Al (TM = Co, Ni, Cu)—toughness generally increases with ν or a lower μ, but, for a given ν (or μ), alloy chemistry exerts an additional influence [93].

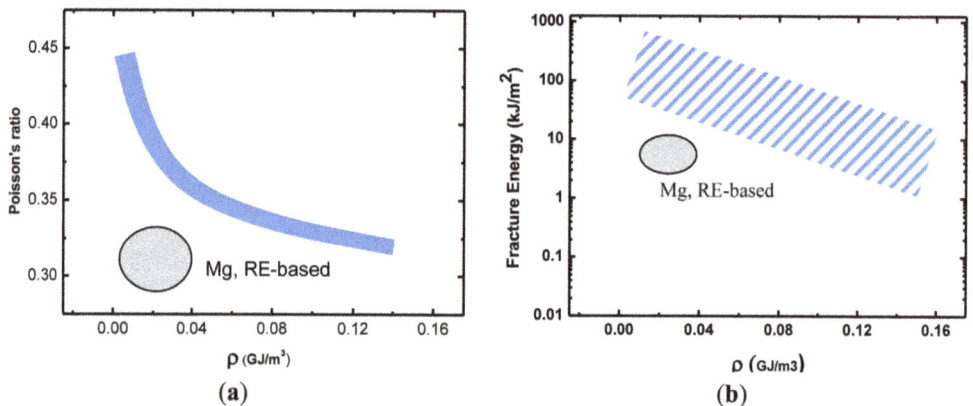

Figure 10. Schematic plots based on the work in [55], showing inverse correlation between STZ barrier energy density and (**a**) Poisson's ratio and (**b**) fracture energy. The Mg- and rare earth (RE)-based BMGs are outliers, probably because of their low glass transition temperature.

3.4. Toughness and the Critical Fictive Temperature

In glass physics, the fictive temperature (T_f) is defined as the temperature where a thermodynamic property (e.g., specific volume) extrapolated on the glass line intersects the line of the equilibrium liquid [94,95]. As mentioned by Badrinarayanan *et al.* [95], the fictive temperature correlates with the structural state of a glass, and can be measured upon heating, whereas the glass transition temperature (T_g) is strictly measured on cooling. However, as clearly shown in their work [95], T_f is only about 1.5 °C lower than T_g and thus, the two are practically the same and both depend on the imposed cooling rate [95,96]. Certainly, the calorimetric T_g measured upon heating in a differential scanning calorimeter (DSC) can be safely taken to be T_f. Faster cooling of the undercooled liquid leads to a higher T_g and a higher fraction of free volume [96]. More slowly cooled alloys possess a lower T_g and the glass so obtained is more structurally relaxed, *i.e.*, has less fraction of free volume. Structural relaxation is well known to lead to a reduction in toughness [65,66]. So, it can be imagined that if the freezing/glass transition temperature happens to be too low (assuming crystallization does not intervene), the glass obtained will be so relaxed, that it can be brittle. This freezing temperature corresponding to the onset of brittleness is defined as the critical fictive temperature (T_{fc}) by Kumar *et al.* [97] and it is a material property, linked to the structural configuration of the glass.

Kumar *et al.* [97] studied annealing-induced embrittlement in Zr-, Pd- and a Pt-based BMG and explained their findings using the concept of T_{fc}. They annealed specimens at various temperatures around the calorimetric T_g, in order to fully relax them and attain a structural configuration characteristic of the annealing temperature. In other words, they obtained glassy specimens having a fictive temperature that equals the annealing temperature. Such specimens were tested in bending and specimens showing 2.5% strain to failure were considered brittle; those with larger strain were called ductile. Through these experiments, they identified the annealing temperature that leads to embrittlement (2.5% failure strain), and this annealing temperature is the T_{fc} for the particular BMG. Based on these experiments, it was proposed in [97] that the intrinsic plasticity of BMGs in their as-cast state depends on how the T_{fc} of the glass compares with the glass transition temperature (T_g). For glasses like Pt-BMG, T_g–T_{fc} ~57 K, and the glass is always ductile in the as-cast state; further it is also resistant to annealing-induced embrittlement. This is because the supercooled liquid congeals into a glass before it ever reaches the structural state (free volume fraction) characteristic of T_{fc}. For Zr-BMGs, T_g–T_{fc} ~25 K, so the alloy is ductile in the as-cast state, but is susceptible to relaxation-induced embrittlement upon sub-T_g annealing. Importantly, it was also hypothesized that Mg- or Fe-based BMGs are brittle in the as-cast state because their T_{fc} is much higher than T_g, and during cooling at usual rates to form bulk glasses, the liquid already acquires a structural configuration characteristic of T_{fc}, and it is thus difficult to avoid embrittlement.

They also argued that all the Zr, Pd and Pt-BMGs in their experiments, even after relaxation, have Poisson's ratios well above the critical 0.32 discussed earlier, suggesting that embrittlement cannot be explained in terms of a change in the Poisson's ratio, in contrast to the theory advanced in [69]. However, it must be stressed that in [69], brittleness is defined in terms of fracture energy, whereas Kumar *et al.* [97] considered a glass to have embrittled once it showed a failure strain of 2.5%. The concomitant drop in fracture energy has not been reported. As stated in Section 2.3, Gu *et al.* [30] showed that a reduction in the failure strain may not always mean a drop in fracture toughness/energy. It would be worthwhile to also correlate the toughness drop with T_{fc}.

From the perspective of alloy development, knowing the T_{fc} of a new composition, *a priori*, is difficult, but the concept of a critical fictive temperature is appealing and poses questions about its microscopic origins, e.g., does it mean a critical free volume for all BMGs at the T_{fc}? Or is it a critical STZ volume? These questions would make for an interesting research topic. Also, a question that comes up is whether the T_{fc} is related to the fragility of the undercooled liquid? Because fragility tells us how the viscosity (free volume) changes with temperature [98,99], and since T_{fc} is related to a structural state, it is reasonable to expect a link.

4. Concluding Remarks

The toughness of BMGs has now been measured using a range of techniques and various parameters like Poisson's ratio (ν), barrier energy density of STZs (ρ), STZ volume and critical fictive temperature have been proposed to explain toughness. Such parameters are broadly consistent with each other, e.g., tougher alloys have higher ν, higher STZ volume and a lower ρ. Some key issues can be summarized as follows:

(1) The testing technique can influence toughness values, e.g., notch tests *versus* fatigue pre-cracked fracture toughness tests, and in comparing different BMGs, it would be useful to have valid K_{Ic} data for a broad range of alloys. The very reactive Mg- or RE-based glasses often contain oxide inclusions (more so than other glasses), which impair toughness and make comparison between various BMGs difficult. Hence one theme of research is to design processing routes to obtain alloys free from oxides. This will be challenging, but better filtration techniques, or electrolytic reduction of alloy melts [100] could offer possibilities.

(2) Although mode II fracture energy increases only gradually with Poisson's ratio (without an abrupt brittle-to-ductile transition), the stress state may well affect this correlation. It is of clear interest to examine the correlation with K_{Ic} data, obtained from samples free of extrinsic effects like oxides.

(3) Some BMGs show unusually high toughness arising from multiple shear banding in K_{Ic} tests, e.g., $Zr_{61}Ti_2Cu_{25}Al_{12}$ [32]. Why only certain compositions show profuse shear banding and enhanced toughness is not clear and uncovering the underlying mechanisms is of much interest.

(4) A related issue is that of local fluctuations in shear modulus, arising from ordering in the glass, which can affect toughness significantly. The question now must be how can these fluctuations be predicted and described in a tangible form so as to aid alloy design?

(5) Using the STZ volume to describe ductile–brittle transition appears to be quite successful for Zr-BMGs and it would be worthwhile investigating other, less tough alloys like Fe- or Mg-based glasses in this framework.

(6) The microscopic origins of the critical fictive temperature are not entirely clear and thus represent an interesting research topic.

(7) Toughness of metallic glasses in thin film form has shown interesting results, such as the suppression of shear fracture, and this can be an area of further research, since it will be relevant to applications in micro-electro-mechanical systems (MEMS).

Acknowledgments: The author thanks the anonymous reviewers for their comments and Mr Shravan Kumar (Monash University, Australia), for his help with the manuscript.

Conflicts of Interest: The author declares no conflict of interest.

References

1. Inoue, A. Stabilization of metallic supercooled liquid and bulk amorphous alloys. *Acta Mater.* **2000**, *48*, 279–306. [CrossRef]
2. Peker, A.; Johnson, W.L. A highly processable metallic glass: $Zr_{41.2}Ti_{13.8}Cu_{12.5}Ni_{10}Be_{22.5}$. *Appl. Phys. Lett.* **1993**, *63*, 2342–2344. [CrossRef]
3. Conner, R.D.; Dandliker, R.B.; Johnson, W.L. Mechanical properties of tungsten and steel fiber reinforced $Zr_{41.25}Ti_{13.75}Cu_{12.5}Ni_{10}Be_{22.5}$ metallic glass matrix composites. *Acta Mater.* **1998**, *46*, 6089–6102. [CrossRef]
4. Choi-Yim, H.; Busch, R.; Köster, U.; Johnson, W.L. Synthesis and characterization of particulate reinforced $Zr_{57}Nb_5Al_{10}Cu_{15.4}Ni_{12.6}$ bulk metallic glass composites. *Acta Mater.* **1999**, *47*, 2455–2462. [CrossRef]
5. Tang, X.-P.; Geyer, U.; Busch, R.; Johnson, W.L.; Wu, Y. Diffusion mechanisms in metallic supercooled liquids and glasses. *Nature* **1999**, *402*, 160–162.
6. Zumkley, T.; Naundorf, V.; Macht, M.-P.; Frohberg, G. Effect of reversible structural relaxation on diffusion in a ZrTiCuNiBe bulk glass. *Scr. Mater.* **2001**, *45*, 471–477. [CrossRef]
7. Gangopadhyay, A.K.; Croat, T.K.; Kelton, K.F. The effect of phase separation on subsequent crystallization in $Al_{88}Gd_6La_2Ni_4$. *Acta Mater.* **2000**, *48*, 4035–4043. [CrossRef]

8. Madge, S.V.; Alexander, D.T.L.; Greer, A.L. An EFTEM study of compositional variations in Mg–Ni–Nd bulk metallic glasses. *J. Non-Cryst. Solids* **2003**, *317*, 23–29. [CrossRef]
9. Park, E.S.; Kim, D.H. Phase separation and enhancement of plasticity in Cu–Zr–Al–Y bulk metallic glasses. *Acta Mater.* **2006**, *54*, 2597–2604. [CrossRef]
10. Greer, A.L.; Rutherford, K.L.; Hutchings, I.M. Wear resistance of amorphous alloys and related materials. *Int. Mater. Rev.* **2002**, *47*, 87–112. [CrossRef]
11. Madge, S.V.; Caron, A.; Gralla, R.; Wilde, G.; Mishra, S.K. Novel W-based metallic glass with high hardness and wear resistance. *Intermetallics* **2014**, *47*, 6–10. [CrossRef]
12. Xu, T.; Pang, S.; Li, H.; Zhang, T. Corrosion resistant Cr-based bulk metallic glasses with high strength and hardness. *J. Non-Cryst. Solids* **2015**, *410*, 20–25. [CrossRef]
13. Schuh, C.A.; Hufnagel, T.C.; Ramamurty, U. Mechanical behavior of amorphous alloys. *Acta Meter.* **2007**, *55*, 4067–4109. [CrossRef]
14. Conner, R.D.; Rosakis, A.J.; Johnson, W.L.; Owen, D.M. Fracture toughness determination for a beryllium-bearing bulk metallic glass. *Scr. Mater.* **1997**, *37*, 1373–1378. [CrossRef]
15. Lewandowski, J.J.; Shazly, M.; Shamimi Nouri, A. Intrinsic and extrinsic toughening of metallic glasses. *Scr. Mater.* **2006**, *54*, 337–341. [CrossRef]
16. Nishiyama, N.; Amiya, K.; Inoue, A. Novel applications of bulk metallic glass for industrial products. *J. Non-Cryst. Solids* **2007**, *353*, 3615–3621. [CrossRef]
17. Inoue, A.; Takeuchi, A. Recent development and application products of bulk glassy alloys. *Acta Mater.* **2011**, *59*, 2243–2267. [CrossRef]
18. Liquidmetal Technologies. Available online: http://liquidmetal.com/ (accessed on 14 May 2015).
19. Nanosteel. Available online: https://nanosteelco.com/ (accessed on 14 May 2015).
20. Greer, A.L.; Cheng, Y.Q.; Ma, E. Shear bands in metallic glasses. *Mater. Sci. Eng. R* **2013**, *74*, 71–132. [CrossRef]
21. Xu, J.; Ramamurty, U.; Ma, E. The fracture toughness of bulk metallic glasses. *JOM* **2010**, *62*, 10–18. [CrossRef]
22. Greer, J.R.; de Hosson, J.T.M. Plasticity in small-sized metallic systems: Intrinsic *versus* extrinsic size effect. *Prog. Mater. Sci.* **2011**, *56*, 654–724. [CrossRef]
23. Gilbert, C.J.; Schroeder, V.; Ritchie, R.O. Mechanisms for fracture and fatigue-crack propagation in a bulk metallic glass. *Metall. Mater. Trans. A* **1999**, *30*, 1739–1753. [CrossRef]
24. Lowhaphandu, P.; Lewandowski, J.J. Fracture toughness and notched toughness of bulk amorphous alloy: Zr–Ti–Cu–Ni–Be. *Scr. Mater.* **1998**, *38*, 1811–1817. [CrossRef]
25. Kim, C.P.; Suh, J.Y.; Wiest, A.; Lind, M.L.; Conner, R.D.; Johnson, W.L. Fracture toughness study of new Zr-based Be-bearing bulk metallic glasses. *Scr. Mater.* **2009**, *60*, 80–83. [CrossRef]
26. Launey, M.E.; Busch, R.; Kruzic, J.J. Effects of free volume changes and residual stresses on the fatigue and fracture behavior of a Zr–Ti–Ni–Cu–Be bulk metallic glass. *Acta Mater.* **2008**, *56*, 500–510. [CrossRef]
27. Keryvin, V.; Nadot, Y.; Yokoyama, Y. Fatigue pre-cracking and toughness of the $Zr_{55}Cu_{30}Al_{10}Ni_5$ bulk metallic glass for two oxygen levels. *Scr. Mater.* **2007**, *57*, 145–148. [CrossRef]
28. Flores, K.M.; Dauskardt, R.H. Fracture and deformation of bulk metallic glasses and their composites. *Intermetallics* **2004**, *12*, 1025–1029. [CrossRef]
29. Flores, K.M.; Dauskardt, R.H. Mode II fracture behavior of a Zr-based bulk metallic glass. *J. Mech. Phys. Solids* **2006**, *54*, 2418–2435. [CrossRef]
30. Gu, X.J.; Poon, S.J.; Shiflet, G.J.; Lewandowski, J.J. Compressive plasticity and toughness of a Ti-based bulk metallic glass. *Acta Mater.* **2010**, *58*, 1708–1720. [CrossRef]
31. Demetriou, M.D.; Launey, M.E.; Garrett, G.; Schramm, J.P.; Hofmann, D.C.; Johnson, W.L.; Ritchie, R.O. A damage-tolerant glass. *Nat. Mater.* **2011**, *10*, 123–128. [CrossRef] [PubMed]
32. He, Q.; Shang, J.K.; Ma, E.; Xu, J. Crack-resistance curve of a Zr–Ti–Cu–Al bulk metallic glass with extraordinary fracture toughness. *Acta Mater.* **2012**, *60*, 4940–4949. [CrossRef]
33. Varadarajan, R.; Thurston, A.K.; Lewandowski, J.J. Increased toughness of zirconium-based bulk metallic glasses tested under mixed mode conditions. *Metall. Mater. Trans. A* **2010**, *41*, 149–158. [CrossRef]
34. Xu, J.; Ma, E. Damage-tolerant Zr–Cu–Al bulk metallic glasses with record-breaking fracture toughness. *J. Mater. Res.* **2014**, *29*, 1489–1499. [CrossRef]
35. Gludovatz, B.; Naleway, S.E.; Ritchie, R.O.; Kruzic, J.J. Size-dependent fracture toughness of bulk metallic glasses. *Acta Mater.* **2014**, *70*, 198–207. [CrossRef]

36. Conner, R.D.; Li, Y.; Nix, W.D.; Johnson, W.L. Shear band spacing under bending of Zr-based metallic glass plates. *Acta Mater.* **2004**, *52*, 2429–2434. [CrossRef]
37. Argon, A.S.; Salama, M. The mechanism of fracture in glassy materials capable of some inelastic deformation. *Mater. Sci. Eng.* **1976**, *23*, 219–230. [CrossRef]
38. Narasimhan, R.; Tandaiya, P.; Singh, I.; Narayan, R.L.; Ramamurty, U. Fracture in metallic glasses: Mechanics and mechanisms. *Int. J. Fract.* **2015**, *191*, 53–75. [CrossRef]
39. Jiang, M.Q.; Ling, Z.; Meng, J.X.; Dai, L.H. Energy dissipation in fracture of bulk metallic glasses via inherent competition between local softening and quasi-cleavage. *Philos. Mag.* **2008**, *88*, 407–426. [CrossRef]
40. Wang, G.; Chan, K.C.; Xu, X.H.; Wang, W.H. Instability of crack propagation in brittle bulk metallic glass. *Acta Mater.* **2008**, *56*, 5845–5860. [CrossRef]
41. Narayan, R.L.; Tandaiya, P.; Narasimhan, R.; Ramamurty, U. Wallner lines, crack velocity and mechanisms of crack nucleation and growth in a brittle bulk metallic glass. *Acta Mater.* **2014**, *80*, 407–420. [CrossRef]
42. Murali, P.; Guo, T.F.; Zhang, Y.W.; Narasimhan, R.; Li, Y.; Gao, H.J. Atomic scale fluctuations govern brittle fracture and cavitation behaviour in metallic glasses. *Phys. Rev. Lett.* **2011**, *107*, 215501. [CrossRef] [PubMed]
43. Kimura, H.; Masumoto, T. Deformation and fracture of an amorphous Pd–Cu–Si alloy in V-notch bending tests-II. Ductile-brittle transition. *Acta Metall.* **1980**, *28*, 1677–1693. [CrossRef]
44. Xi, X.K.; Zhao, D.Q.; Pan, M.X.; Wang, W.H.; Wu, Y.; Lewandowski, J.J. Fracture of Brittle metallic glasses: Brittleness or plasticity. *Phys. Rev. Lett.* **2005**, *94*, 125510. [CrossRef] [PubMed]
45. Madge, S.V.; Louzguine-Luzgin, D.V.; Lewandowski, J.J.; Greer, A.L. Toughness, extrinsic effects and Poisson's ratio of bulk metallic glasses. *Acta Mater.* **2012**, *60*, 4800–4809. [CrossRef]
46. Ghidelli, M.; Gravier, S.; Blandin, J.-J.; Raskin, J.-P.; Lani, F.; Pardoen, T. Size-dependent failure mechanisms in ZrNi thin metallic glass films. *Scr. Mater.* **2014**, *89*, 9–12. [CrossRef]
47. Ghidelli, M.; Gravier, S.; Blandin, J.-J.; Djemia, P.; Mompiou, F.; Abadias, G.; Raskin, J.-P.; Pardoen, T. Extrinsic mechanical size effects in thin ZrNi metallic glass films. *Acta Mater.* **2015**, *90*, 232–241. [CrossRef]
48. Han, Z.; Wu, W.F.; Li, Y.; Wei, Y.J.; Gao, H.J. An instability index of shear band for plasticity in metallic glasses. *Acta Mater.* **2009**, *57*, 1367–1372. [CrossRef]
49. Mondal, K.; Kumar, G.; Ohkubo, T.; Oishi, K.; Mukai, T.; Hono, K. Large apparent compressive strain of metallic glasses. *Philos. Mag. Lett.* **2007**, *87*, 625–635. [CrossRef]
50. Li, G.; Jiang, M.Q.; Jiang, F.; He, L.; Sun, J. The ductile to brittle transition behavior in a Zr-based bulk metallic glass. *Mater. Sci. Eng. A* **2015**, *625*, 393–402. [CrossRef]
51. Madge, S.V.; Wada, T.; Louzguine-Luzgin, D.V.; Greer, A.L.; Inoue, A. Oxygen embrittlement in a Cu–Hf–Al bulk metallic glass. *Scr. Mater.* **2009**, *61*, 540–543. [CrossRef]
52. Hess, P.A.; Poon, S.J.; Shiflet, G.J.; Dauskardt, R.H. Indentation fracture toughness of amorphous steel. *J. Mater. Res.* **2005**, *20*, 783–786. [CrossRef]
53. Keryvin, V.; Hoang, V.H.; Shen, J. Hardness, toughness, brittleness and cracking systems in an iron-based bulk metallic glass by indentation. *Intermetallics* **2009**, *17*, 211–217. [CrossRef]
54. Kruzic, J.J.; Kim, D.K.; Koester, K.J.; Ritchie, R.O. Indentation techniques for evaluating the fracture toughness of biomaterials and hard tissues. *J. Mech. Behav. Biomed. Mater.* **2009**, *2*, 384–395. [CrossRef] [PubMed]
55. Liu, Y.H.; Wang, K.; Inoue, A.; Sakurai, T.; Chen, M.W. Energetic criterion on the intrinsic ductility of bulk metallic glasses. *Scr. Mater.* **2010**, *62*, 586–589. [CrossRef]
56. Hsieh, P.J.; Lin, S.C.; Su, H.C.; Jang, J.S.C. Glass forming ability and mechanical properties characterization on $Mg_{58}Cu_{31}Y_{11-x}Gd_x$ bulk metallic glasses. *J. Alloys Compd.* **2009**, *483*, 40–43. [CrossRef]
57. Lewandowski, J.J.; Gu, X.J.; Shamimi Nouri, A.; Poon, S.J.; Shiflet, G.J. Tough Fe-based bulk metallic glasses. *Appl. Phys. Lett.* **2008**, *92*, 091918. [CrossRef]
58. Wang, W.H. The elastic properties, elastic models and elastic perspectives of metallic glasses. *Prog. Mater. Sci.* **2012**, *57*, 487–656. [CrossRef]
59. Zhang, Y.; Greer, A.L. Correlations for predicting plasticity or brittleness of metallic glasses. *J. Alloys. Compd.* **2007**, *434–435*, 2–5. [CrossRef]
60. Nagendra, N.; Ramamurty, U.; Goh, T.T.; Li, Y. Effect of crystallinity on the impact toughness of a La-based bulk metallic glass. *Acta Mater.* **2000**, *48*, 2603–2615. [CrossRef]
61. Raghavan, R.; Shastry, V.V.; Kumar, A.; Jayakumar, T.; Ramamurty, U. Toughness of as-cast and partially crystallized composites of a bulk metallic glass. *Intermetallics* **2009**, *17*, 835–839. [CrossRef]

62. Yokoyama, Y.; Yamasaki, T.; Liaw, P.K.; Inoue, A. Study of the structural relaxation-induced embrittlement of hypoeutectic Zr–Cu–Al ternary bulk glassy alloys. *Acta Mater.* **2008**, *56*, 6097–6108. [CrossRef]

63. Yokoyama, Y.; Yamasaki, T.; Nishijima, M.; Inoue, A. Drastic increase in the toughness of structural relaxed hypoeutectic $Zr_{59}Cu_{31}Al_{10}$ bulk glassy alloy. *Mater. Trans. JIM* **2007**, *48*, 1276–1281. [CrossRef]

64. Madge, S.V. Mg-based Bulk Metallic Glasses. Ph.D. Thesis, University of Cambridge, Cambridge, UK, 2003.

65. Wu, T.; Spaepen, F. The relation between embrittlement and structural relaxation of an amorphous metal. *Philos. Mag. B* **1990**, *61*, 739–750. [CrossRef]

66. Raghavan, R.; Murali, P.; Ramamurty, U. On the factors influencing the ductile-to-brittle transition in a bulk metallic glass. *Acta Mater.* **2009**, *57*, 3332–3340. [CrossRef]

67. Chen, H.S.; Krause, J.T.; Coleman, E. Elastic constants, hardness and their implications to flow properties of metallic glasses. *J. Non-Cryst. Solids* **1975**, *18*, 157–171. [CrossRef]

68. Schroers, J.; Johnson, W.L. Ductile bulk metallic glass. *Phys. Rev. Lett.* **2004**, *93*, 255506. [CrossRef] [PubMed]

69. Lewandowski, J.J.; Wang, W.H.; Greer, A.L. Intrinsic plasticity or brittleness of metallic glasses. *Philos. Mag. Lett.* **2005**, *85*, 77–87. [CrossRef]

70. Greaves, G.N.; Greer, A.L.; Lakes, R.S.; Rouxel, T. Poisson's ratio and modern materials. *Nat. Mater.* **2011**, *10*, 823–837. [CrossRef] [PubMed]

71. Liu, Z.Q.; Wang, W.H.; Jiang, M.Q.; Zhang, Z.F. Intrinsic factor controlling the deformation and ductile-to-brittle transition of metallic glasses. *Philos. Mag. Lett.* **2014**, *94*, 658–668. [CrossRef]

72. Kelly, A.; Tyson, W.R.; Cottrell, A.H. Ductile and brittle crystals. *Philos. Mag.* **1967**, *15*, 567–586. [CrossRef]

73. Gu, X.J.; McDermott, A.G.; Poon, S.J.; Shiflet, S.J. Critical Poisson's ratio for plasticity in Fe–Mo–C–B–Ln bulk amorphous steel. *Appl. Phys. Lett.* **2006**, *88*, 211905. [CrossRef]

74. Jia, P.; Zhu, Z.; Ma, E.; Xu, J. Notch toughness of Cu-based bulk metallic glasses. *Scr. Mater.* **2009**, *61*, 137–140. [CrossRef]

75. Kumar, G.; Rector, D.; Conner, R.D.; Schroers, J. Embrittlement of Zr-based bulk metallic glasses. *Acta Mater.* **2009**, *57*, 3572–3583. [CrossRef]

76. Leonhard, A.; Xing, L.Q.; Heilmaier, M.; Gebert, A.; Eckert, J.; Schultz, L. Effect of crystalline precipitates on the mechanical behavior of bulk glass forming Zr-based alloys. *Nanostr. Mater.* **1998**, *10*, 805–817. [CrossRef]

77. Keryvin, V.; Bernard, C.; Sanglebœuf, J.-C.; Yokoyama, Y.; Rouxel, T. Toughness of $Zr_{55}Cu_{30}Al_{10}Ni_5$ bulk metallic glass for two oxygen levels. *J. Non-Cryst. Solids* **2006**, *352*, 2863–2868. [CrossRef]

78. Madge, S.V.; Sharma, P.; Louzguine-Luzgin, D.V.; Greer, A.L.; Inoue, A. Mechanical behaviour of Zr-La-Cu-Ni-Al glass-based composites. *Intermetallics* **2011**, *19*, 1474. [CrossRef]

79. Granata, D.; Fischer, E.; Wessels, V.; Loeffler, J.F. Fluxing of Pd–Si–Cu bulk metallic glass and the role of cooling rate and purification. *Acta Mater.* **2014**, *71*, 145–152. [CrossRef]

80. Kinaka, M.; Kato, H.; Hasegawa, M.; Inoue, A. High specific strength Mg-based bulk metallic glass matrix composite highly ductilized by Ti dispersoid. *Mater. Sci. Eng. A* **2008**, *494*, 299–303. [CrossRef]

81. Madge, S.V.; Sharma, P.; Louzguine-Luzgin, D.V.; Greer, A.L.; Inoue, A. New La-based glass-crystal *ex situ* composites with enhanced toughness. *Scr. Mater.* **2010**, *62*, 210–213. [CrossRef]

82. Madge, S.V.; Louzguine-Luzgin, D.V.; Inoue, A.; Greer, A.L. Large compressive plasticity in a la-based glass-crystal composite. *Metals* **2013**, *3*, 41–48. [CrossRef]

83. Lee, M.L.; Li, Y.; Schuh, C.A. Effect of a controlled volume fraction of dendritic phases on tensile and compressive ductility in La-based metallic glass matrix composites. *Acta Mater.* **2004**, *52*, 4121–4131. [CrossRef]

84. Lewandowski, J.J. Modern fracture mechanics. *Philos. Mag.* **2013**, *93*, 3893–3906. [CrossRef]

85. Poon, S.J.; Zhu, A.; Shiflet, G.J. Poisson's ratio and intrinsic plasticity of metallic glasses. *Appl. Phys. Lett.* **2008**, *92*, 261902. [CrossRef]

86. Shi, Y.; Luo, J.; Yuan, F.; Huang, L. Intrinsic ductility of solids. *J. Appl. Phys.* **2014**, *115*, 043528. [CrossRef]

87. Spaepen, F. A microscopic mechanism for steady state inhomogeneous flow in metallic glasses. *Acta Metall.* **1977**, *25*, 407–415. [CrossRef]

88. Pan, D.; Inoue, A.; Sakurai, T.; Chen, M.W. Experimental characterization of shear transformation zones for plastic flow of bulk metallic glasses. *Proc. Natl. Acad. Sci.* **2008**, *105*, 14769–14772. [CrossRef] [PubMed]

89. Johnson, W.L.; Samwer, K. A universal criterion for plastic yielding of metallic glasses with a $(T/T_g)^{2/3}$ temperature dependence. Shear transformation zone volume dertermining ductile-brittle transition of bulk metallic glasses. *Phys. Rev. Lett.* **2005**, *95*, 195501. [CrossRef] [PubMed]

Metals **2015**, *5*, 1279–1305

90. Jiang, F.; Jiang, M.Q.; Wang, H.F.; Zhao, Y.L.; He, L.; Sun, J. Shear transformation zone volume determining ductile-brittle transition of bulk metallic glasses. *Acta Mater.* **2011**, *59*, 2057–2068. [CrossRef]

91. Jiang, M.Q.; Wilde, G.; Chen, J.H.; Qu, C.B.; Fu, S.Y.; Jiang, F.; Dai, L.H. Cryogenic-temperature-induced transition from shear to dilatational failure in metallic glasses. *Acta Mater.* **2014**, *77*, 248–257. [CrossRef]

92. Demetriou, M.D.; Kaltenboech, G.; Suh, J.; Garrett, G.; Floyd, M.; Crewdson, C.; Hofmann, D.C.; Kozachkov, H.; Wiest, A.; Schramm, J.P.; *et al.* Glassy steel optimized for glass-forming ability and toughness. *Appl. Phys. Lett.* **2009**, *95*, 041907. [CrossRef]

93. He, Q.; Cheng, Y.Q.; Ma, E.; Xu, J. Locating bulk metallic glasses with high fracture toughness: Chemical effects and composition optimization. *Acta Mater.* **2011**, *59*, 202–215. [CrossRef]

94. Egami, T. Formation and deformation of metallic glasses: Atomistic theory. *Intermetallics* **2006**, *14*, 882–887. [CrossRef]

95. Badrinarayanan, P.; Zheng, W.; Li, Q.; Simon, S.L. The glass transition temperature *versus* the fictive temperature. *J. Non-Cryst. Solids* **2007**, *353*, 2603–2612. [CrossRef]

96. Cahn, R.W.; Greer, A.L. Metastable states of alloys. In *Physical Metallurgy*, 4th ed.; Cahn, R.W., Haasen, P., Eds.; Elsevier Science: Amsterdam, The Netherlands, 1996; Volume 2, pp. 1724–1830.

97. Kumar, G.; Neibecker, P.; Liu, Y.H.; Schroers, J. Critical fictive temperature for plasticity in metallic glasses. *Nat. Commun.* **2013**, *4*. [CrossRef]

98. Angell, C.A. Formation of glasses from liquids and biopolymers. *Science* **1995**, *267*, 1924–1935. [CrossRef] [PubMed]

99. Busch, R.; Liu, W.; Johnson, W.L. Thermodynamics and kinetics of the $Mg_{65}Cu_{25}Y_{10}$ bulk metallic glass forming liquid. *J. Appl. Phys.* **1998**, *83*, 4134–4141. [CrossRef]

100. Bossuyt, S.; Madge, S.V.; Chen, G.Z.; Castellero, A.; Deledda, S.; Eckert, J.; Fray, D.J.; Greer, A.L. Electrochemical removal of oxygen for processing glass-forming alloys. *Mater. Sci. Eng. A* **2004**, *375–377*, 240–243. [CrossRef]

metals

MDPI

Review

Stress-Corrosion Interactions in Zr-Based Bulk Metallic Glasses

Petre Flaviu Gostin [1],*, Dimitri Eigel [1], Daniel Grell [2], Margitta Uhlemann [1], Eberhard Kerscher [2], Jürgen Eckert [1,3] and Annett Gebert [1]

[1] Leibniz-Institute for Solid State and Materials Research IFW Dresden, Dresden D-01171, Germany; d.eigel@ifw-dresden.de (D.E.); m.uhlemann@ifw-dresden.de (M.U.); j.eckert@ifw-dresden.de (J.E.); a.gebert@ifw-dresden.de (A.G.)
[2] Materials Testing, TU Kaiserslautern, Gottlieb-Daimler-Strasse, Kaiserslautern D-67663, Germany; daniel.grell@mv.uni-kl.de (D.G.); kerscher@mv.uni-kl.de (E.K.)
[3] Institute of Materials Science, Faculty of Mechanical Science and Engineering, TU Dresden, Dresden D-01062, Germany
*   Author to whom correspondence should be addressed; f.p.gostin@ifw-dresden.de; Tel.: +49-351-465-9767; Fax: +49-351-465-9452.

Academic Editors: K. C. Chan and Jordi Sort Viñas

Received: 7 May 2015; Accepted: 10 July 2015; Published: 15 July 2015

Abstract: Stress-corrosion interactions in materials may lead to early unpredictable catastrophic failure of structural parts, which can have dramatic effects. In Zr-based bulk metallic glasses, such interactions are particularly important as these have very high yield strength, limited ductility, and are relatively susceptible to localized corrosion in halide-containing aqueous environments. Relevant features of the mechanical and corrosion behavior of Zr-based bulk metallic glasses are described, and an account of knowledge regarding corrosion-deformation interactions gathered from *ex situ* experimental procedures is provided. Subsequently the literature on key phenomena including hydrogen damage, stress corrosion cracking, and corrosion fatigue is reviewed. Critical factors for such phenomena will be highlighted. The review also presents an outlook for the topic.

Keywords: metallic glasses; amorphous alloys; zirconium alloys; corrosion; stress corrosion cracking; corrosion fatigue; hydrogen damage; shear bands; mechanical behavior

1. Introduction

In the last few decades a remarkably growing understanding of the glass-forming ability (GFA) of metallic alloy systems and the optimization of casting techniques have enabled the production of bulk metallic glasses with up to several centimetres in thickness. This makes this new class of material very interesting for engineering applications, e.g., in electronic packaging, pressure sensors, sportive equipment, biomedical devices, or MEMS [1]. Compared to crystalline alloys, metallic glasses exhibit unusual chemical compositions and unique short- and medium-range ordered (SRO/MRO) structures. Similar to inorganic glasses, with their ideally single-phase nature, they lack defects like grain boundaries, dislocations, or second phases. In consequence, this particular alloy state yields some exceptional properties. Most importantly, the topology of SRO and MRO controls the material's response to mechanical stress [2]. The room temperature mechanical performance is characterized by superior hardness and ultimate strength as well as large elastic strain limits enabling large reversible mechanical energy storage, for instance. But upon straining beyond the elastic limit metallic glasses fail with only limited macroscopic plastic deformation. In absence of the possibility for dislocation mediated crystallographic slip, the deformation mode of metallic glasses is very heterogeneous; it is based on the formation of localized shear bands, which result in catastrophic failure due to uninhibited shear band propagation. Meanwhile, the underlying deformation processes have been intensively studied in dependence on a wide parameter field and detailed mechanistic descriptions have been provided [3,4]. New concepts refer to improvement of plasticity by retarding shear band propagation, e.g., by development of "ductile glasses" and "glass-matrix composites" with different length-scales of constituent phases [5] or by mechanical pre-treatments like shot-peening [6] or cold rolling [7] causing shear-band multiplication.

For metallic glasses the resistance against crack propagation determining fracture toughness and fatigue resistance is a critical aspect [8]. Due to their single-phase, quasi-brittle nature, behaviour similar to that of oxide or silicate glasses is expected, but several metallic glasses behave nearly like crystalline alloys. Studies dealing with those phenomena are still limited and the results obtained so far are not very consistent. Besides limited availability of material and use of non-standardized testing methods this is mostly attributed to the large number of influencing parameters [9]. Not only materials properties were identified to be important, but also in particular environmental influences causing chemical reactions in the crack regions. Contradicting traditional opinions, recent studies have clearly demonstrated that bulk metallic glasses are prone to corrosion degradation mainly due to localized dissolution reactions [10]. Moreover, in particular the early-late transition metal glass formers are good hydrogen absorbers based on occupation of the interstitial sites of the SRO units by hydrogen atoms [11]. The superposition of those environmentally induced reactions to stress-driven degradation processes may give rise to environmentally induced cracking phenomena like hydrogen damage, stress corrosion cracking, and corrosion fatigue.

Studies of corrosion-deformation interactions in metallic glasses have been reviewed in several publications [10,12–15]. However, those review papers focused mostly on Fe- and Ni-based metallic glasses. The present review is dedicated to Zr-based metallic glasses, which have gained increasing attention in the last two decades and which form at present the most prominent alloy family of metallic glasses.

2. Mechanical Properties at Room Temperature

Macroscopic mechanical properties of multi-component Zr-based alloys mostly obtained from uniaxial compression tests are in typical ranges of 70–96 GPa Young's modulus, 1400–1900 MPa ultimate strength and around 2% elastic limit, given for selected standard alloys $Zr_{55}Cu_{30}Al_{10}Ni_5$, $Zr_{57}Cu_{15.4}Ni_{12.6}Al_{10}Nb_5$ (Vitreloy 106) and $Zr_{41.2}Ti_{13.8}Cu_{12.5}Ni_{10}Be_{22.5}$ (Vitreloy 1) [1]. Zr-based alloys have been the main targets for principal analyses of deformation mechanisms of metallic glasses with special emphasis on the processes of shear band nucleation and propagation and for development of concepts for property improvements [3–7,16]. Recent studies focused on multi-axial loading conditions in bending tests. Wang *et al.* [17] described the inhomogeneous distribution of residual stresses in bent Zr–Ti–Al–Cu–Ni glass specimen and identified a high sensitivity of hardness to it, which is more pronounced in regions under tension. Zhao *et al.* [18] analyzed shear deformation mechanisms. The inhomogeneous stress distribution in a bent glass specimen led to regions which were still elastic whereby for others at the tension surface the critical shear stress was reached resulting in a regular pattern of shear bands. Introduction of significant stress gradients in a metallic glass specimen yields a larger plasticity than under uniaxial tension. Following this concept the surface imprinting technology was developed as new approach for making glasses more ductile [19].

Intrinsic Plasticity and Toughness of Metallic Glasses

Plasticity plays a key role in environmentally induced cracking of metals/alloys, in contrast to inorganic glasses and ceramics, which are brittle and crack by reaction of the environment with highly stressed bonds at an atomically sharp crack tip [20,21]. Metallic glasses may be considered as quasi-brittle materials because they do not possess sufficient intrinsic micro-mechanisms to mitigate high stress concentrations at crack tips. This is accompanied by lack of strain hardening and crack propagation barriers like grain boundaries and should result in "ideal brittle fracture" behavior.

The same fact was derived for inorganic glasses based on a series of chemical bond rupture events and absence of plastic deformation giving rise to low toughness [8]. However, Lewandowski *et al.* [22] revealed that similar as for polycrystalline metals, for metallic glasses a correlation exists between the fracture energy G and the μ/B ratio (μ—elastic shear modulus, B—bulk modulus) with a critical ratio μ/B_{crit} = 0.41–0.43, which determines their intrinsic plasticity (low μ/B) or brittleness (high μ/B). This relation can also be expressed in terms of the Poisson's ratio ν, with ν_{crit} = 0.31–0.32 and $\nu > \nu_{crit}$ indicating certain glass toughness. For example Mg-based glasses are "brittle", whereas Zr-based glasses are "tough" with some variations among the particular compositions. Cast Zr-based glasses exhibit capability for high plasticity strongly localized in narrow shear bands. A high strength level combined with capacity for local plastic flow can give rise to high fracture toughness. But macroscopic plasticity in tension remains limited since fracture occurs under tension only in one dominant shear band. The fracture surface shows the characteristic vein-like morphology as result of the viscosity drop in the shear band. Annealing below the glass transition causing structural relaxation and densification can reduce the toughness of certain Zr-based BMGs.

3. Corrosion Aspects

The corrosion behavior of metallic glasses is determined by their composition, *i.e.*, the reactivity of their constituent elements, by their thermodynamic metastability and their ideal single-phase chemically and structurally homogeneous nature. Zr–Cu-based bulk glass formers have been a main target for fundamental corrosion studies in different aqueous media. In a wide pH value range excellent passivation ability was detected [10]. This is based on the competitive processes of Cu dissolution and formation of barrier-type thin valve-metal (e.g., Zr, Al, Ti, Nb) oxide films which is determined by the Cu:valve-metal atomic ratio of the alloy and by the surface finishing state [23].

Zr-based glassy alloys exhibit in halide-containing solutions a high sensitivity for pitting which comprises local dissolution of valve-metal components and enrichment of Cu-complexes. TEM studies revealed Cu-rich nanocrystals at pit walls which govern the pit morphology evolution by (i) protecting capped areas against dissolution and (ii) speeding the dissolution of uncapped areas by local galvanic cells. This gives rise to rapid pit growth and low repassivation. Their pitting susceptibility and repassivation ability can be significantly influenced by alloying elements [24–29]. In a recent study, both these properties were found to depend strongly on Cu concentration [30]. High pitting resistance is achieved by decreasing the Cu and increasing the valve metals contents. The pitting susceptibility of samples prepared under real casting conditions is related with the presence of chemical defects, *i.e.*, crystalline inclusions, or physical defects, e.g., air pockets or pores [10,31,32].

3.1. Pitting at Local Surface Defects

It is well known that stress corrosion cracking often initiates at preexisting surface discontinuities/defects or corrosion-induced features, e.g., pits [33]. Before attempting to understand the role of such surface defects on the stress corrosion cracking initiation process, it is useful to first evaluate how these influence the corrosion behavior of Zr-based BMGs, especially pitting. In a recent study, Vickers micro-indents were used as model surface defects to study their impact on the pitting susceptibility of a typical Zr-based BMG, *i.e.*, $Zr_{59}Ti_3Cu_{20}Al_{10}Ni_8$ [34]. Multiple micro-indents were created by applying various loads from 25 to 2000 g on a mechanically polished surface. The pitting susceptibility was evaluated by anodic potentiodynamic polarization in 0.01 M Na_2SO_4 + 0.01 M NaCl. As shown in Figure 1, for the reference as-polished state and for 200 g micro-indents no pitting takes place, whereas for 2000 g micro-indents pitting clearly takes place at a potential of about 0.44 V *vs.* SCE. Subsequent SEM analysis revealed pits were located preferentially inside or at the rim of micro-indents (see inset in Figure 1). This demonstrates that isolated mechanical surface defects can have a significant detrimental effect on the pitting susceptibility of Zr-based BMGs.

As mentioned above, one of the limitations of metallic glasses is their inhomogeneous plastic deformation behavior in shear bands and lack of strain hardening, which hinders their use in structural applications. Shot-peening induces surface compressive residual stresses and favors nucleation of shear bands near the surface, which in turn translates into larger macroscopic plasticity in compression and in bending [6]. The process of shot-peening severely alters not only the stress field and the surface topography, but also the structure of the metallic glass to a depth of up to 150 μm [35]. This in turn causes a slight improvement of spontaneous passivity in halide-free environments [36]. However, after prolonged shot-peening a decrease of the pitting resistance was detected. Figure 2 shows a SEM image of a shot-peened bulk glassy $Zr_{59}Ti_3Cu_{20}Al_{10}Ni_8$ alloy sample surface after exposure to 6 M HCl for several seconds. One large pit and several smaller ones can be seen. Craters formed by shot-peening are preferentially attacked and are fully corroded. It was found that the degradation patterns are driven by local stress fields surrounding the craters and scratches. In conclusion, shot-peening is expected to have antagonistic effects on the environmentally induced cracking of Zr-based BMGs: it increases plasticity which should be beneficial for mitigating stress concentration at crack tips, but at the same time it decreases the resistance to pitting corrosion which can lead to initiation of cracking.

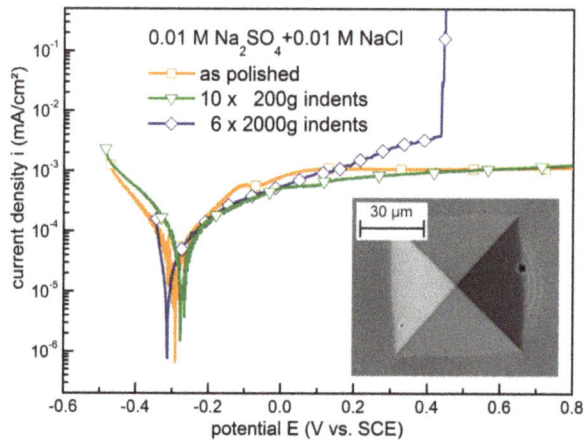

Figure 1. Anodic potentiodynamic polarization curves of bulk glassy $Zr_{59}Ti_3Cu_{20}Al_{10}Ni_8$ alloy in the as polished state and after applying Vickers micro-indent arrays with two different loads.

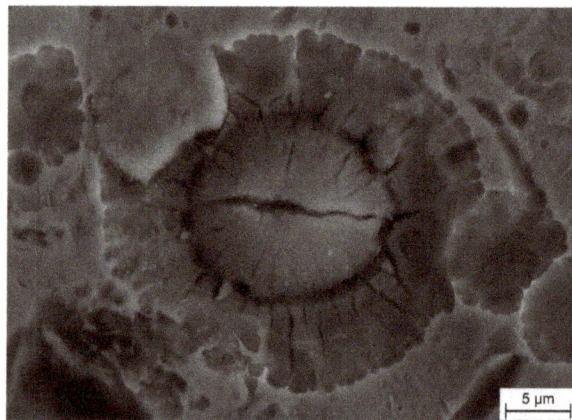

Figure 2. SEM image of a bulk glassy $Zr_{59}Ti_3Cu_{20}Al_{10}Ni_8$ alloy sample surface after shot-peening for 30 s and subsequent exposure to 6 M HCl for several seconds.

3.2. Pitting at Shear Steps and Shear Bands

Although shear steps at a surface created by shear banding can be regarded as surface defects, considering the focus of this paper, they are treated separately in this subsection. Already early in the history of metallic glass research, it has been recognized that shear bands in metallic glasses can be preferentially etched [37]. More recent studies demonstrated that shear steps at the surface produced by shear banding are preferred initiation sites for pitting corrosion. Gebert *et al.*, carried out uniaxial compression loading to generate multiple shear bands in a bulk glassy $Zr_{59}Ti_3Cu_{20}Al_{10}Ni_8$ alloy samples [34]. These samples were subsequently exposed to an aggressive 12 M HCl solution for 1 to 2 min. As shown exemplarily in Figure 3, pits preferentially initiate at shear steps. This may be explained in terms of the passive film being mechanically broken down along the shear step. Additionally, pits have a tendency to grow along the shear steps creating elongated attack features with a width comparable to that of the heat affected zone of shear bands, *i.e.*, 4–10 μm [38]. In the vicinity of shear bands chemical and structural changes and sometimes crystallization can occur due to the shearing process and the associated heating [39]. Moreover, near shear bands local residual stresses may be expected, which can accelerate dissolution [40]. Similar preferential pitting corrosion at shear steps was observed for the $Zr_{64.13}Cu_{15.75}Ni_{10.12}Al_{10}$ BMG [41]. The shear bands/steps were generated by cold rolling. However, preferential pitting at shear steps was observed only in chronopotentiometric tests. No clear preferential pitting occurred under free corrosion conditions, *i.e.*, at the open circuit potential. In this study, structural changes in shear bands, more than the surface shear offset or the residual stresses, were assumed to be the main factor for preferential pitting at shear bands. On the contrary, Wang *et al.* [42] found for the $(Zr_{62}Cu_{23}Fe_5Al_{10})_{97}Ag_3$ BMG that the shear offset (step), rather than the structural changes in the shear bands, is responsible for pitting initiation. Supporting this conclusion was the observation that preferential pitting does not occur at shear bands when the surface was polished. In conclusion, it is clear that shear steps at the surface of Zr-based BMGs are preferential sites for initiation of pitting corrosion. However, there is disagreement regarding the main reason and further work is necessary to explain this observation.

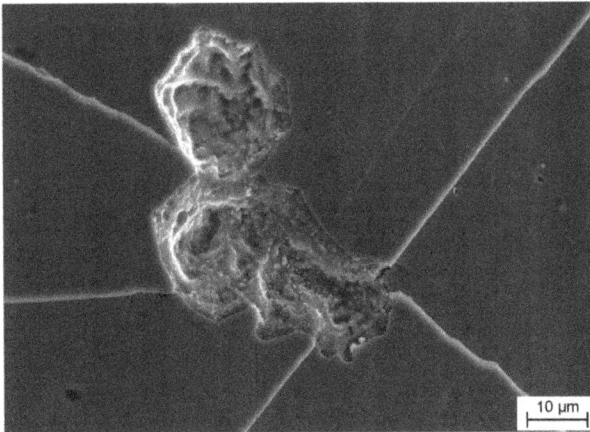

Figure 3. SEM image of a pit on the lateral surface of a bulk glassy $Zr_{59}Ti_3Cu_{20}Al_{10}Ni_8$ alloy sample compressed to fracture.

In the above paragraph only *ex situ* studies were presented, *i.e.*, first shear steps were created at the surface of metallic glass specimens by mechanical loading in air and their corrosion behavior was subsequently analyzed in electrolyte. In a very recent work, Grell *et al.* [43] carried out *in situ* electrochemical experiments during quasistatic three-point bending of flat bulk glassy

$Zr_{52.5}Cu_{17.9}Al_{10}Ni_{14.6}Ti_5$ alloy (Vitreloy 105) samples in 0.01 M Na_2SO_4. Open circuit potential as well as potentiostatic measurements reveal clear transients corresponding to the exposure of bare alloy when shear bands create a step at the surface causing breakdown of the passive film followed by repassivation. However, to the authors' knowledge *in situ* tests of pitting susceptibility during shear step formation have not yet been carried out.

4. Environmentally Induced Cracking

Most work related to environmentally induced cracking of Zr-based BMGs was done on hydrogen effects (e.g., [44–52]) and on corrosion fatigue [53–60]. Studies on stress corrosion cracking of Zr-based BMGs are rather scarce [9,10,13,53,61].

According to general knowledge gained from studies on conventional crystalline metals and alloys, as the yield strength increases, the resistance to environmentally induced cracking decreases [62]. If this holds also for metallic glasses, they should have lower resistance to the various types of environmentally induced cracking, e.g., stress corrosion cracking or hydrogen damage.

4.1. Hydrogen Effects

Zr-based metallic glasses exhibit significant capability for absorption of hydrogen. Hydrogen atoms occupy interstitial sites in the SRO structure with different energy levels being determined by the local atomic environment and the hydrogen affinity of the constituents. Preferential occupation of interstitial sites of polyhedra formed by Zr atoms was verified [11,63]. The incorporation of hydrogen leads to a reduction of free volume and the buildup of local strain which retards atomic mobility causing increased internal friction [64,65]. When the hydrogen concentration reaches a threshold locally, phase separation into regions of ZrH_x and Cu-rich phases occurs [66,67]. For Zr–Al–Cu–Ni glasses H/M (M = metal) ratios of up 1.9 were measured and a significant effect on the thermal behavior was detected [67,68]. Based on these principal findings strong effects of absorbed hydrogen on crack initiation and propagation processes in Zr-based glasses can be predicted.

Absorbed (pre-charged) hydrogen has significant effects on the mechanical properties of Zr-based BMGs. For different alloy compositions an increase of hardness with increasing hydrogen content in the amorphous structure was evidenced and attributed to a restriction of shear band formation. This was compared with thermal relaxation effects causing embrittlement [34,46,69]. Wang *et al.* [70] investigated glassy I-shaped notched $Zr_{57}Cu_{15.4}Ni_{12.6}Al_{10}Nb_5$ (Vitreloy 106) specimen (made from $1.5 \times 10 \times 12$ mm^3 plates) under sustained tensile load and severe hydrogen charging. Enhanced local shear band nucleation at the notch, but limited growth was observed with increasing charging time and interpreted as enhanced localized plastic flow. Shan *et al.* [48] observed the damage evolution during 3-point bending of notched bulk glassy $Zr_{41.2}Ti_{13.8}Cu_{12.5}Ni_{10}Be_{22.5}$ (Vitreloy 1) samples ($2 \times 4 \times 15$ mm^3) under sustained loading conditions and different hydrogen charging rates. Below a critical current density of 20 $mA \cdot cm^{-2}$, hydrogen is incorporated into the amorphous structure without hydride formation or blistering. Initiation of hydrogen-induced cracking and hydrogen-induced delayed fracture takes place at a stress value representing 63% of that required for fracture of uncharged specimens. Above the critical charging rate, blistering and micro-cracking occurred causing a relative stress for crack initiation of 26%. Suh and Dauskardt [44] prepared fatigue pre-cracked CT specimen of this bulk glass (from plates with 3 mm thickness) and investigated the effect of pre-charged hydrogen on fracture toughness and fatigue crack growth. Increased charging time was found to have a degrading effect on initiation fracture toughness. In contrast, fatigue crack growth rates were found to be retarded with increasing charging time. Microscopic fatigue crack analysis revealed a tortuous and deflected crack path of pre-charged samples in comparison to a straight crack path of uncharged ones. The contradicting results were interpreted as a competition between degradation of the inherent resistance to crack extension and increased crack tip shielding.

In summary, Zr-based metallic glasses exhibit significant capability for absorption of hydrogen, which in turn severely affects their mechanical properties depending on the concentration of absorbed hydrogen. On the one hand, increasing hydrogen concentration leads to increased hardness. On the other hand, hydrogen appears to enhance nucleation of shear bands at notch tips and to decrease the threshold intensity factor. Furthermore, hydrogen has a retarding effect on the crack growth rate under sustained or cyclic loading.

4.2. Chloride-Induced Stress Corrosion Cracking

Kawashima *et al.* [71] conducted slow strain rate (5×10^{-6} s^{-1}) tensile tests on glassy Zr–Cu–Al(Ni) specimens ($3.5 \div 8$ mm diameter rods) in various aqueous electrolytes under free corrosion conditions. A significant reduction in fracture stress from more than 1800 to 880 MPa in NaCl solution was detected, while this effect was dependent on alloy composition. Cracks were found to initiate at pits. Brittle fracture features were observed close to the crack origin. These were attributed to presumed hydrogen-induced subcritical crack growth. In another study, Schroeder *et al.* [55] carried out crack velocity measurements on compact tension specimens of the bulk glass forming $Zr_{41.2}Ti_{13.8}Cu_{12.5}Ni_{10}Be_{22.5}$ alloy in 0.5 M NaCl at constant stress and free corrosion conditions. Above a K_{ISCC}, a sharp increase in crack velocity of several orders of magnitude was observed. This was followed by a plateau at $10^{-5} \div 10^{-4}$ m·s^{-1}. It was assumed that crack propagation took place by an anodic dissolution assisted mechanism. Almost identical results were reported by Nakai and Yoshioka [58] for a different Zr-based BMG, *i.e.*, $Zr_{55}Cu_{30}Ni_5Al_{10}$, in 3.5 wt. % NaCl.

Gostin *et al.* [61] recently investigated stress corrosion cracking phenomena in bulk glass forming $Zr_{52.5}Cu_{17.9}Al_{10}Ni_{14.6}Ti_5$ alloy (Vitreloy 105) in 0.01 M Na_2SO_4 + 0.01 M NaCl solution. Experiments were carried out on $2 \times 2.5 \times 27$ mm^3 samples at constant deflection and constant anodic potential. Applied stress values were in the elastic regime, and applied potential values were in the interval between the repassivation and the pitting potential. Deflection, stress, and the current density were monitored *in situ* during the stress corrosion cracking process. Figure 4 shows exemplarily the evolution of deflection, stress, and current density for one experiment carried out at an initial applied stress of 30% of the yield strength at 0.2% offset, and an applied anodic potential of 50 mV *vs.* SCE. It can be seen that initially the current density is very low corresponding to a passive state. After a certain incubation time, *i.e.*, in this case 133 s, the current density suddenly increases by several orders of magnitude. Since at this moment there is no significant change in the deflection and the stress signals, this sudden current increase is attributed to pitting corrosion. After another 1353 s, the stress starts to decrease signalling the initiation of the cracking process. As the crack propagates, the measured stress level continuously and rapidly decreases down to a value of 645 MPa, when catastrophic fracture by unstable crack growth occurs. Experiments performed at various stress and potential levels revealed two clear trends: (1) the time to fracture is shorter (SCC resistance is lower) for higher potentials or higher stresses; (2) the time required by a pit/a group of pits to turn into a crack is shorter at higher stresses.

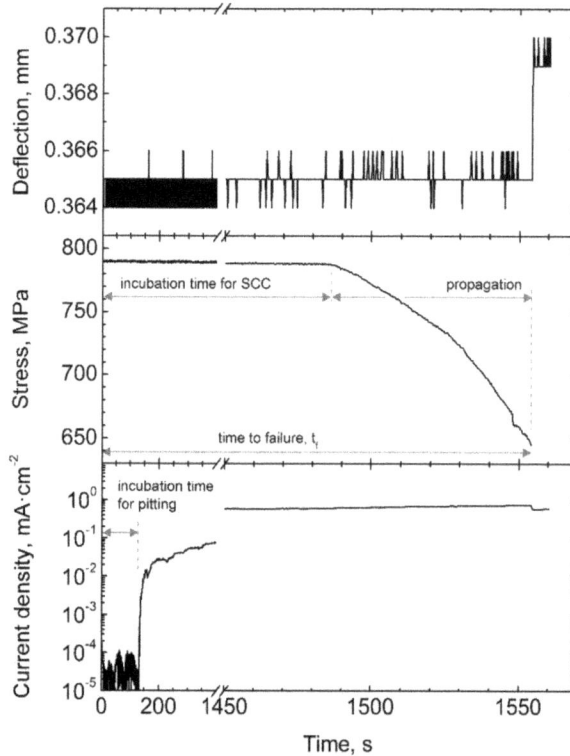

Figure 4. The evolution of deflection, stress, and current density of a bulk glass forming $Zr_{52.5}Cu_{17.9}Al_{10}Ni_{14.6}Ti_5$ alloy (Vitreloy 105) sample in 0.01 M Na_2SO_4 + 0.01 M NaCl solution under three-point bending at an initial stress of 30% of the yield strength at 0.2% offset, and an applied anodic potential of 50 mV *vs.* SCE.

Fractography analysis revealed the presence of multiple pits typically located at sample edges. In all cases, one or more pits were found at the crack initiation location confirming that cracks originate at pits. This is shown in Figure 5a. In the stable crack propagation region of the fracture surface, two different zones were observed. As shown in Figure 5b, close to the lateral side of the fractured bending sample, the crack propagated mostly in a mechanical brittle manner: this zone is characterized by cleavage features and by the absence of corrosion signs. The other zone, which is close to the tension side of the fractured bending sample and shown in Figure 5c, is dominated by fine striations and crack branching. Observation of finer topographical features was impeded due to corrosion. One possible mechanism (explaining the striations and crack branches) could comprise alternate crack sharpening by anodic dissolution at the crack tip and crack blunting by formation of shear bands and/or crack branching. Anodic dissolution may take place preferentially along shear bands. As shown in the previous section, pre-formed shear bands can be preferential sites for pitting initiation and continued pit growth.

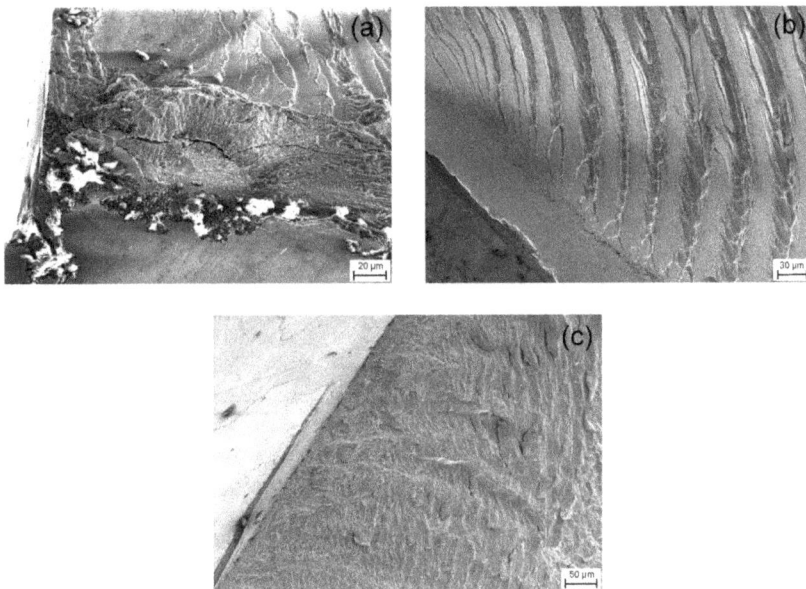

Figure 5. Fracture surface of a bulk glass forming $Zr_{52.5}Cu_{17.9}Al_{10}Ni_{14.6}Ti_5$ alloy (Vitreloy 105) sample fractured by stress corrosion cracking in 0.01 M Na_2SO_4 + 0.01 M NaCl under three point bending and anodic polarization. (**a**) Crack initiation location; (**b**) Stable crack propagation region close to lateral side (perpendicular to tension side) of fractured bending sample; (**c**) Stable crack propagation region close to tension side of fractured bending sample.

In conclusion, few Zr-based BMGs were investigated regarding their sensitivity to SCC. Those first works demonstrated clearly high susceptibility to SCC in chloride-containing aqueous environments. Pitting is a necessary initial stage and cracking by stress-corrosion interactions initiates at pits. The crack growth rate is usually high and is attributed to active anodic dissolution at the crack tip. However, one study claims hydrogen damage to be responsible for crack propagation. One possible SCC mechanism could consist of alternate crack sharpening by anodic dissolution at the crack tip and crack blunting by formation of shear bands and/or crack branching, but more detailed investigations are needed for a comprehensive description of the SCC mechanism in Zr-based BMGs.

4.3. Corrosion Fatigue

Schroeder and Ritchie [54,56] prepared fatigue pre-cracked CT specimen ($4.4 \times 20 \times 20$ mm^3) of the $Zr_{41.2}Ti_{13.8}Cu_{12.5}Ni_{10}Be_{22.5}$ BMG to investigate fatigue crack growth processes under the influence of aqueous media (de-ionized water, 0.5 M Na_2SO_4 or $NaClO_4$ solution, 0.005–0.5 M NaCl solution). Fatigue experiments were conducted under increasing and decreasing range of stress intensity factor ΔK and under constant cyclic loading. A particular sensitivity of the fatigue crack growth to NaCl solutions under OCP conditions was detected with a crack growth rate being three orders of magnitude higher than that in air. The fatigue crack growth behavior in the da/dN vs. ΔK plot revealed for the salt solutions a medium-controlled characteristic of an abrupt threshold, whereupon the growth rate increased by 3–5 orders of magnitude to reach a plateau regime (region II) where the crack growth is relatively independent from the stress intensity range. Compared to the fatigue threshold ΔK_{th} in air of 3 MPa·m$^{1/2}$, the value was strongly reduced to ~0.9 MPa·m$^{1/2}$ in salt solutions. The constant fatigue crack growth rates in region II were under OCP conditions directly proportional to the NaCl concentration, whereby ΔK_{th} remained nearly constant. Under cathodic control the growth rates decreased, whereby no changes were observed under anodic control in a limited potential window. The poor environmental cracking resistance of this particular BMG was mainly attributed to stress enhanced anodic dissolution at the crack tip.

Morrison *et al.* [57] employed the $Zr_{52.5}Cu_{17.9}Ni_{14.6}Al_{10}Ti_5$ (Vitreloy 105) BMG to analyze the corrosion-fatigue behavior in four-point bending tests. They used rectangular specimens ($3.5 \times 3.5 \times 30$ mm^3) and tested them in 0.6 M NaCl solution. While at stress amplitudes of $\geq$800 MPa the corrosion-fatigue life was similar to that in air, an increasingly deleterious effect of the salt solution was observed as the stress amplitude decreased due to increasing length of exposure time. At 425 MPa the corrosion-fatigue life in NaCl solution was more than one order of magnitude lower than that in air. The fatigue-endurance limit reached in the salt solution ~50 MPa which relates to a decrease of ~88%. This behavior of the Zr-based BMG was found to be similar to that of crystalline Al. With SEM no significant effect of the test medium on fracture morphology was noticed. Crack initiation occurred on the machined and polished samples preferentially at the outer pins of the tensile surface. The corrosion-fatigue mechanism was concluded to be driven by anodic dissolution with no indication for hydrogen embrittlement.

Wiest *et al.* [59] used results of this study as reference data for their corrosion-fatigue studies on Zr–Ti(–Co)–Be rod samples (3 mm diam. $\times$ 6 mm, 2 mm diam. $\times$ 4 mm) in compression loading. These alloys with no or low late TM-content exhibit very low corrosion rates and strong barrier-type passive behavior, but a lower pitting resistance than Vitreloy 105. The latter explains their poor corrosion fatigue performance. Huang *et al.* [60] prepared $Zr_{55}Cu_{30}Al_{10}Ni_5$ + 1 at. % Y rectangular glass samples ($3 \times 3 \times 25$ mm^3) for 4-point-bending in an artificial body fluid. They confirmed a low impact of the medium on fatigue lifetime at high stress ranges ($\geq$450 MPa), but a detrimental effect in the lower range. The fatigue strength decreased by 40% compared to air tests. A degradation mechanism was proposed considering the random formation of a corrosion pit at the tension side, from where crack initiation takes place due to stress concentration and exposure of the bare metal surface. Crack propagation was stated to be driven by anodic dissolution resulting in a rough fracture surface.

All the above-described studies characterized the corrosion fatigue behavior in aqueous environments. However, for Zr-based BMGs ambient air can also cause corrosion fatigue phenomena leading to significant reduction of their fatigue resistance. The extent of degradation of fatigue properties by ambient air is strongly dependent on alloy chemistry. Philo *et al.* [72] analyzed the fatigue crack growth behavior of $Zr_{58.5}Cu_{15.6}Ni_{12.8}Al_{10.3}Nb_{2.8}$ BMG (Vitreloy 106a), and found a regime in the da/dN *vs.* ΔK curve, where the crack growth was relatively independent of ΔK, implying a controlling effect of the ambient air. This plateau occurs at $2.5 \cdot 10^{-8}$ m·cycle^{-1} and is much higher than for $Zr_{44}Ti_{11}Ni_{10}Cu_{10}Be_{25}$ BMG (Vitreloy 1b), *i.e.*, 10^{-9} m·cycle^{-1} [73]. It is not yet clear what the mechanisms are involved in the observed corrosion fatigue effect in ambient air. Surface oxidation and/or hydrogen embrittlement were suggested as the main reason [9,74]. Studies regarding effects of hydrogen on the fatigue behavior of Zr-based BMGs were presented in the subsection *Hydrogen Effects*.

In summary, the corrosion fatigue behavior of several Zr-based BMGs was studied and all showed high susceptibility/poor resistance to corrosion fatigue cracking in chloride containing aqueous solutions. Some Zr-based BMGs exhibit a corrosion fatigue effect in ambient air, and this is strongly dependent on alloy chemistry. In chloride containing aqueous solutions, corrosion fatigue crack propagation appears to be controlled by anodic dissolution. It is not yet clear what the main reasons and the underlying mechanisms are for the observed corrosion fatigue phenomena in ambient air.

5. Outlook

Those first reported studies reviewed here clearly reveal a high sensitivity of Zr-based bulk metallic glasses to corrosion-deformation interactions in halide-containing aqueous environments. However, tests on these aspects were conducted so far only for a few selected alloy compositions with high GFA, e.g., on special Be-containing glasses, which enable the preparation of glassy samples in larger dimensions needed for tests with CT specimens or bending plates. Moreover, different mechanical loading procedures and conditions were applied under various environmental influences which do not permit a direct comparison of the results and which have led in part to contradicting conclusions. Therefore, more systematic studies are needed for a comprehensive mechanistic description of those environmentally assisted mechanical degradation phenomena. Considering general corrosion and hydrogenation studies, besides structural particularities the glassy alloy composition is expected to play a significant role in environmentally assisted crack initiation and propagation. This refers in particular to the Cu:valve-metal ratio which controls anodic and cathodic reactions rates. A reliable analysis of those compositional effects is still missing. Furthermore, introduction of finely dispersed heterogeneities into an amorphous phase like clusters/nanocrystals was demonstrated to be a successful strategy to increase macroscopic plasticity under compression or tensile load in air. But there is no clear understanding yet how this affects the environmental-assisted fracture and fatigue behavior. Also, the role of different types of surface defects, e.g., of notches or of micro-indents, is mostly unclear. Altogether, more detailed microscopic and spectroscopic studies of crack regions at different propagation states and of fracture surfaces are needed. These should be focused on analyzing processes on the sub-micron scale like shear band evolution or chemical reactions under the special conditions of highly localized stresses, crevice geometry, and varying media and polarization states. Such knowledge is the pre-requisite for mechanistic models of stress corrosion cracking and corrosion fatigue phenomena and their dependence on structural and compositional particularities of Zr-based bulk metallic glasses.

Acknowledgments: A. Davenport, J.R. Scully, and N. Birbilis are acknowledged for fruitful discussions. The authors are grateful to M. Frey, S. Donath and M. Johne for technical assistance. Financial support for this work was provided by the German Research Society (DFG) under grants GE 1106/11-1 and KE 1426/4-1 in the frame of the SPP 1594 collaborative project.

Author Contributions: P.F.G., M.U., A.G. and E.K. wrote the text. D.E. and D.G. carried out experiments. J.E. supervised the research project.

Conflicts of Interest: The authors declare no conflict of interest.

References

1. Suryanarayana, C.; Inoue, A. *Bulk Metallic Glasses*; CRC Press: Boca Raton, FL, USA, 2011; p. 548.
2. Wondraczek, L.; Mauro, J.C.; Eckert, J.; Kühn, U.; Horbach, J.; Deubener, J.; Rouxel, T. Towards ultrastrong glasses. *Adv. Mater.* **2011**, *23*, 4578–4586. [CrossRef] [PubMed]
3. Trexler, M.M.; Thadhani, N.N. Mechanical properties of bulk metallic glasses. *Prog. Mater. Sci.* **2010**, *55*, 759–839. [CrossRef]
4. Cheng, Y.Q.; Cao, A.J.; Sheng, H.W.; Ma, E. Local order influences initiation of plastic flow in metallic glass: Effects of alloy composition and sample cooling history. *Acta Mater.* **2008**, *56*, 5263–5275. [CrossRef]
5. Eckert, J.; Das, J.; Pauly, S.; Duhamel, C. Processing routes, microstructure and mechanical properties of metallic glasses and their composites. *Adv. Eng. Mater.* **2007**, *9*, 443–453. [CrossRef]
6. Zhang, Y.; Wang, W.H.; Greer, A.L. Making metallic glasses plastic by control of residual stress. *Nat. Mater.* **2006**, *5*, 857–860. [CrossRef] [PubMed]
7. Song, K.K.; Pauly, S.; Zhang, Y.; Scudino, S.; Gargarella, P.; Surreddi, K.B.; Kühn, U.; Eckert, J. Significant tensile ductility induced by cold rolling in $Cu_{47.5}Zr_{47.5}Al_5$ bulk metallic glass. *Intermetallics* **2011**, *19*, 1394–1398. [CrossRef]
8. Schuh, C.A.; Hufnagel, T.C.; Ramamurty, U. Mechanical behavior of amorphous alloys. *Acta Mater.* **2007**, *55*, 4067–4109. [CrossRef]
9. Kruzic, J.J. Understanding the Problem of Fatigue in Bulk Metallic Glasses. *Metall. Mater. Trans. A* **2011**, *42*, 1516–1523. [CrossRef]
10. Scully, J.R.; Gebert, A.; Payer, J. Corrosion and related mechanical properties of bulk metallic glasses. *J. Mater. Res.* **2007**, *22*, 302–313. [CrossRef]
11. Bankmann, J.; Pundt, A.; Kirchheim, R. Hydrogen loading behaviour of multi-component amorphous alloys: Model and experiment. *J. Alloy. Compd.* **2003**, *356–357*, 566–569. [CrossRef]

12. Archer, M.D.; Corke, C.C.; Harji, B.H. The Electrochemical Properties of Metallic Glasses. *Electrochim. Acta* **1987**, *32*, 13–26. [CrossRef]

13. Scully, J.R.; Lucente, A. Corrosion of Amorphous Metals. In *ASM Handbook Volume 13B, Corrosion: Materials*; Cramer, S.D., Covino, B.S., Jr., Eds.; ASM International: Materials Park, OH, USA, 2005; Volume 13B, pp. 476–489.

14. Hashimoto, K. Chemical properties. In *Amorphous Metallic Alloys*; Luborsky, F.E., Ed.; Butterworths: London, UK, 1983; pp. 471–486.

15. Waseda, Y.; Aust, K.T. Corrosion behaviour of metallic glasses. *J. Mater. Sci.* **1981**, *16*, 2337–2359. [CrossRef]

16. Wilde, G.; Rosner, H. Nanocrystallization in a shear band: An *in situ* investigation. *Appl. Phys. Lett.* **2011**. [CrossRef]

17. Wang, L.; Bei, H.; Gao, Y.F.; Lu, Z.P.; Nieh, T.G. Effect of residual stresses on the hardness of bulk metallic glasses. *Acta Mater.* **2011**, *59*, 2858–2864. [CrossRef]

18. Zhao, J.X.; Wu, F.F.; Zhang, Z.F. Analysis on shear deformation mechanism of metallic glass under confined bending test. *Mater. Sci. Eng. A* **2010**, *527*, 6224–6229. [CrossRef]

19. Scudino, S.; Jerliu, B.; Pauly, S.; Surreddi, K.B.; Kühn, U.; Eckert, J. Ductile bulk metallic glasses produced through designed heterogeneities. *Scr. Mater.* **2011**, *65*, 815–818. [CrossRef]

20. Newman, R.C. Stress-Corrosion Cracking Mechanisms. In *Corrosion Mechanisms in Theory and Practice*, 3rd ed.; Marcus, P., Ed.; CRC Press: Boca Raton, FL, USA, 2011; pp. 499–544.

21. Clarke, D.R.; Faber, K.T. Fracture of ceramics and glasses. *J. Phys. Chem. Solids* **1987**, *48*, 1115–1157. [CrossRef]

22. Lewandowski, J.J.; Wang, W.H.; Greer, A.L. Intrinsic plasticity or brittleness of metallic glasses. *Philos. Mag. Lett.* **2005**, *85*, 77–87. [CrossRef]

23. Gebert, A.; Gostin, P.F.; Schultz, L. Effect of surface finishing of a Zr-based bulk metallic glass on its corrosion behaviour. *Corros. Sci.* **2010**, *52*, 1711–1720. [CrossRef]

24. Huang, L.; Yokoyama, Y.; Wu, W.; Liaw, P.K.; Pang, S.J.; Inoue, A.; Zhang, T.; He, W. Ni-free Zr–Cu–Al–Nb–Pd bulk metallic glasses with different Zr/Cu ratios for biomedical applications. *J. Biomed. Mater. Res. B* **2012**, *100B*, 1472–1482. [CrossRef] [PubMed]

25. Lu, H.B.; Zhang, L.C.; Gebert, A.; Schultz, L. Pitting corrosion of Cu–Zr metallic glasses in hydrochloric acid solutions. *J. Alloys Compd.* **2008**, *462*, 60–67. [CrossRef]

26. Pang, S.J.; Zhang, T.; Kimura, H.; Asami, K.; Inoue, A. Corrosion Behavior of Zr–(Nb–)Al–Ni–Cu Glassy Alloys. *Mater. Trans. JIM* **2000**, *41*, 1490–1494. [CrossRef]

27. Pang, S.J.; Zhang, T.; Asami, K.; Inoue, A. Formation, corrosion behavior, and mechanical properties of bulk glassy Zr–Al–Co–Nb alloys. *J. Mater. Res.* **2003**, *18*, 1652–1658. [CrossRef]

28. Raju, V.R.; Kühn, U.; Wolff, U.; Schneider, F.; Eckert, J.; Reiche, R.; Gebert, A. Corrosion behaviour of Zr-based bulk glass-forming alloys containing Nb or Ti. *Mater. Lett.* **2002**, *57*, 173–177. [CrossRef]

29. Li, Y.H.; Zhang, W.; Dong, C.; Qiang, J.B.; Fukuhara, M.; Makino, A.; Inoue, A. Effects of Ni addition on the glass-forming ability, mechanical properties and corrosion resistance of Zr–Cu–Al bulk metallic glasses. *Mater. Sci. Eng. A* **2011**, *528*, 8551–8556. [CrossRef]

30. Gostin, P.F.; Eigel, D.; Grell, D.; Eckert, J.; Kerscher, E.; Gebert, A. Comparing the pitting corrosion behavior of prominent Zr-based bulk metallic glasses. *J. Mater. Res.* **2015**, *30*, 233–241. [CrossRef]

31. Gebert, A.; Buchholz, K.; Leonhard, A.; Mummert, K.; Eckert, J.; Schultz, L. Investigations on the electrochemical behaviour of Zr-based bulk metallic glasses. *Mater. Sci. Eng. A* **1999**, *267*, 294–300. [CrossRef]

32. Paillier, J.; Mickel, C.; Gostin, P.F.; Gebert, A. Characterization of corrosion phenomena of Zr–Ti–Cu–Al–Ni metallic glass by SEM and TEM. *Mater. Charact.* **2010**, *61*, 1000–1008. [CrossRef]

33. Jones, R.H. Stress-Corrosion Cracking. In *ASM Handbook Volume 13A, Corrosion: Fundamentals, Testing and Protection*; Cramer, S.D., Covino, B.S., Jr., Eds.; ASM International: Materials Park, OH, USA, 2003; Volume 13A, pp. 346–366.

34. Gebert, A.; Gostin, P.F.; Uhlemann, M.; Eckert, J.; Schultz, L. Interactions between mechanically generated defects and corrosion phenomena of Zr-based bulk metallic glasses. *Acta Mater.* **2012**, *60*, 2300–2309. [CrossRef]

35. Mear, F.O.; Vaughan, G.; Yavari, A.R.; Greer, A.L. Residual-stress distribution in shot-peened metallic-glass plate. *Philos. Mag. Lett.* **2008**, *88*, 757–766. [CrossRef]

36. Gebert, A.; Concustell, A.; Greer, A.L.; Schultz, L.; Eckert, J. Effect of shot-peening on the corrosion resistance of a Zr-based bulk metallic glass. *Scr. Mater.* **2010**, *62*, 635–638. [CrossRef]

37. Pampillo, C.A.; Chen, H.S. Comprehensive Plastic-Deformation of a Bulk Metallic Glass. *Mater. Sci. Eng.* **1974**, *13*, 181–188. [CrossRef]
38. Guo, H.; Wen, J.; Xiao, N.M.; Zhang, Z.F.; Sui, M.L. The more shearing, the thicker shear band and heat-affected zone in bulk metallic glass. *J. Mater. Res.* **2008**, *23*, 2133–2138. [CrossRef]
39. Greer, A.L.; Cheng, Y.Q.; Ma, E. Shear bands in metallic glasses. *Mat. Sci. Eng. R* **2013**, *74*, 71–132. [CrossRef]
40. Gutman, E.M. *Mechanochemistry of Materials*; Cambridge International Science Publishing: Cambridge, UK, 1998.
41. Nie, X.P.; Cao, Q.P.; Wu, Z.F.; Ma, Y.; Wang, X.D.; Ding, S.Q.; Jiang, J.Z. The pitting corrosion behavior of shear bands in a Zr-based bulk metallic glass. *Scr. Mater.* **2012**, *67*, 376–379. [CrossRef]
42. Wang, Y.M.; Zhang, C.; Liu, Y.; Chan, K.C.; Liu, L. Why does pitting preferentially occur on shear bands in bulk metallic glasses? *Intermetallics* **2013**, *42*, 107–111. [CrossRef]
43. Grell, D.; Gostin, P.F.; Eckert, J.; Gebert, A.; Kerscher, E. *In situ* electrochemical analysis during deformation of a Zr-based bulk metallic glass: A sensitive tool revealing early shear banding. *Adv. Eng. Mater.* **2015**. accepted.
44. Suh, D.; Dauskardt, R.H. Hydrogen effects on the mechanical and fracture behavior of a Zr–Ti–Ni–Cu–Be bulk metallic glass. *Scr. Mater.* **2000**, *42*, 233–240. [CrossRef]
45. Suh, D.; Dauskardt, R.H. The effects of hydrogen on deformation and fracture of a Zr–Ti–Ni–Cu–Be bulk metallic glass. *Mater. Sci. Eng. A* **2001**, *319*, 480–483. [CrossRef]
46. Suh, D.; Dauskardt, R.H. Effects of pre-charged hydrogen on the mechanical and thermal behavior of Zr–Ti–Ni–Cu–Be bulk metallic glass alloys. *Mater. Trans.* **2001**, *42*, 638–641. [CrossRef]
47. Daewoong, S.; Asoka-Kumar, P.; Dauskardt, R.H. The effects of hydrogen on viscoelastic relaxation in Zr–Ti–Ni–Cu–Be bulk metallic glasses: Implications for hydrogen embrittlement. *Acta Mater.* **2002**, *50*, 537–551.
48. Shan, G.B.; Wang, Y.W.; Chu, W.Y.; Li, J.X.; Hui, X.D. Hydrogen damage and delayed fracture in bulk metallic glass. *Corros. Sci.* **2005**, *47*, 2731–2739. [CrossRef]
49. Shan, G.B.; Wei, B.C.; Li, J.X.; Qiao, L.J.; Chu, W.Y. Effects of hydrogen on shear bands and cracking in Zr base bulk amorphous alloy. *Acta Metall. Sin.* **2006**, *42*, 689–693.
50. Jayalakshmi, S.; Kim, K.B.; Fleury, E. Effect of hydrogenation on the structural, thermal and mechanical properties of $Zr_{50}Ni_{27}Nb_{18}Co_5$ amorphous alloy. *J. Alloy. Compd.* **2006**, *417*, 195–202. [CrossRef]
51. Shan, G.B.; Li, J.X.; Yang, Y.Z.; Qiao, L.J.; Chu, W.Y. Hydrogen-enhanced plastic deformation during indentation for bulk metallic glass of $Zr_{65}Al_{7.5}Ni_{10}Cu_{17.5}$. *Mater. Lett.* **2007**, *61*, 1625–1628. [CrossRef]
52. Dong, F.Y.; Su, Y.Q.; Luo, L.S.; Wang, L.; Wang, S.J.; Guo, L.J.; Fu, H.Z. Enhanced plasticity in Zr-based bulk metallic glasses by hydrogen. *Int. J. Hydrog. Energy* **2012**, *37*, 14697–14701. [CrossRef]
53. Ritchie, R.O.; Schroeder, V.; Gilbert, C.J. Fracture, fatigue and environmentally-assisted failure of a Zr-based bulk amorphous metal. *Intermetallics* **2000**, *8*, 469–475. [CrossRef]
54. Schroeder, V.; Gilbert, C.J.; Ritchie, R.O. A comparison of the mechanisms of fatigue-crack propagation behavior in a Zr-based bulk amorphous metal in air and an aqueous chloride solution. *Mater. Sci. Eng. A* **2001**, *317*, 145–152. [CrossRef]
55. Schroeder, V.; Gilbert, C.J.; Ritchie, R.O. Effect of aqueous environment on fatigue-crack propagation behavior in a Zr-based bulk amorphous metal. *Scr. Mater.* **1999**, *40*, 1057–1061. [CrossRef]
56. Schroeder, V.; Ritchie, R.O. Stress-corrosion fatigue-crack growth in a Zr-based bulk amorphous metal. *Acta Mater.* **2006**, *54*, 1785–1794. [CrossRef]
57. Morrison, M.L.; Buchanan, R.A.; Liaw, P.K.; Green, B.A.; Wang, G.Y.; Liu, C.T.; Horton, J.A. Corrosion-fatigue studies of the Zr-based Vitreloy 105 bulk metallic glass. *Mater. Sci. Eng. A* **2007**, *467*, 198–206. [CrossRef]
58. Nakai, Y.; Yoshioka, Y. Stress Corrosion and Corrosion Fatigue Crack Growth of Zr-Based Bulk Metallic Glass in Aqueous Solutions. *Metall. Mater. Trans. A* **2010**, *41A*, 1792–1798. [CrossRef]
59. Wiest, A.; Wang, G.Y.; Huang, L.; Roberts, S.; Demetriou, M.D.; Liaw, P.K.; Johnson, W.L. Corrosion and corrosion fatigue of Vitreloy glasses containing low fractions of late transition metals. *Scr. Mater.* **2010**, *62*, 540–543. [CrossRef]
60. Huang, L.; Wang, G.Y.; Qiao, D.C.; Liaw, P.K.; Pang, S.J.; Wang, J.F.; Zhang, T. Corrosion-fatigue study of a Zr-based bulk-metallic glass in a physiologically relevant environment. *J. Alloy. Compd.* **2010**, *504*, S159–S162. [CrossRef]

61. Gostin, P.F.; Eigel, D.; Grell, D.; Uhlemann, M.; Kerscher, E.; Eckert, J.; Gebert, A. Stress corrosion cracking of a Zr-based bulk metallic glass. *Mater. Sci. Eng. A* **2015**, *639*, 681–690. [CrossRef]

62. Craig, B. Introduction to Environmentally Induced Cracking. In *ASM Handbook*; Cramer, S.D., Covino, B.S.J., Eds.; ASM International: Materials Park, OH, USA, 2003; Volume 13A, p. 345.

63. Mattern, N.; Gebert, A. Hydrogenation of $Zr_{60}Ti_2Cu_{20}Al_{10}Ni_8$ bulk metallic glass. *Appl. Phys. Lett.* **2003**, *83*, 1134–1135. [CrossRef]

64. Mizubayashi, H.; Shibasaki, M.; Murayama, S. Local strain around hydrogen in amorphous $Cu_{50}Zr_{50}$ and $Cu_{50}Ti_{50}$. *Acta Mater.* **1999**, *47*, 3331–3338. [CrossRef]

65. Hasegawa, M.; Kotani, K.; Yamaura, S.; Kato, H.; Kodama, I.; Inoue, A. Hydrogen-induced internal friction of Zr-based bulk glassy alloys in a rod shape above 90 K. *J. Alloys Compd.* **2004**, *365*, 221–227. [CrossRef]

66. Rodmacq, B.; Maret, M.; Laugier, J.; Billard, L.; Chamberod, A. Hydrogen-Induced Phase-Separation in Amorphous $Cu_{0.5}Ti_{0.5}$ Alloys. 1. Room-Temperature Experiments. *Phys. Rev. B* **1988**, *38*, 1105–1115. [CrossRef]

67. Ismail, N.; Gebert, A.; Uhlemann, M.; Eckert, J.; Schultz, L. Effect of hydrogen on $Zr_{65}Cu_{17.5}Al_{7.5}Ni_{10}$ metallic glass. *J. Alloys Compd.* **2001**, *314*, 170–176. [CrossRef]

68. Ismail, N.; Uhlemann, M.; Gebert, A.; Eckert, J. Hydrogenation and its effect on the crystallisation behaviour of $Zr_{55}Cu_{30}Al_{10}Ni_5$ metallic glass. *J. Alloys Compd.* **2000**, *298*, 146–152. [CrossRef]

69. Yoo, B.G.; Oh, J.H.; Kim, Y.J.; Jang, J.I. Effect of hydrogen on subsurface deformation during indentation of a bulk metallic glass. *Intermetallics* **2010**, *18*, 1872–1875. [CrossRef]

70. Wang, Y.W.; Chu, W.Y.; Li, J.X.; Hui, X.D.; Wang, Y.; Gao, K.W.; Su, Y.J.; Qiao, L.J. In situ observation of hydrogen-enhanced localized plastic flow in Zr-based bulk metallic glass. *Mater. Lett.* **2004**, *58*, 2393–2396. [CrossRef]

71. Kawashima, A.; Yokoyama, Y.; Inoue, A. Zr-based bulk glassy alloy with improved resistance to stress corrosion cracking in sodium chloride solutions. *Corros. Sci.* **2010**, *52*, 2950–2957. [CrossRef]

72. Philo, S.L.; Heinrich, J.; Gallino, I.; Busch, R.; Kruzic, J.J. Fatigue crack growth behavior of a $Zr_{58.5}Cu_{15.6}Ni_{12.8}Al_{10.3}Nb_{2.8}$ bulk metallic glass-forming alloy. *Scr. Mater.* **2011**, *64*, 359–362. [CrossRef]

73. Launey, M.E.; Busch, R.; Kruzic, J.J. Effects of free volume changes and residual stresses on the fatigue and fracture behavior of a Zr–Ti–Ni–Cu–Be bulk metallic glass. *Acta Mater.* **2008**, *56*, 500–510. [CrossRef]

74. Philo, S.L.; Kruzic, J.J. Fatigue crack growth behavior of a Zr–Ti–Cu–Ni–Be bulk metallic glass: Role of ambient air environment. *Scr. Mater.* **2010**, *62*, 473–476. [CrossRef]

metals

MDPI

Review
Dynamics and Geometry of Icosahedral Order in Liquid and Glassy Phases of Metallic Glasses

Masato Shimono * and Hidehiro Onodera

National Institute for Materials Science, 1-2-1 Sengen, Tsukuba 305-0047, Japan; onodera.hidehiro@nims.go.jp
* Author to whom correspondence should be addressed; shimono.masato@nims.go.jp; Tel.: +81-29-860-4965;
  Fax: +81-02-860-4960.

Academic Editors: Jordi Sort Viñas and K.C. Chan
Received: 22 May 2015; Accepted: 25 June 2015; Published: 2 July 2015

Abstract: The geometrical properties of the icosahedral ordered structure formed in liquid and glassy phases of metallic glasses are investigated by using molecular dynamics simulations. We investigate the Zr-Cu alloy system as well as a simple model for binary alloys, in which we can change the atomic size ratio between alloying components. In both cases, we found the same nature of icosahedral order in liquid and glassy phases. The icosahedral clusters are observed in liquid phases as well as in glassy phases. As the temperature approaches to the glass transition point T_g, the density of the clusters rapidly grows and the icosahedral clusters begin to connect to each other and form a medium-range network structure. By investigating the geometry of connection between clusters in the icosahedral network, we found that the dominant connecting pattern is the one sharing seven atoms which forms a pentagonal bicap with five-fold symmetry. From a geometrical point of view, we can understand the mechanism of the formation and growth of the icosahedral order by using the Regge calculus, which is originally employed to formulate a theory of gravity. The Regge calculus tells us that the distortion energy of the pentagonal bicap could be decreased by introducing an atomic size difference between alloying elements and that the icosahedral network would be stabilized by a considerably large atomic size difference.

Keywords: metallic glasses; icosahedral order; medium-range order; molecular dynamics simulation; glass-forming ability; Regge calculus

---

1. Introduction

Icosahedral symmetry is considered to play an important role in the atomic scale structure of glassy phases in spherically interacting systems. The Dense Random Packing (DRP) model was proposed by Bernal for liquid phases in his pioneering work [1], and was later applied to a structure of amorphous metals [2]. In this model, the icosahedral cluster formed by 13 atoms located at the center and the 12 vertices of an icosahedron is a key building block and is basis of the icosahedral short-range order in liquid and glassy phases. The early works of computer simulation have shown that the icosahedral order would exist in both liquid and glassy phases [3,4]. After the discovery of good metallic glass-formers [5,6], experimental observations [7–12] has reported that the icosahedral short-range order does exist in metallic glasses and that some medium-range order may also exist beyond the icosahedral short-range order. However, it was little known how the icosahedral short-range order is arranged and extended to form a medium-range order in glassy phases. In other words, the interrelation between the icosahedral clusters was not clear. In recent years, two milestone models for icosahedral medium-range structure have been proposed: One is the fcc packing of icosahedral clusters proposed by Miracle [13] and the other is the icosahedral packing of icosahedral clusters and the strings of connected of icosahedra proposed by Sheng *et al.* [14]. Being enlightened to these models, a family of network-type models has been proposed with a special attention on the bonding topology between icosahedral clusters, such as the "bicap sharing" (or "interpenetrating" depending on authors) network [15–17], the string-like backbone network [18,19], and the superclusters [20,21]. Especially, Ding *et al.* have elegantly shown [22] that the icosahedral order can be generally understood as the polyhedral packing by the Frank-Kasper polyhedra [23], which are formulated based on the notion of DRP and include the icosahedron as a member. In addition, recent experimental observations [24] and simulation studies [25,26] have revealed that the icosahedral network formation is closely related to the "slowing down" of the relaxation dynamics in supercooled liquids near the glass transition. Despite the understanding of the nature of icosahedral order is gradually deepened in both structural and dynamical aspects, the physics behind the formation of the icosahedral medium-range order is not fully understood yet. In the present study, we investigate geometrical and dynamical properties of the icosahedral order formed in both liquid and glassy phases of metallic glasses by using molecular dynamics (MD) simulations.

MD simulation is a method to calculate the moving trajectories of atoms by solving the Newtonian equations of motion numerically. It is a powerful tool to investigate the atomic scale structure because all information of atomic configurations can be drawn at any time in the course of calculations. Since the purpose of our study is to clarify the geometrical nature underlying the icosahedral medium-range ordered structure formed in glassy phases and supercooled liquid phases of metallic glasses, the MD technique is highly useful. This article is planned as follows. The methods of MD simulation are given in Section 2. Section 3 is devoted to the simulation results and discussion. The results for the Zr-Cu system are shown in Section 3.1, the results for a model alloy system are shown in Section 3.2, and the Regge calculus is introduced in Section 3.3 to discuss the geometrical properties of the icosahedral ordered structure found in the simulations. The conclusion is given in Section 4.

2. Methods

2.1. Interatomic Potentials

In the classical molecular dynamics simulations, interatomic potentials between constituent atoms play a decisive role in the calculation. In this study, we use two different types of potentials: One is a many-body potential [27] based on the electron density theory for the Zr-Cu system, and the other is a two-body potential [28] for a model alloy system.

For the Zr-Cu system, we use a many-body Finnis-Sinclare type potential [27] developed by Rosato *et al.* [29], which has a functional form as

$$V_i = -\sqrt{\rho_i} + \sum_j \varphi_{ij} \tag{1}$$

$$\varphi_{iji}(r_{ij}) = A_{ij} \exp p_{ij}(1 - r_{ij}/r_{ij}^0) \tag{2}$$

$$\rho_i(r_{ij}) = \sum_j \xi_{ij}^2 \exp 2q_{ij}(1 - r_{ij}/r_{ij}^0) \tag{3}$$

where r_{ij} is the distance between atoms i and j and the parameters p_{ij}, q_{ij}, A_{ij}, ξ_{ij} and r_{ij}^0 are determined by us [30] to reproduce the mixing enthalpy of the Zr-Cu system and the lattice parameters, elastic moduli and the cohesive energies of hcp-Zr, fcc-Cu, and the B2-ZrCu phase.

It is well known that the atomic size ratio between the constituent elements plays an important role in the formation of metallic glasses [31]. Therefore, for a simple model for binary alloys, we assume the interaction between atoms i and j to be described by the 8-4 type Lennard-Jones potential [28] V^{ij} as

$$V^{ij}(r) = e_0^{ij}\left\{ \left(r_0^{ij}/r\right)^8 - 2\left(r_0^{ij}/r\right)^4 \right\} \tag{4}$$

The merit of using this potentials is that we can independently vary the atomic size and the chemical bond strength by changing the parameters r_0^{ij} and e_0^{ij}, respectively. In this study, to focus on the atomic size effect, we assume for a binary system composed of elements A and B as $r_0^{AA} = 1$, $r_0^{BB} \leq 1$, $r_0^{AB} = (r_0^{AA} + r_0^{BB})/2$, and $e_0^{AA} = e_0^{BB} = e_0^{AB} = 1$. Thus, we can vary the atomic size ratio r_0^{BB} of the element B to A, and the concentration x_B of the smaller element B. The atomic masses of both elements are also supposed to be the same unit mass. All physical quantities are expressed in the above units for the model system.

2.2. Simulation Procedure

The simulation system consists of 4000 to 16,000 atoms, and is confined in a cubic simulation cell with periodic boundary conditions imposed in all three directions. The temperature of the system is controlled by scaling the atomic momenta. The pressure of the system is kept zero by changing the size of the simulation cell according to the constant pressure formalism [32].

In the simulation the alloy system is started from a liquid phase above the melting point and quenched down to solidify, which brings us a glassy phase or a crystalline phase depending on the cooling rate. After a glassy phase is produced, the system is heated up again and kept at a constant temperature for isothermal annealing, if needed. By monitoring the volume, energy, radial distribution of atoms, and the atomic diffusion, we can detect the phase properties at any stage.

2.3. Icosahedral Symmetry

We shall investigate the local atomic structure of liquid and glassy phases with paying a special attention to the icosahedral symmetry. For this purpose, we use the Voronoi tessellation analysis [1], in which the local symmetry around each atom is indexed by a set of integers (n_3, n_4, n_5, n_6), where n_i is the number of i-edged faces of the Voronoi cell. By calculating this index for each atom, we define the icosahedral cluster by the atom that has the Voronoi index $(0, 0, 12, 0)$ and its neighbors.

3. Results and Discussion

3.1. Icosahedral Order in Zr-Cu System

3.1.1. Icosahedral Medium-range Order in Zr-Cu Metallic Glasses

Firstly we examine the icosahedral order in glassy phases of the Zr-Cu alloys. Figure 1 shows all icosahedral clusters picked up by the Voronoi analysis found in an as-quenched $N = 8000$ $Zr_{40}Cu_{60}$ glassy phase. To investigate the medium-range order or the interrelation between the icosahedral clusters, we also calculate the Voronoi indices between the icosahedral clusters. Unfortunately, we could not detect any signature of fcc order or icosahedral order of the clusters, but found an inhomogeneous and string-like [14,18] network structure as shown in Figure 1. To characterize the topological nature of this network, we check the connecting pattern between adjacent clusters.

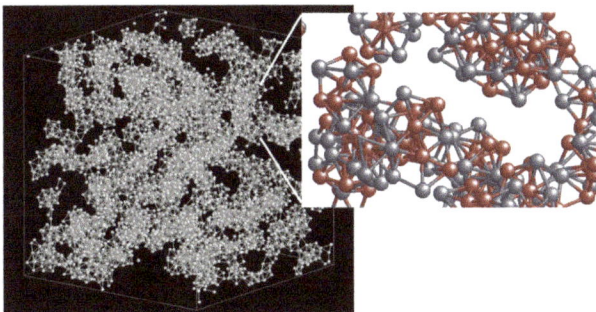

Figure 1. Snapshot of icosahedral clusters formed in an as-quenched $Zr_{40}Cu_{60}$ glassy phase. In the inset, the gray and brown spheres denote the Zr and Cu atoms.

3.1.2. Geometrical Feature of Icosahedral Network

When two icosahedral clusters are linked together, the linking patterns can be classified into the following four types [15] as illustrated in Figure 2: (1) Vertex sharing, where one atom is shared by two clusters; (2) edge sharing, where two atoms forming a link are shared; (3) face sharing, where three atoms forming a triangle are shared; and (4) bicap sharing, where seven atoms forming a pentagonal bicap are shared or two icosahedra are interpenetrating each other. We have counted the population of these four linking patterns for an as-quenched $N = 8000$ $Zr_{40}Cu_{60}$ glassy phase. The results are shown in Figure 2, where the population of the isolated and the linked icosahedral clusters are also shown as well as the population of vertex sharing, the edge sharing, the face sharing, and the bicap sharing connection. We can see that the network mainly consists of bicap sharing connection and this type of connection should be a key to understand the medium-range order in metallic glasses. It is consistent with the recent experimental observation [12] by the scanning electron nanodiffraction that suggests a face sharing or bicap sharing model of the icosahedral medium-range order in a $Zr_{36}Cu_{64}$ glass.

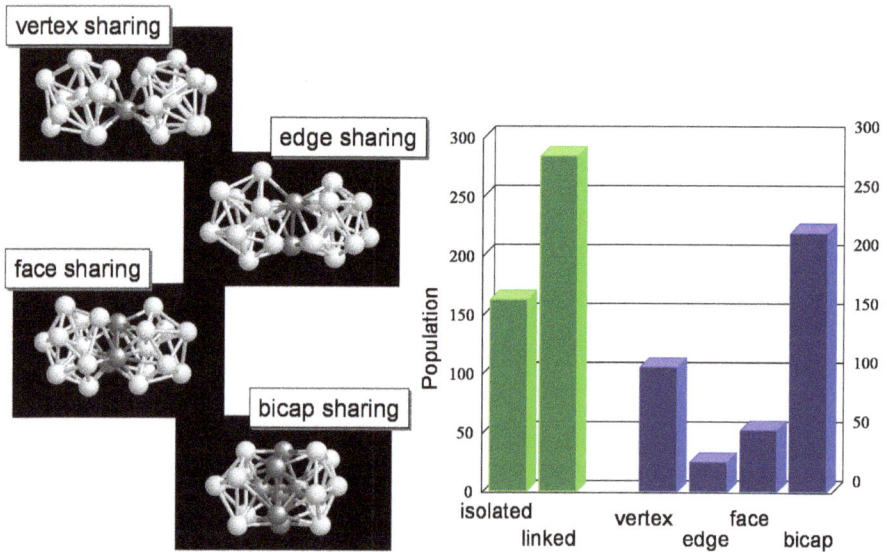

Figure 2. Linking patterns between the icosahedral clusters and their population found in an $N = 8000$ $Zr_{40}Cu_{60}$ glassy phase.

3.1.3. Icosahedral Order in Supercooled Liquids

The icosahedral clusters are also found in liquid phases. Figure 3 shows the temperature dependence of the atomic volume in a quenching process of the $N = 4000$ $Zr_{40}Cu_{60}$ system together with snapshots of icosahedral clusters found in supercooled liquid phases at $T = 1044, 928$, and 814 K, where only the central atoms of the clusters are depicted by white spheres. In higher temperature liquid phases, most of the icosahedral clusters are isolated. As the temperature decreasing, a sign of networking can be found in supercooled liquid phases at near T_g.

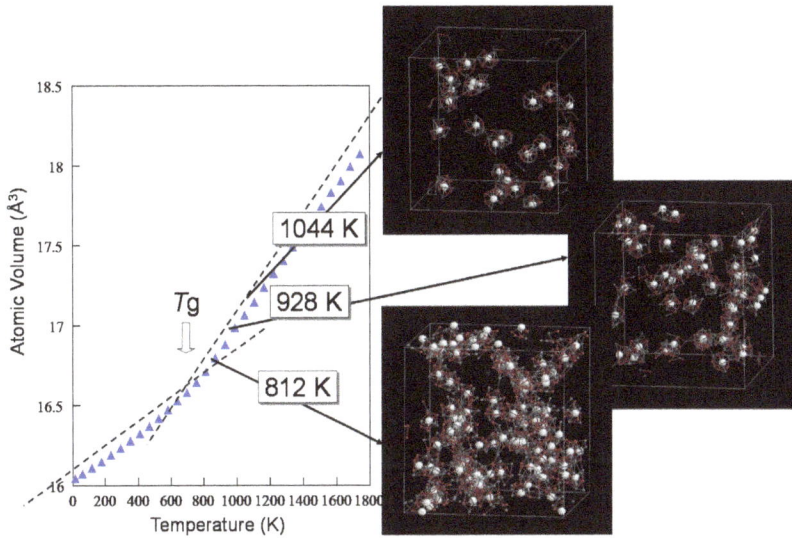

Figure 3. Volume change in a quenching process of the $N = 4000$ $Zr_{40}Cu_{60}$ alloy system and snapshots of icosahedral clusters formed in supercooled liquid phases, where the central atoms of the clusters are depicted by white spheres.

3.1.4. Lifetime of Icosahedral Clusters

The growth of the icosahedral order in supercooled liquid phases is closely related with the stability of the icosahedral clusters. To estimate it, we calculate the lifetime of the icosahedral clusters in supercooled liquids. The lifetime of the icosahedral cluster is defined as follows. Once an icosahedral cluster forms, the cluster is "living" when it keeps both the Voronoi index and the neighboring atoms. Let us show some illustrative examples. In Figure 4, the case (a) and the case (b) are considered to be "decaying" due to the change of symmetry or neighbors, but the case (c) is considered to be still "living" although the arrangement order of the neighbors are changed.

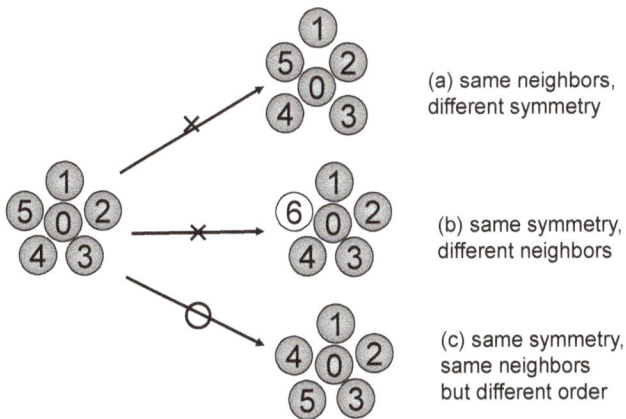

Figure 4. Changing patterns of atomic configuration of the cluster and definition of cluster decaying.

Under this definition, we have calculated the distribution of lifetime of the icosahedral clusters in supercooled liquid phases. Figure 5a shows the histogram of the lifetime of the clusters for 458 decaying events found in an $N = 4000$ $Zr_{40}Cu_{60}$ supercooled liquid phase at $T = 812$ K. More than 65% of the clusters has decayed within the first 2 ps after their formation, but the distribution of the lifetime has a long tail. Figure 5b shows the temperature dependence of the average lifetime in supercooled liquids of the $Zr_{40}Cu_{60}$ alloy system, where the average was taken from 458, 276, and 171 decaying events for the T = 812, 928, and 1044 K cases, respectively. The average lifetime goes longer at lower temperature and its temperature dependence indicates some activation process in the cluster decay. However, when we try to fit the lifetime distribution shown in Figure 5a to a single exponential function, we always failed at any temperature. Therefore, we try to fit the distributions by a stretched exponential function as $N(t) = N_0 \exp\{-(t/\tau)^\beta\}$. The results of the fitted values are $\beta = 0.74$ and $\tau = 0.72$ ps for $T = 812$ K. The fact that the exponent β is not unity indicates that the icosahedral cluster decaying cannot be described by a single process, and that there are two or more different processes with different time scales. Thus, we should investigate the time evolution of the decaying process of the icosahedral clusters more closely.

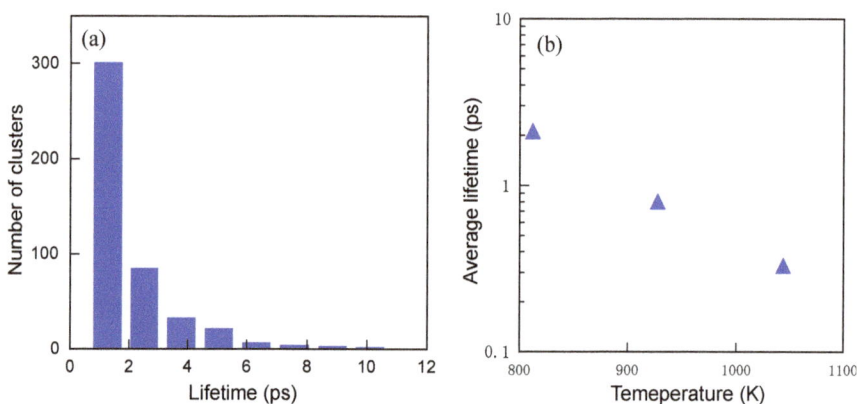

Figure 5. (**a**) Distribution of lifetime of the icosahedral clusters in an $N = 4000$ $Zr_{40}Cu_{60}$ supercooled liquid phase at $T = 812$ K and (**b**) temperature dependence of the average lifetime of the icosahedral clusters in liquid phases of the $Zr_{40}Cu_{60}$ alloy system.

3.1.5. Lifetime and Cluster Bonding

Let us show some examples of the cluster decaying process by depicting a series of snapshots of atomic configurations. Figure 6a is a snapshot of a supercooled liquid phase of the $N = 4000$ $Zr_{40}Cu_{60}$ system at just above T_g. More than 50 icosahedral clusters exist at this moment. Among them, we have picked up some clusters as shown in Figure 6b, where deep green atoms and deep red atoms denote isolated clusters, and light green ones and dark yellow ones denote connected clusters by bicap sharing. The colors of central atoms of the clusters are changed from those of their neighbors for eye guide, and the splitting of the dark yellow ones is due to the periodic boundary condition of the simulation cell. In the first frame at $t = 0$ (Figure 6b), all clusters keep the icosahedral symmetry. As the time goes on at $t = 0.5, 1.0, 1.5, 2.0$ ps (Figure 6c–f, respectively), the isolated clusters changed their neighbors and lost their initial structure. On the other hand, the two-connected clusters (the light green atoms) and the three-connected clusters (the dark yellow atoms) keep their initial configurations even in the last frame at $t = 2.0$ ps (Figure 6f).

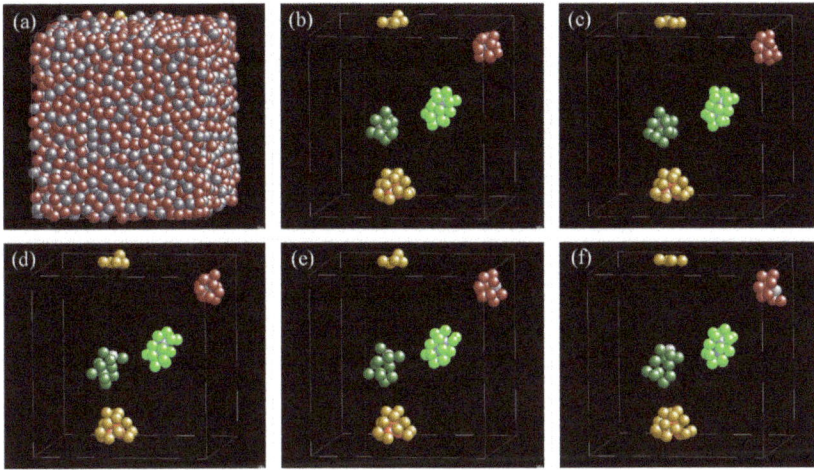

Figure 6. Snapshots of change of atomic configuration of icosahedral clusters in a supercooled liquid phase of the $Zr_{40}Cu_{60}$ system at just above T_g. (**a**) All atoms at $t = 0$; (**b**) two isolated clusters and two connected "superclusters" picked up at $t = 0$; (**c**) configuration change at $t = 0.5$ ps; (**d**) at $t = 1.0$ ps; (**e**) at $t = 1.5$ ps; and (**f**) at $t = 2.0$ ps.

This behavior indicates that the cluster lifetime would be elongated by connecting to each other. Thus, we check the relation between lifetime and cluster networking. We calculated the average lifetime separately for isolated clusters, two-connected clusters, and multi-connected clusters with three or more connections in an $N = 4000$ $Zr_{40}Cu_{60}$ supercooled liquid phase at $T = 812$ K. The results are shown in Figure 7, where the average was taken from 30, 21, and 14 decaying events for the isolated, two-connected, and multi-connected clusters, respectively. We can see that the lifetime becomes longer if clusters are connected via bicap sharing, as already reported by Malins *et al.* [26] in a simulation study of a model binary glass former. Therefore, the icosahedral order in supercooled liquids is stabilized by the network formation of clusters connected via bicap sharing.

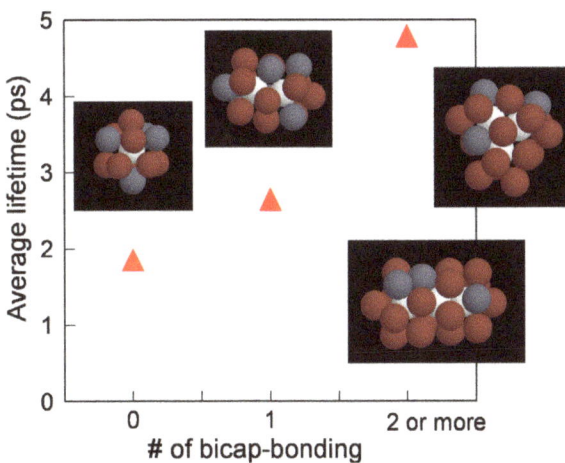

Figure 7. Relation between the average lifetime of the icosahedral clusters and the number of the bicap sharing bonds belonging to the clusters.

3.2. Icosahedral Order in Model System

3.2.1. Atomic Size Effect on Glass-forming Ranges of Model Binary System

To focus on a geometrical aspects of icosahedral order in metallic glasses, we proceed to analysis for a model alloy system interacting with the Lennard-Jones type potential, where we can freely change the atomic size ratio between constituent elements. Here we consider a model A-B binary alloy systems, where the element A has a unit size and the size of element B changes from 0.8 to 1, and the composition x_B of B is ranging from 0 to 1. The heat of mixing between the elements A and B is fixed to be zero to focus on a geometrical effect.

The glass-forming range by rapid quenching from liquid phases in this model system has been investigated in the previous studies [33–35] and the results are shown in Figure 8. The color of the glass-forming region indicates the cooling rate, where a darker region corresponds to a lower cooling rate. We can find that the glass-forming ability goes up as the atomic size difference increases.

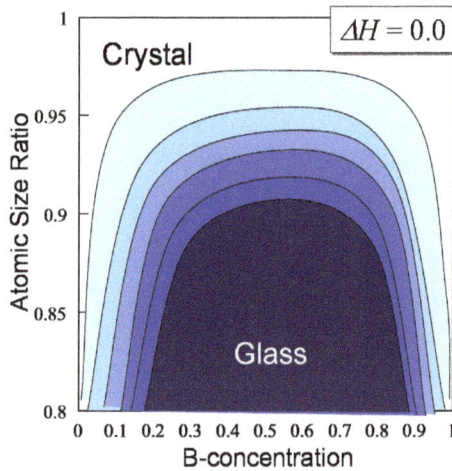

Figure 8. Dependence of glass-formation range on the cooling rate mapped on the atomic size ratio *vs.* composition plane. Darker region corresponds to lower cooling rate.

3.2.2. Geometrical Feature of Icosahedral Network

As found in the Zr-Cu system, we have found the network structure formed by icosahedral clusters mainly connected by bicap sharing in the glassy phases also in the model system. Figure 9 shows the properties of the cluster networks found in as-quenched glassy phases in the $N = 4000$ $A_{80}B_{20}$ and the $N = 4000$ $A_{40}B_{60}$ systems with the atomic size ratio 1:0.8, which approximately agrees with that between Zr and Cu atoms. The dominant connecting pattern is bicap sharing in both cases as found in the Zr-Cu system and the network is more extended in the $A_{40}B_{60}$ case than in the $A_{80}B_{20}$ case. If we pick up the cap sharing bonds and draw them as sticks, we can see that the network extends all over the system as shown in Figure 9b for the $A_{40}B_{60}$ case.

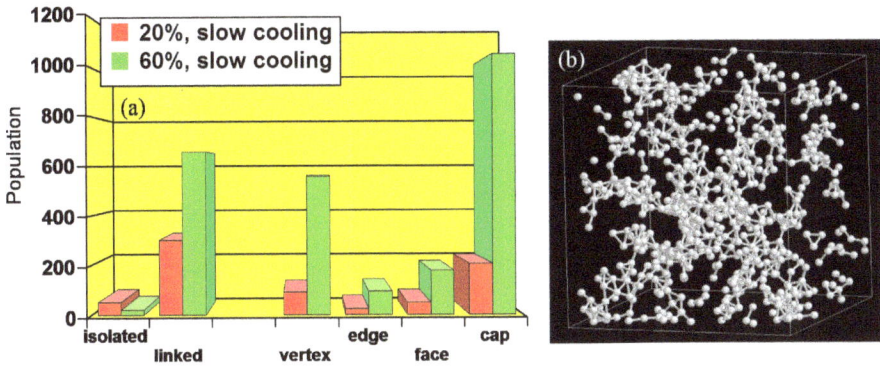

Figure 9. (a) Population of isolated/linked icosahedral clusters and those of linking patterns between clusters found in glassy phases of the $N = 4000$ $A_{80}B_{20}$ system and the $N = 4000$ $A_{40}B_{60}$ system; (b) Network structure formed by the bicap sharing bonds between the icosahedral clusters found in a glassy phase of the $A_{40}B_{60}$ system, where spheres and sticks denote the central atoms of the icosahedral clusters and the bicap sharing bonds, respectively.

3.2.3. Atomic Size Ratio Dependence of Icosahedral Order

The relation between the icosahedral order formation and the atomic size ratio is confirmed by investigating the icosahedral order in supercooled liquid phases. For a fixed composition $x_B = 0.5$, the dependence of the density of the icosahedral clusters on the atomic size ratio of the supercooled phases was calculated, and the results are shown in Figure 10. The number of icosahedral clusters in the supercooled liquids increases as the size difference increases to the atomic size ratio 1:0.8. On the other hand, it begins to decrease beyond the atomic size difference 0.2 because the relative stability of the icosahedral clusters to other types of clusters (e.g., the trigonal prisms [36]) decreases due to the large atomic size difference. It means that the proper size difference between constituent elements brings higher icosahedral order and higher stability of the supercooled liquid phases.

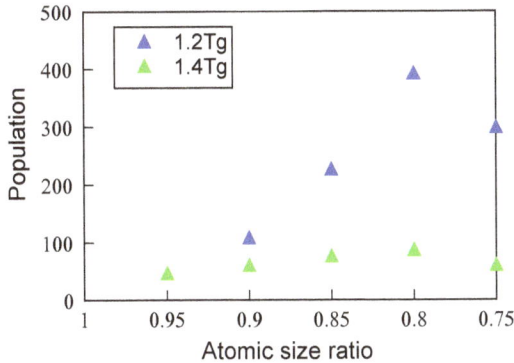

Figure 10. Dependence of the population of the icosahedral clusters in the supercooled liquid phases of the $N = 4000$ $A_{50}B_{50}$ system at $T = 1.4T_g$ and $T = 1.2T_g$ on the atomic size ratio.

3.2.4. Cluster Lifetime and Atomic Size difference

To estimate the stability of the icosahedral order in supercooled liquids in the model system, we have also calculated [35] the dependence of the average lifetime of the icosahedral clusters on the atomic size ratio. The results for supercooled liquids of the N = 4000 $A_{50}B_{50}$ system at T = 1.2T_g are shown in Figure 11, where the average was taken from 45, 54, and 50 decaying events for the $A_{50}B_{50}$ system with the atomic size ratio 1:0.9, 1:0.85, and 1:0.8, respectively. The average lifetime of the clusters increases as the size difference increases in this range. The enhanced stability of the icosahedral clusters due to the atomic size difference would be mainly originated from two factors: One is enhancement of stability of single cluster due to atomic size difference and the other is enhancement of connectivity between the clusters due to the increase of cluster density. Since the stability of a sole cluster has been discussed in the previous study [35], we shall investigate the enhancement of the stability of the icosahedral order due to cluster connection by focusing on the geometry of cluster networking.

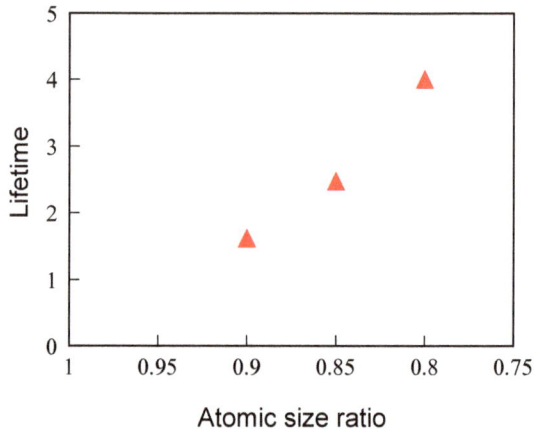

Figure 11. Dependence of the average lifetime of the icosahedral clusters in a supercooled liquid phase of the N = 4000 $A_{50}B_{50}$ system at T = 1.2T_g on the atomic size ratio. The time unit normalized by the model parameters described in the Section 2.1 is used.

3.2.5. Icosahedral Network in Supercooled Liquids

The network structure of connected icosahedral clusters via bicap sharing is also observed in supercooled liquid phases. Figure 12 shows snapshots of the cluster networks found in supercooled liquid phases of the N = 4000 $A_{50}B_{50}$ system with the atomic size ratio 1:0.9 and 1:0.8, and the dependence of the number of cluster connection via bicap sharing in a supercooled liquid phase at T = 1.2Tg on the atomic size ratio of the N = 4000 $A_{50}B_{50}$ system. Comparing with the cluster density shown in Figure 10, the number of bicap sharing bonds has larger dependence on the atomic size ratio. That is another reason why the stability of icosahedral structure in supercooled liquids is strongly enhanced by a large atomic size difference between alloying elements.

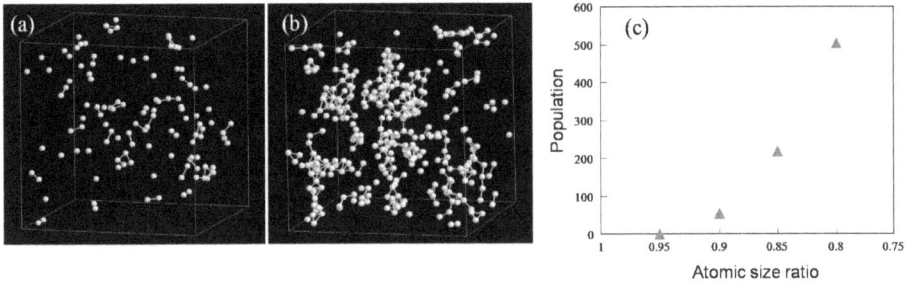

Figure 12. (**a**) Network structure formed by the bicap sharing bonds between the icosahedral clusters found in supercooled liquid phases of the N = 4000 $A_{50}B_{50}$ system with the atomic size ratio 1:0.9 and (**b**) the atomic size ratio 1:0.8, where spheres and sticks denote the central atoms of the icosahedral clusters and the bicap sharing bonds between clusters, respectively. (**c**) Dependence of the number of cluster connection via bicap sharing in a supercooled liquid phase of the N = 4000 $A_{50}B_{50}$ system at $T = 1.2T_g$ on the atomic size ratio.

3.2.6. Unit for 5-Fold Symmetry

The property of the network formed by bicap sharing bonds should depend on the properties of the sharing part of the bonding, that is, the pentagonal bicap formed by seven atoms. Therefore we shall change our focus from the icosahedral cluster to the pentagonal bicap or the bond surrounded by a 5-membered ring, which we call "5-ring bond", as illustrated in Figure 13. Similarly we also call the bonds which are surrounded by 4 or 6 neighbors as 4-ring or 6-ring bonds, respectively.

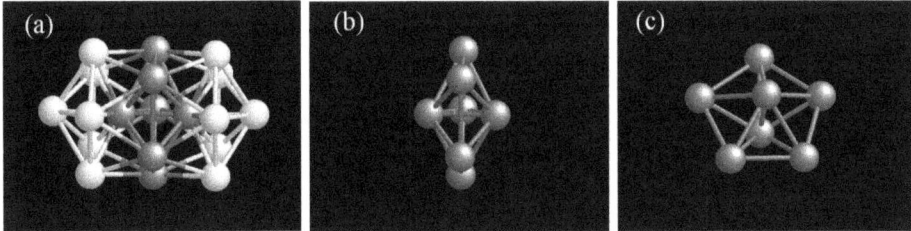

Figure 13. (**a**) Snapshot of two icosahedral clusters connected by bicap sharing; (**b**) Side view of a pentagonal bicap or a 5-ring bond formed by shared 7 atoms; and (**c**) perspective view of a 5-ring bond.

Here we add a note on the definition of "atomic bonding" in this study. We define that two atoms are "bonding" if the two atoms are sharing a common Voronoi face. This definition has the shortcomings that unnatural bonds might be generated when a large atomic size difference exists between constituent elements [37], because the atomic size is not taken into account in the Voronoi tessellation procedure. Therefore, one should use a kind of weighted Voronoi tessellation technique [37] for more detailed analyses.

3.2.7. Geometric Change in Solidification

To investigate the icosahedral order formation in solidifying stage, we calculate the temperature dependence of the population of 4-, 5- and 6-ring bonds together with that of icosahedral clusters in a quenching process of the N = 4000 $A_{50}B_{50}$ system with an atomic size ratio 1:0.8. The results are shown in Figure 14. The number of 5-ring bonds rapidly grows just above T_g as well as that of icosahedral clusters, which indicates that the icosahedral network by the bicap sharing connection would grow in this temperature range.

3.2.8. Geometric Change in Relaxation

We have also calculated the change of population of the 4-ring, 5-ring, and 6-ring bonds in a relaxation process of a glassy phase of the $A_{50}B_{50}$ system with the atomic size ratio 1:0.8 annealed just below T_g. The results are shown in Figure 15. We find that the 5-ring bonds increase and 4-ring bonds decrease in the relaxation stage. It indicates that the relaxation goes on with the growth of the 5-ring network supplied by transformation from 4-rings into 5-rings.

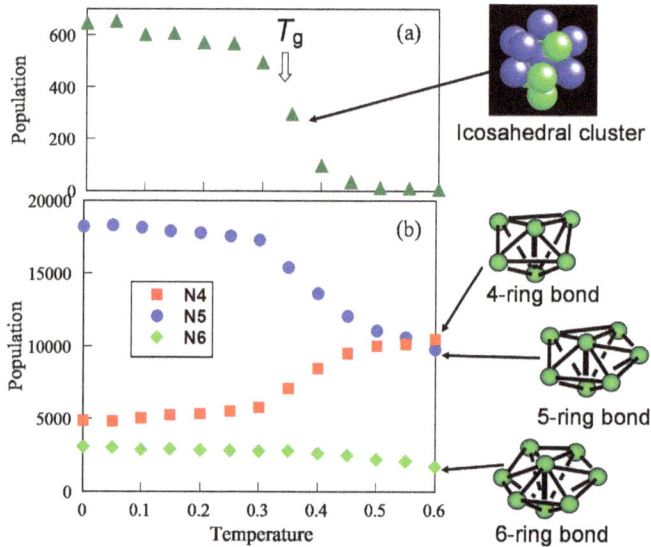

Figure 14. (a) Temperature dependence of population of the icosahedral clusters in a quenching process of the $N = 4000$ $A_{50}B_{50}$ system with an atomic size ratio 1:0.8; (b) Temperature dependence of population of the 4-ring, 5-ring, and 6-ring bonds in the same process.

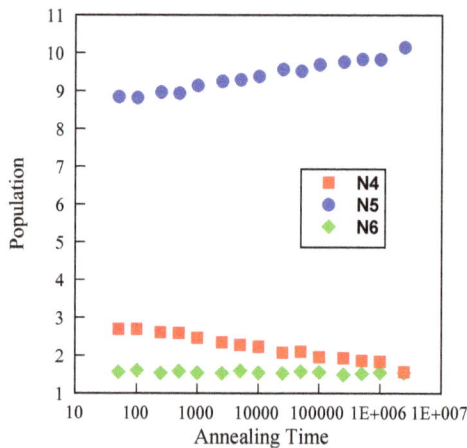

Figure 15. Time dependence of population of the 4-ring, 5-ring, and 6-ring bonds in an annealing process at just below T_g of the $N = 4000$ $A_{50}B_{50}$ system with an atomic size ratio 1:0.8.

3.3. Icosahedral Order and Regge Calculus

3.3.1. Geometric Consideration Based on DRP Model

The idea of the Dense Random Packing (DRP) [1] has given rise to various studies on the atomic structure of liquids and glasses [38–40]. The role of the 5-ring bond can be also understood by a simple geometric consideration based on the DRP model. In the DRP model, the basic building block is mutually bonded tetrahedral cluster of 4-atoms. In other words, the DRP structure is a space-filling with the tetrahedra in three dimensions. The fact that the regular tetrahedron has a dihedral angle around 70.5°, which cannot completely fit to 360° is the reason why the DRP structure cannot fill the whole three dimensional space as crystalline structures do. Therefore, the DRP structure is always accompanied with frustration. As illustrated in Figure 16, the 4-ring bond is too few, the 5-ring bond is a little few, and the 6-ring bond is too many. Among them, the 5-ring has the lowest frustration or distortion, which is why 5-rings dominate in DRP structure.

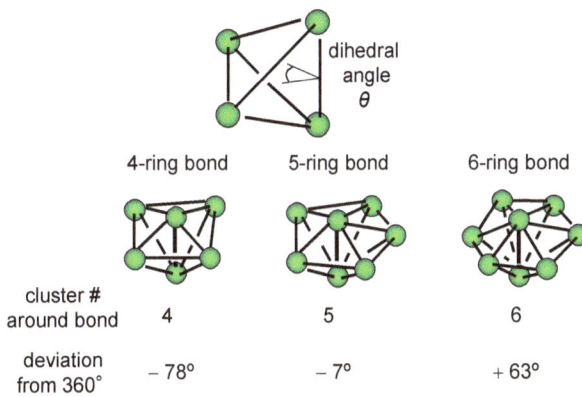

Figure 16. Tetrahedral cluster and 4-, 5-, and 6-ring bonds and the deficit angles around each bonds.

3.3.2. Regge Calculus

To estimate this type of distortion energy, the Regge calculus [41,42] is an appropriate formalism, which was originally proposed as a model of theory of gravitation. In Einstein's theory of gravity, the energy of continuum space-time is expressed by a scalar curvature. On the other hand, in the Regge calculus, the space-time is discretized into simplices (*i.e.*, triangles in two dimensions and tetrahedra in three dimensions) and the energy is estimated by the deviation from 360° or the "deficit angle" located at "hinge". As illustrated in Figure 17, the deficit angle is located at each vertex in two dimensions.

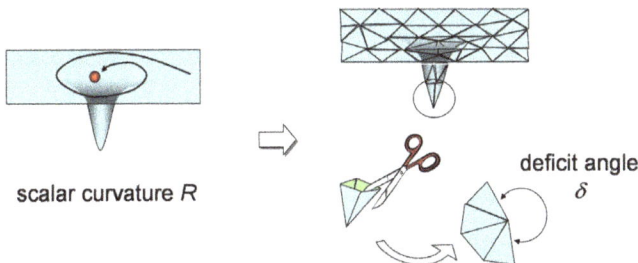

Figure 17. Illustrative view between the continuum theory and the discrete theory of curvature in two-dimensional manifold.

In three dimensions, the space is divided by tetrahedra, and the deficit angle is located at each link or bond, as illustrated in Figure 18. Thus, the hinges are vertices in two dimensions and links in three dimensions.

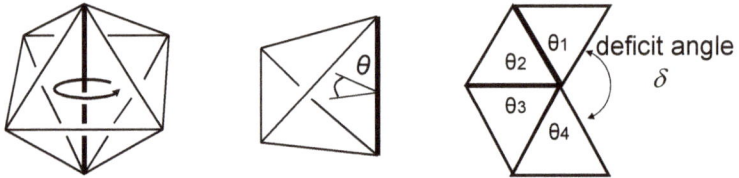

Figure 18. Illustrative view of definition of deficit angle around a bond in three dimensions.

3.3.3. Variety of Tetrahedral Building Blocks: Binary Case

For a monoatomic system, there is no variation of shape of tetrahedra but the regular tetrahedron, so the lowest energy bond for pure elements is the 5-ring bond. On the other hand, if we introduce the different sized elements, the shape variation of tetrahedra increases and the variation of dihedral angles also goes up, as illustrated in Figure 19 for a binary AB system with an atomic size ratio 1:0.8.

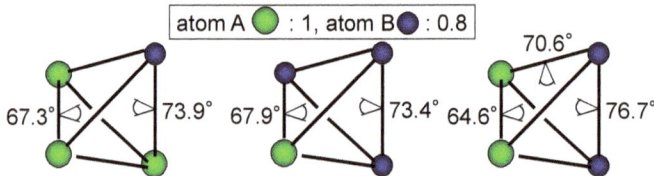

Figure 19. Variety of dihedral angles of tetragonal clusters found in a binary system with the atomic size ratio 1:0.8.

In this binary system, we have 24 different configurations of 5-rings. Among them we can find six configurations which has a lower distortion or a smaller deficit angle than that of monoatomic 5-rings as shown in Figure 20. These configurations of 5-rings with lower distortion would be the key structure to construct an icosahedral ordered network with low frustration in the alloy system.

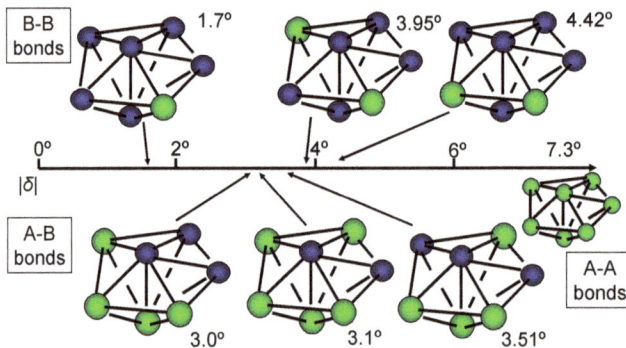

Figure 20. Configurations of 5-ring bonds which has a smaller deficit angle found in a binary system with the atomic size ratio 1:0.8 than that of the 5-ring bond in monoatomic systems.

3.3.4. Icosahedral Network and Low Distortion 5-Rings

The frustration in the network formed by bicap sharing bonds would get lower and the connectivity of the network would be enhanced by the existence of these types of low distortion 5-rings. Therefore, we examined the distribution and population of these types of 5-rings in the icosahedral network via bicap sharing in glassy phases of the AB system. Figure 21 shows snapshots of the icosahedral networks formed by bicap sharing bonds and their portions consisting the top 6 types of low distortion 5-rings shown in Figure 20 found in an as-quenched glassy phase of the $N = 4000$ $A_{50}B_{50}$ system with the atomic size ratio 1:0.8. The top 6 configurations of 5-rings distribute over all icosahedral networks of bicap sharing and its fraction in the network is 50.9%, which is considerably greater than the expectation value 45/128 = 35.2% of finding the above six configurations when we make 5-rings by randomly choosing seven atoms from A or B atoms with equal probability from the $A_{50}B_{50}$ composition. Figure 21c,d shows a bonding topology and the atomic configuration of a portion of the icosahedral network connecting solely the top six configurations of 5-rings.

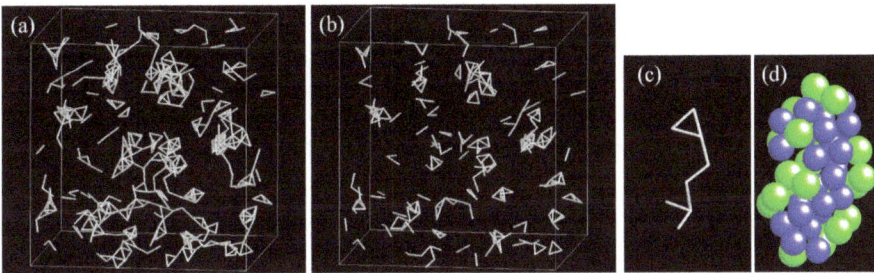

Figure 21. (**a**) Snapshot of the icosahedral networks linked by bicap sharing and (**b**) their portions consisting of the top six configurations with low distortion as depicted in Figure 20 found in an as-quenched glassy phase of the $N = 4000$ $A_{50}B_{50}$ system with the atomic size ratio 1:0.8. (**c**) Snapshot of bonding topology of a fragment of the network shown (**b**), and (**d**) atomic configuration of this portion formed by eight icosahedral clusters, where green and blue spheres denote the A and B atoms, respectively.

3.3.5. Geometry Change in Relaxation

The top six configurations with lower distortion of 5-ring bonds are already dominated in the as-quenched glassy phases as expected. It is likely that their fraction in the icosahedral network would grow in the course of structural relaxation. The results are shown in Figure 22 for an annealing process at just below T_g of a glassy phase of the $N = 4000$ $A_{50}B_{50}$ system with an atomic size ratio 1:0.8. The population of these top six configurations has grown in the annealing process, as well as the total number of bicap 5-ring bonds. However, the fraction of them little changed and simply fluctuated within 50% ± 3% during the annealing process. It means that the relaxation goes mainly by 5-rings formation supplied by 4-rings decay into 5-rings as found in Figure 15, and that the contribution from the configuration change in 5-rings is small, due to low mobility in glassy phases. In other words, the average alloy composition is important to form a lower distortion network of the icosahedral clusters.

3.3.6. Low Distortion 5-Rings and Alloy Compositions of Good Glass-Former

From that viewpoint, we discuss the relation between the composition dependence of icosahedral order and the atomic configurations of 5-rings for the model binary system with the atomic size ratio 1:0.8. As an index for the icosahedral order formation, we calculate the number of the bicap sharing bonds in the networks of icosahedral clusters formed in supercooled liquid phases at $T = 1.2T_g$ of the $N = 4000$ AB system with the atomic size ratio 1:0.8. The results are shown in Figure 23. In this viewgraph, we have indicated the average composition of the 5-rings which have lower distortion as illustrated in Figure 20. Based on the composition dependence of icosahedral order, the predicted composition of the highest icosahedral order and the best glass-former is around $x_B = 0.60$. On the other hand, the composition of the 5-ring of the lowest distortion and the average composition of top 6 configurations are 0.86 and 0.57, respectively, the latter of which is not so bad prediction for highest icosahedral order and good glass-former. It means that the alloy composition which has the biggest chance to form the low distortion 5-rings would be a good glass-former.

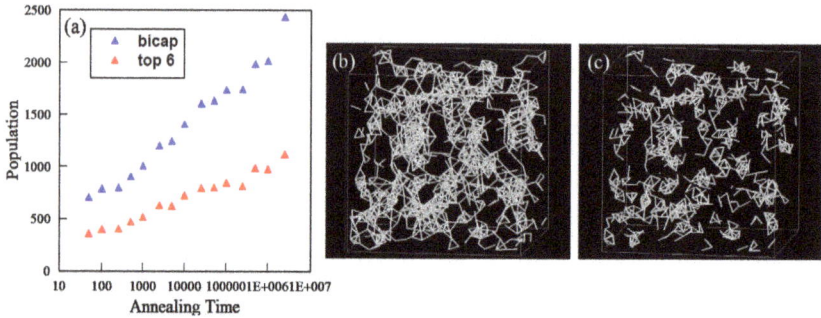

Figure 22. (**a**) Time evolution of the number of the bicap sharing bonds between icosahedral clusters and that of the top 6 configurations of 5-rings in the bicap connections in an annealing process at just below T_g of a glassy phase of the $N = 4000$ $A_{50}B_{50}$ system with an atomic size ratio 1:0.8; (**b**) Snapshot of the icosahedral network via bicap sharing formed in a relaxed glassy phase (annealing time: 2.5×10^6); and (**c**) portion of the top six configurations in the icosahedral network.

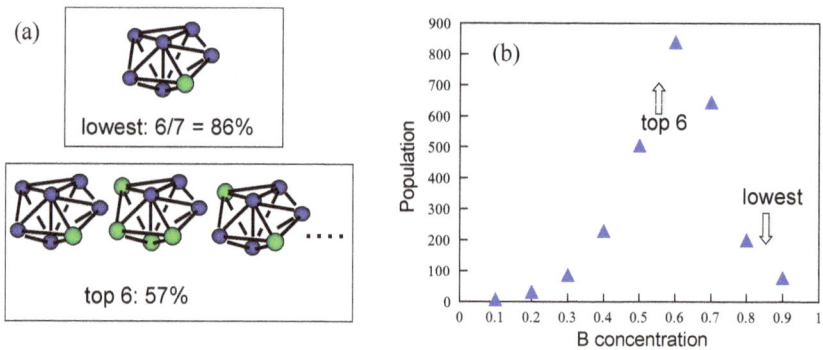

Figure 23. Comparison between the alloy composition of clusters with lower distortion and that of higher density of icosahedral order. (**a**) Atomic configuration of 5-ring bonds with lower distortion; (**b**) Composition dependence of the number of the bicap sharing bonds between the icosahedral clusters in supercooled liquid phases at $T = 1.2T_g$ of the $N = 4000$ AB system with the atomic size ratio 1:0.8. The arrows indicate the composition of the clusters shown in Figure 23a.

3.3.7. Variety of Tetrahedral Building Blocks: Ternary Case

For ternary system with size ratio 1:0.9:0.8, we have further variety of tetrahedron shape. Consequently, we can find many lower distortion configurations in the ternary system than those in binary systems, as shown in Figure 24. That is why the glass-forming ability goes up in ternary or more components metallic systems.

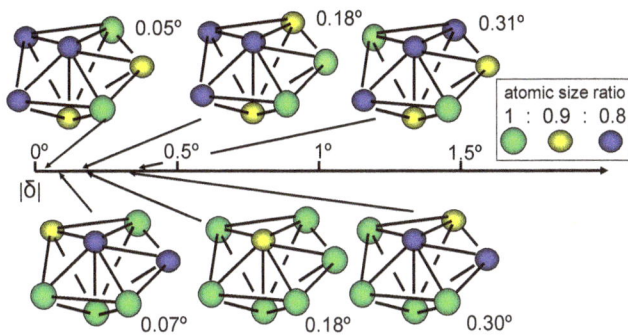

Figure 24. Configurations of 5-ring bonds which have a smaller deficit angle than 0.5 degree found in a ternary system with the atomic size ratio 1:0.9:0.8.

The well-known three empirical rules to acquire a high glass-forming ability in metallic systems proposed by Inoue [31] are the following:

(1) Multicomponent alloy systems consisting of more than three components,
(2) Significantly different atomic size ratios among the main constituent elements,
(3) Large negative heats of mixing among their elements.

The first one (multi-component) and the second one (large atomic size difference) would realize the 5-ring configurations with very low distortion just as shown in Figure 24. Moreover, the third one (negative heat of mixing) makes the probability of forming such low distortion 5-rings greater, because such configurations can be achieved only by mixing different components as shown in Figure 24. In this sense, we can understand the above rules as criteria for lower distortion network formation of icosahedral order.

From a similar point of view, we can understand the geometrical origin of medium-range order structure in metallic glasses by constructing a structural model of icosahedral order by connecting lower strain 5-rings step by step. In the course of construction, we should take not only the icosahedral clusters but also the distorted icosahedra, which are not indexed (0, 0, 12, 0), into account. This type of construction would offer a geometrical basis of the "low frustrated disclination line" [42] or the "icosahedral backbone" [18,19]. This task is complicated and will be the subject of further studies.

3.3.8. Relation between Icosahedral Clusters and Other Types of Clusters

As a structural unit, not only the icosahedral clusters indexed as (0 0 12 0) but other types of clusters also play an important role in liquid and glassy phases of metallic glasses. Lee *et al.* has shown [19] in their simulation study that the interconnection between the icosahedral network and distorted icosahedra (for example, indexed as (0 2 8 2)) form a medium-range structure in the Cu-Zr glassy phases. Ding *et al.* has shown [22] that icosahedral clusters are only dominated in the Cu-centered clusters, but another type of Frank-Kasper polyhedra indexed as (0 0 12 4) called "Z16 cluster" are dominated in the Zr-centered clusters and both types of clusters contribute to the icosahedral order formation in the Cu-Zr supercooled liquids. It means that the icosahedral cluster are only important around "smaller" atoms, but other type of clusters is important around "larger" atoms. Here we should note that the central atom of the Z16 cluster has twelve 5-ring bonds and four 6-ring bonds, so an increase of the Z16 clusters may also contribute to the formation of icosahedral order or 5-ring networks as in the case of the icosahedral clusters. In addition, recent experimental observations [24] and simulation studies [26] have shown that the crystal-like clusters also play an important role in the dynamics at the glass transition, where the time scale of structural relaxation blows up. Therefore, we discuss the interrelation between the icosahedral clusters and other types of clusters for the model binary system shortly.

For the $N = 4000$ $A_{50}B_{50}$ system with an atomic size ratio 1:0.8, the fraction of Voronoi polyhedra with various indices was calculated for an as quenched glassy phase and an relaxed glassy phase annealed at just below T_g for $\Delta t = 2.5 \times 10^6$. The results are shown in Figure 25a, where some of the most abundant polyhedra are shown in each symmetry, that is, the icosahedron-like, the Z16 cluster-like, and crystal-like symmetry. In both cases, the most populous polyhedron is the icosahedron and the next one is the distorted icosahedron indexed by (0 1 10 2). In the relaxed glassy phase, the fraction of the icosahedron-like atoms and Z16 cluster-like atoms have increased due to the structural relaxation. Among them, the fraction of the Z16 cluster showed a remarkable growth (2.75 times higher) [22]. On the other hand, the fraction of crystal-like atoms has a little decreased and it indicates that the crystal-like clusters would transform into the icosahedral ones by annealing, which is consistent with a recent experimental observation in $Zr_{50}Cu_{45}Al_5$ bulk metallic glasses [11]. In the same relaxed phase, 96.7% of the icosahedron-like clusters are the (smaller) B atom-centered, while 99.7% of the Z16 cluster-like clusters are the (larger) A atom-centered.

To get a hint of the structural roles of various types of clusters, we have shown a snapshot of spatial distribution of them in Figure 25b, where a 25% portion of the $N = 4000$ relaxed glassy $A_{50}B_{50}$ alloys is sliced out and only the atoms centered at the icosahedral clusters, the distorted icosahedra, the Z16 cluster, the Z16 cluster-like ones, and the crystal-like ones are depicted by the white, yellow, dark green, light green, and blue spheres. From the figure, the distorted icosahedral atoms and the Z16 cluster-like atoms seem to interconnect to the icosahedral network and form a larger network structure, while the crystal-like atoms seem to occupy a little separated region from the icosahedral network. Clarifying the individual roles of the icosahedral, the distorted icosahedral, and Z16 cluster-like atoms in forming the icosahedral medium-range order is an important issue to be investigated but will be another subject of our future studies.

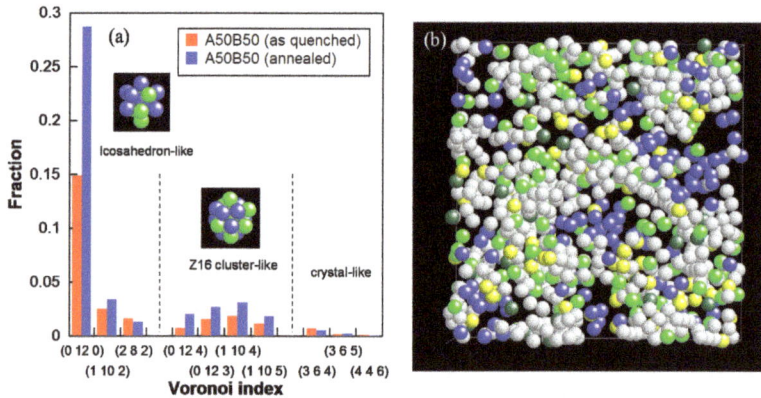

Figure 25. (a) Fractions of various Voronoi polyhedra found in an as quenched glassy phase (**green**) and a relaxed glassy phase (**blue**) of the $N = 4000$ $A_{50}B_{50}$ system with an atomic size ratio 1:0.8; (b) Snapshot of atomic configuration of a 25% portion of an $N = 4000$ relaxed glassy $A_{50}B_{50}$ phase. Only the atoms centered at the icosahedral clusters, the distorted icosahedra, the Z16 clusters, the Z16 cluster-like ones, and the crystal-like ones are depicted by the white, yellow, dark green, light green, and blue spheres.

4. Conclusions

Icosahedral ordered structures formed in liquid and glassy phases of metallic glasses are investigated by using molecular dynamics simulations. In both the Zr-Cu alloy system and a model binary alloy systems, the same feature of the icosahedral order is observed. Formation and decay

Metals **2015**, *5*, 1163–1197

processes of the icosahedral clusters are found in liquid phases. As the temperature decreases, the lifetime of the cluster becomes longer and the density of the cluster grows in supercooled liquid phases. Near T_g, the clusters begin to connect to each other and the network structure of clusters connected via bicap sharing forms, which is the origin of the icosahedral medium-range order. Network formation of icosahedral clusters via cap-sharing in supercooled liquids enhances stability of clusters. In glassy phases, the icosahedral network grows and extends over the whole system in structural relaxation by annealing. The frustration or distortion in the icosahedral network formed by bicap or 5-ring sharing connections can be estimated by using the Regge calculus. The analysis of the distortion in the connected part or the 5-ring bond shows that atomic size difference gives a variety in shape of tetrahedral clusters and generates lower frustrated configurations of 5-ring bonds and icosahedral clusters. The empirical rules between alloying elements for achieving high glass-forming ability in metallic systems can be understood by the analysis of the frustration of the 5-rings.

Author Contributions: The authors contributed equally to this work.

Conflicts of Interest: The authors declare no conflict of interest.

References

1. Bernal, J.D. A Geometrical Approach to the Structure of Liquids. *Nature* **1959**, *183*, 141–147.
2. Cargill, G.S., III. Dense random packing of hard spheres as a structural model for noncrystalline metallic solids. *J. Appl. Phys.* **1970**, *41*, 2248–2250. [CrossRef]
3. Finney, J.L. Modelling the structures of amorphous metals and alloys. *Nature* **1977**, *266*, 309–314. [CrossRef]
4. Yamamoto, R.; Doyama, M. The polyhedron and cavity analyses of a structural model of amorphous iron. *J. Phys. F Metal Phys.* **1979**, *9*, 617–627. [CrossRef]
5. Inoue, A.; Zhang, T.; Masumoto, T. Zr-Al-Ni amorphous alloys with high glass transition temperature and significant supercooled liquid region. *Mater. Trans. JIM* **1990**, *31*, 177–183. [CrossRef]
6. Parker, A.; Johnson, W.L. A highly processable metallic glass: $Zr_{41.2}Ti_{13.8}Cu_{12.5}Ni_{10}Be_{22.5}$. *Appl. Phys. Lett.* **1993**, *63*, 2342–2344.
7. Saida, J.; Sanada, T.; Sato, S.; Imafuku, M.; Matsubara, E.; Inoue, A. Local structure study in Zr-based metallic glasses. *Mater. Trans.* **2007**, *48*, 1703–1707. [CrossRef]
8. Hui, X.; Gao, R.; Chen, G.L.; Shang, S.L.; Wang, Y.; Liu, Z.K. Short-to-medium-range order in $Mg_{65}Cu_{25}Y_{10}$ metallic glass. *Phys. Lett. A* **2008**, *372*, 3078–3084. [CrossRef]
9. Shen, Y.T.; Kim, T.H.; Gangopadhy, A.K.; Kelton, K.F. Icosahedral order, frustration, and the glass transition: Evidence from time-dependent nucleation and supercooled liquid structure studies. *Phys. Rev. Lett.* **2009**, *102*, 057801. [CrossRef] [PubMed]
10. Hirata, A.; Guan, P.-F.; Fujita, T.; Hirotsu, Y.; Inoue, A.; Yavari, A.R.; Sakurai, T.; Chen, M.-W. Direct observation of local atomic order in a metallic glass. *Nat. Mater.* **2011**, *10*, 28–32. [CrossRef] [PubMed]
11. Hwang, J.; Melgarejo, Z.H.; Kalay, Y.E.; Kalay, I.; Kramer, M.J.; Stone, D.S.; Voyles, P.M. Nanoscale Structure and Structural Relaxation in $Zr_{50}Cu_{45}Al_5$ Bulk Metallic Glass. *Phys. Rev. Lett.* **2012**, *108*, 195505. [CrossRef] [PubMed]
12. Liu, A.C.Y.; Neish, M.J.; Stokol, G.; Buckley, G.A.; Smillie, L.A.; de Jonge, M.D.; Ott, R.T.; Kramer, M.J.; Bourgeois, L.J. Systematic Mapping of Icosahedral Short-Range Order in a Melt-Spun $Zr_{36}Cu_{64}$ Metallic Glass. *Phys. Rev. Lett.* **2013**, *110*, 205505. [CrossRef] [PubMed]
13. Miracle, D.B. A structural model for metallic glasses. *Nat. Mater.* **2004**, *3*, 697–702. [CrossRef] [PubMed]
14. Sheng, H.W.; Luo, W.K.; Alamgir, F.M.; Bai, J.M.; Ma, E. Atomic packing and short-to-medium-range order in metallic glasses. *Nature* **2006**, *439*, 419–425. [CrossRef] [PubMed]
15. Shimono, M.; Onodera, H. Icosahedral order in supercooled liquids and glassy alloys. *Mat. Sci. Forum* **2007**, *539–543*, 2031–2035. [CrossRef]
16. Wakeda, M.; Shibutani, Y. Icosahedral clustering with medium-range order and local elastic properties of amorphous metals. *Acta Mater.* **2010**, *58*, 3963–3969. [CrossRef]
17. Xie, Z.-C.; Gao, T.-H.; Guo, X.-T.; Qin, X.-M.; Xie, Q. Growth of icosahedral mediumrange order in liquid TiAl alloy during rapid solidification. *J. Non-Cryst. Solids* **2014**, *394*, 16–21. [CrossRef]

18. Li, M.Z.; Wang, C.Z.; Hao, S.G.; Kramer, M.J.; Ho, K.M. Structural heterogeneity and medium-range order in Zr_xCu_{100-x} metallic glasses. *Phys. Rev. B* **2009**, *80*, 184201. [CrossRef]
19. Lee, M.; Kim, H.-K.; Lee, J.-C. Icosahedral medium-range orders and backbone formation in an amorphous alloy. *Met. Mater. Int.* **2010**, *16*, 877–881. [CrossRef]
20. Cheng, Y.Q.; Ma, E.; Sheng, H.W. Atomic level structure in multicomponent bulk metallic glass. *Phys. Rev. Lett.* **2009**, *102*, 245501. [CrossRef] [PubMed]
21. Lekka, Ch.E.; Evangelakis, G.A. Bonding characteristics and strengthening of CuZr fundamental clusters upon small Al additions from density functional theory calculations. *Scr. Mater.* **2009**, *61*, 974–977. [CrossRef]
22. Ding, J.; Cheng, Y.-Q.; Ma, E. Full icosahedra dominate local order in $Cu_{64}Zr_{34}$ metallic glass and supercooled liquid. *Acta Mater.* **2014**, *69*, 343–354. [CrossRef]
23. Frank, F.C.; Kasper, J.S. Complex alloy structures regarded as sphere packings. I. Definitions and basic principles. *Acta Cryst.* **1958**, *11*, 184–190. [CrossRef]
24. Leocmach, M.; Tanaka, H. Roles of icosahedral and crystal-like order in the hard spheres glass transition. *Nat. Commun.* **2012**, *3*, 974. [CrossRef] [PubMed]
25. Pedersen, U.R.; Schrøder, T.B.; Dyre, J.C.; Harrowell, P. Geometry of Slow Structural Fluctuations in a Supercooled Binary Alloy. *Phys. Rev. Lett.* **2010**, *104*, 105701. [CrossRef] [PubMed]
26. Malins, A.; Eggers, J.; Royall, C.P.; Williams, S.R.; Tanaka, H. Identification of long-lived clusters and their link to slow dynamics in a model glass former. *J. Chem. Phys.* **2013**, *138*, 12A535. [CrossRef] [PubMed]
27. Finnis, M.W.; Sinclair, J.E. A Simple Empirical N-Body Potential for Transition Metals. *Pilos. Mag. A* **1984**, *50*, 45–55. [CrossRef]
28. Sanchez, J.M.; Barefoot, J.R.; Jarrett, R.N.; Tien, J.K. Modeling of γ/γ' phase equilibrium in the Nickel-Aluminum system. *Acta Metall.* **1984**, *32*, 1519–1525. [CrossRef]
29. Rosato, V.; Guillope, M.; Legrand, B. Thermodynamical and structural properties of f.c.c. transition metals using a simple tight-binding model. *Philos. Mag. A* **1989**, *59*, 321–336. [CrossRef]
30. Shimono, M.; Onodera, H. Molecular dynamics study on structural relaxation of metallic glasses. *Mat. Sci. Forum* **2010**, *638–642*, 1648–1652. [CrossRef]
31. Inoue, A. High strength bulk amorphous alloys with low critical cooling rates. *Mater. Trans. JIM* **1995**, *36*, 866–875. [CrossRef]
32. Andersen, H.C. Molecular dynamics simulations at constant pressure and/or temperature. *J. Chem. Phys.* **1980**, *72*, 2384–2393. [CrossRef]
33. Shimono, M.; Onodera, H. Geometrical and chemical factors in the glass-forming ability. *Scripta Mater.* **2001**, *44*, 1595–1598. [CrossRef]
34. Shimono, M.; Onodera, H. Structural Relaxation in Supercooled Liquids. *Mater. Trans.* **2005**, *46*, 2830–2837. [CrossRef]
35. Shimono, M.; Onodera, H. Icosahedral symmetry, fragility and stability of supercooled liquid state of metallic glasses. *Rev. Métallurgie* **2012**, *109*, 41–46. [CrossRef]
36. Ohkubo, T.; Kai, H.; Hirotsu, Y. Structural modeling of Pd–Si and Fe–Zr–B amorphous alloys based on the microphase separation model. *Mater. Sci. Eng. A* **2001**, *304–306*, 300–304. [CrossRef]
37. Park, J.; Shibutani, Y. Weighted Voronoi tessellation technique for internal structure of metallic glasses. *Intermetallics* **2007**, *15*, 187–192. [CrossRef]
38. Hansen, J.P.; McDonald, I.R. *Theory of Simple Liquids*, 3rd ed.; Academic Press: New York, NY, USA, 2006.
39. Miracle, D.B. The efficient cluster packing model—An atomic structural model for metallic glasses. *Acta Mater.* **2006**, *54*, 4317–4336. [CrossRef]
40. Hopkins, A.B.; Stillinger, F.H.; Torquato, S. Densest local sphere-packing diversity. II. Application to three dimensions. *Phys. Rev. E* **2011**, *83*, 011304. [CrossRef]
41. Regge, T. General relativity without coordinates. *Nuovo Cimento* **1961**, *19*, 558–571. [CrossRef]
42. Nelson, D.R. Order, frustration, and defects in liquids and glasses. *Phys. Rev. B* **1983**, *28*, 5515–5535. [CrossRef]

metals

MDPI

Review

Mechanical Relaxation of Metallic Glasses: An Overview of Experimental Data and Theoretical Models

Chaoren Liu [1], Eloi Pineda [1,2,*] and Daniel Crespo [1,2]

[1] Department of Physics, Universitat Politècnica Catalunya—BarcelonaTech, Esteve Terradas 7–8, Castelldefels 08860, Spain; chaorenliu@gmail.com (C.L.); daniel.crespo@upc.edu (D.C.)
[2] Centre for Research in NanoEngineering, Universitat Politècnica Catalunya—BarcelonaTech, Esteve Terradas 7–8, Castelldefels 08860, Spain
* Author to whom correspondence should be addressed; eloi.pineda@upc.edu; Tel.: +34-935-521-141; Fax: +34-935-521-122.

Academic Editors: K. C. Chan and Jordi Sort Viñas
Received: 20 May 2015; Accepted: 12 June 2015; Published: 19 June 2015

Abstract: Relaxation phenomena in glasses are a subject of utmost interest, as they are deeply connected with their structure and dynamics. From a theoretical point of view, mechanical relaxation allows one to get insight into the different atomic-scale processes taking place in the glassy state. Focusing on their possible applications, relaxation behavior influences the mechanical properties of metallic glasses. This paper reviews the present knowledge on mechanical relaxation of metallic glasses. The features of primary and secondary relaxations are reviewed. Experimental data in the time and frequency domain is presented, as well as the different models used to describe the measured relaxation spectra. Extended attention is paid to dynamic mechanical analysis, as it is the most important technique allowing one to access the mechanical relaxation behavior. Finally, the relevance of the relaxation behavior in the mechanical properties of metallic glasses is discussed.

Keywords: metallic glass; relaxation; secondary relaxation; mechanical spectroscopy; amorphous alloy; aging; dynamic mechanical analysis; internal friction; viscoelasticity; anelasticity

1. Introduction

Relaxation is a universal phenomenon driving a system from an excited state towards a more stable one. In this work we will focus on relaxations implying changes of the atomic or molecular positions of a substance, generally known as structural relaxations. Relaxations involving changes of the electronic states are not discussed here. There are different techniques capable of exploring the relaxation response under different stimuli. These techniques comprehend mechanical and dielectric spectroscopy, nuclear magnetic resonance, neutron scattering and various electromagnetic radiation scattering. They probe the relaxation dynamics through the time evolution of macroscopic variables such as density, enthalpy, stress, strain and electric polarization, or by monitoring microscopic parameters such as nuclear spin orientation and mean square or rotational angle displacements.

The relaxation times obtained from different techniques coincide in some systems and time-temperature windows, but not necessary in others as illustrated in Figure 1 [1]. The relaxation time τ_l (liquid structural relaxation-longitudinal) determined from Brillouin scattering, $\tau_{\text{reorientation}}$ obtained from vibrational spectroscopy, τ_σ determined from electrical conductivity, τ_s from viscosity and τ_H from differential scanning calorimetry of $0.4Ca(NO_3)_2 0.6KNO_3$ (CKN glass) split off from each other when temperature decreases, this indicating a decoupling of the relaxation times associated to different structural movements as the supercooled liquid approaches the glass transition temperature

(T_g). In the case of metallic glasses some of these experimental techniques cannot be used, as for example dielectric spectroscopy. This opens a hole in the frequency window usually probed in other substances, which is partially filled with the information obtained by mechanical relaxation techniques. This work is focused on mechanical relaxation of metallic glasses, where experimental data is obtained in terms of macroscopic stress and strain variables in experiments involving relatively long times (or low frequencies). The high-frequency (vibrational) dynamics is not within the scope of this work.

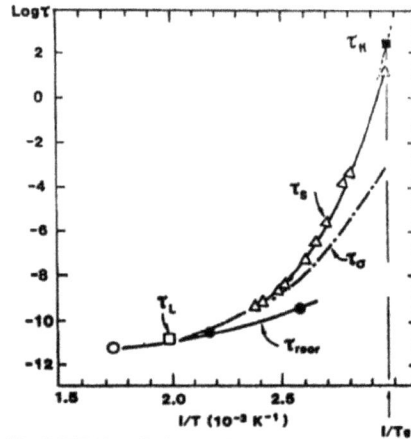

Figure 1. Relaxation times associated to different probed properties in CKN glass. Reprinted from Reference [1] with permission from Elsevier.

Although there is not yet a comprehensive theory, it is generally accepted that glass formation is not a thermodynamic phase transformation but a kinetic process that freezes the system in an out-of-equilibrium configuration at temperatures below T_g [2–4]. Therefore, glasses are permanently prone to change its configuration towards a more stable state through irreversible atomic movements. This process is called physical aging and it may have time scales much larger than the experimental ones, seeming that the system is stable from the macroscopic point of view. This process is also called structural relaxation, in the sense that the system is relaxing form a higher free-energy configuration towards a lower one. Nevertheless, the structural relaxation understood as the response of the system to an applied external stimulus, is an intrinsic process present both in equilibrium (liquid) and out-of-equilibrium (glassy) states. For the sake of clarity, in this work, we denote as relaxation only the latter kind of process while the former one will be always referred as aging. Of course, in many cases, similar structural movements are responsible for mechanical relaxation and physical aging. However, we would like to emphasize here that the relaxation response is a well-defined property dependent only on temperature and pressure for a particular glassy configuration while physical aging is a history dependent process driving the system through different glassy states.

This work aims to review early work as well as recent new findings on the relaxation of metallic glasses studied by mechanical spectroscopy. It is also intended as an easy introduction for beginners in the field. It is organized in the following way: Firstly, we will review the relaxation dynamics, viscosity and physical aging behaviors of glasses introducing the major models used for their interpretation; Secondly, we will summarize the main characteristics of the mechanical spectroscopy technique, the different kind of processes contributing to the mechanical relaxation of glasses and the model functions used for analyzing the response of glasses both in the time and the frequency domains; Finally, experimental data on mechanical relaxation of metallic glasses are presented, from early works measuring internal friction as a probe to study the physical aging to more recent works characterizing the presence and effects of secondary relaxations in bulk metallic glasses (BMGs). Eventually, recent progress on fitting the Dynamo-Mechanical-Analysis (DMA) results considering different fitting models and the relationship between the relaxation behavior and other mechanical properties will be also discussed.

2. Basics of Glass Relaxation Dynamics

2.1. Relaxation in the Supercooled Liquid Region

The activation of viscous flow reflects the complete relaxation of the system, accommodating its structure under the application of an external force. Therefore, in structural glasses, viscosity is directly related to the primary relaxation time of the system, the so-called α-relaxation. In the supercooled liquid region, when $T > T_g$, the system is ergodic and the α-relaxation time, τ_α, characterizes how long takes the system to return to internal equilibrium after being excited by an external force or a change in the temperature-pressure conditions. Generally speaking, the viscosity (η) and τ_α deviate from Arrhenius behavior. As summarized by Angell [5], different theories like the simple liquid model, mode coupling theory, random walk model, or theoretical results for random packed spheres or cooperatively rearranging systems propose different dependences on temperature.

Actually, in the relatively narrow temperature region where most of the experimental viscosity data is obtained, all models fit the data with just two or three parameters and it is difficult to determine which model have superior validity. Experimentally, for many liquids above T_g, the relaxation dynamics can be described in the form of the Vogel-Fulcher-Tammann (VFT) equation for the viscosity

$$\eta(T) = \eta_0 \exp\left(\frac{D^* T_0}{T - T_0}\right) \tag{1}$$

D^* and T_0 being the strength parameter and the VFT temperature respectively. The strength parameter D^* is used to distinguish between strong and fragile liquids; strong liquids are defined by a large D^* and an almost Arrhenius-like behavior ($T_0 \rightarrow 0$). On the other hand, fragile liquids are characterized by small D^* and a very rapid breakdown of shear resistance when heating above T_g. Analogous temperature dependence is also found for τ_α above T_g [6]. The strong-fragile nature of liquids near T_g is also usually characterized by the fragility parameter

$$m = \left.\frac{d \log \eta}{d\left(T_g/T\right)}\right|_{T=T_g} = \left.\frac{d \log \tau_\alpha}{d\left(T_g/T\right)}\right|_{T=T_g} \tag{2}$$

In addition to the main α-relaxation, secondary relaxations may be also found in the supercooled liquid region. Below some critical temperature, particular structural movements are decoupled from the main process giving rise to faster relaxations which usually follow an Arrhenius-like temperature behavior. As discussed below, some models predict the presence of a β-relaxation below a critical temperature ($T_c > T_g$) as a universal feature of the glass transition process [5,7].

2.2. Relaxation and Aging below T_g

In an intermediate temperature region near and not too far below the glass transition, the situation is complex. The relaxation dynamics cannot be described by the VFT equation anymore, the system is not in an ergodic state and its properties are not uniquely defined by the temperature-pressure conditions; they depend on the particular glassy state reached during the previous history. Furthermore, the degree of aging determine the physical and mechanical properties like density, elastic constants, diffusivity, Curie temperature (for ferromagnetic glasses) [8], electrical resistivity, enthalpy, *etc.*, as well as the relaxation dynamics of the system. In this intermediate region, physical aging must be considered as occurring continuously on all experimental time scales, but without reaching equilibrium except for very long annealing times (t_a). This complexity might be solved in a first-order approach by introducing a fictive temperature T_f, which is used to define the glassy state of a system [9].

If aging is not considered the glassy dynamics of many systems can be approached as function of T_f using the Adams-Gibbs-Vogel (AGV) model [10,11]

$$\tau_\alpha(T) = \tau_0 \exp\left(\frac{B}{T(1 - T_0/T_f)}\right)$$ (3)

where B and T_0 are empirical parameters, τ_0 the pre-exponential factor and T_f defines the temperature at which liquid (VFT) and glass Arrhenius-like (AGV) dynamics intersect each other. This fictive temperature evolves as a consequence of aging

$$\frac{dT_f}{dt_a} = \frac{T - T_f}{\tau_{aging}}$$ (4)

with limiting condition of $T_f = T$ for a completely aged system attaining internal equilibrium.

At low temperatures ($T \ll T_g$) the aging time, τ_{aging}, is usually long enough to consider T_f constant and the glass structure frozen in an isoconfigurational state, with properties dependent on T, P and T_f. This low temperature region can be defined as the range where the cooperative stress relaxation (α-relaxation) of the viscous liquid is completely frozen and the structural state of the glass does not change in laboratory time-scale; as T_f is constant, the glass properties are defined only by the temperature-pressure conditions. Relaxation in this glassy range involves decoupled, localized motion of easily mobile species; this is usually called secondary relaxations. They are sometimes classified further as β, γ, δ-relaxations in polymers where the stepwise freezing of various local degrees of freedom may be associated with specific molecular groups. While remaining in an isoconfigurational state, relaxations are thermally activated processes with well-defined temperature dependences, $\tau(T)$, usually following Arrhenius-like behaviors.

On the other hand, at intermediate temperatures, the complexity comes from the fact that τ_{aging}, which controls the aging evolution, is at the same time dependent on the degree of aging. Different models have been introduced in order to model this complex behavior. One common approach to model the viscosity, the glassy dynamics and other properties is the Tool-Narayanaswamy-Moynihan (TNM) equation [10,12–14]

$$\tau(T, T_f) = \tau_0 \exp\left(\frac{x E}{k_B T} + \frac{(1 - x) E}{k_B T_f}\right)$$ (5)

where x is a dimensionless non-linearity parameter (the TNM parameter), E is an activation energy and k_B is the Boltzmann constant as usual. Under this approach, the viscosity dependence on time could be described as a function of the change of the fictive temperature with complexity attributed to the non-linearity parameter x. Some models propose different approaches for the time evolution of the α-relaxation time during aging at a given temperature. Lunkenheimer et al. [15] proposed an expression for $\tau_\alpha(t_a)$ assuming that τ_{aging} is equal to τ_α. This assumption is commonly adopted for structural glasses, as it is reasonable to expect that the movements accommodating the structure to an external force should be similar to the ones driving the system towards more stable configurations during aging.

The study of physical aging in metallic glasses has been extensively done by calorimetric techniques. Based on Chen's work [16], the aging process characterized by DSC can show a broad distribution of activation energies. Chen's work shows that the spectrum has two separable broad processes, attributed to β and α relaxations respectively [16]. In $Pd_{48}Ni_{32}P_{20}$ glass the low-energy peak corresponds to an activation energy E = 92.4 kJ/mol (0.96 eV). Tsyplakov [17] obtained similar results on the activation energy spectrum using DSC and mechanical relaxation. He interpreted the data by assuming that aging of metallic glasses is a change in the concentration of frozen defects similar to Dumbbell interstitials in simple crystals [17–19]. In the Dumbbell interstitials model, the activation energy shows a broad distribution of values. Nagel's work on positron annihilation studies of free volume changes during aging of $Zr_{65}Al_{7.5}Ni_{10}Cu_{17.5}$ glass [20] suggests that the isothermal aging kinetics obeys a Kohlrausch-Williams-Watts (KWW) law with β_{KWW} exponent of about 0.3 between 230 and 290 °C. The effective activation energy was found $E \sim 120$ kJ/mol.

Below T_g, there is also a reversible thermal relaxation component, faster than the irreversible aging. By thermal cycling, the annealing induced relaxation can be separated into reversible and irreversible components and has been interpreted by chemical short range ordering (CSRO) and topological short range ordering (TSRO) respectively [21,22]. The former was explained by the activation energy spectrum, while the latter was explained in terms of free volume theory [21]. However, it should be noted that this sharp separation between TSRO and CSRO is criticized because CSRO is unlikely without an accompanying TSRO [23,24]. Borrego *et al.* [25] studied aging by monitoring enthalpy and Curie temperature changes in Fe(Co)-Si-Al-Ga-P-C-B and Finemet glasses. They found that aging can be interpreted as driven by two relaxation times of minutes and hours respectively, in this case they associated the fast process to TSRO and the slow one to CSRO changes.

Khonik [26,27] treated plastic flow below T_g as irreversible structural relaxation with distributed activation energies modified by external stress, developing the so called directional structural relaxation (DSR) model. According to DSR model, relaxation and aging involve structural movements generally anisotropic at the atomic level. In the presence of a mechanical stress, however, the distribution of the local events may become asymmetric producing a net distortion in the direction of energetically favored orientations. The DSR model is a general approach which includes any relaxation mechanism based on the motion of defects, and it even can be applied to relaxation in crystalline materials. At $T \sim T_g$, cooperative atomic motions cause viscous flow, mechanical relaxation and aging. In the DSR model the relaxation centers are divided into irreversible and reversible ones, the former being responsible for viscoplastic low-frequency internal friction, plastic flow and even for reversible strain recovery, whereas the latter cause anelastic processes seen at higher frequencies. In spite of these and other models, the microscopic mechanisms of aging are still far from being understood. Recent work shows that, in the microscopic scale, the aging of metallic glasses is a complex process leaded by the release of internal stresses involving both smooth and sudden (avalanche-like) movements of the structure [28].

The activation energies of the processes controlling physical aging in metallic glasses show typical values around 100 kJ/mol [16,17,29]. Generally, aging is thought as being driven by thermally activated localized structural rearrangements and then controlled by the same molecular movements responsible of the secondary relaxations. Hu [30] performed a survey of the sub-T_g aging and relaxation data of several metallic glasses obtained by DSC or DMA and they found a common relationship of $E_\beta = 26RT_g$. Although enthalpy changes can be the result of many different types of structural rearrangements while mechanical measurements respond only to shear deformation, comparison between as-quenched and relaxed samples using both enthalpy and mechanical techniques suggest that structural relaxation could be characterized by both of these techniques and the results are consistent [17,31]. However, as Chen pointed out [32], the secondary relaxation process observed from calorimetry in metallic glasses does not necessarily correspond to the one observed by shear deformation. The internal friction measurements probe shear relaxations, while enthalpy relaxation samples all sorts of relaxation processes, chemical and topological.

At still lower T, plastic deformation of MG is controlled either by creep or by highly localized shear banding depending on the applied deformation rate. In this range, secondary relaxations are related to anelastic processes concerning easy mobile species with similar behaviors as in crystalline materials. Maddin [33] suggested that the creep behavior of $Pd_{80}Si_{20}$ metallic glass is governed by a single thermally activated process with an activation energy $E = 50$ kJ/mol (0.52 eV). Based on the calculation of the activation volume of 25 Å, close to the volume of one constituent atom, they proposed that the steady state creep of the metallic glass is due to transfer of atoms across a distance of one lattice spacing in order to relax the applied stress.

2.3. Theoretical Frameworks for Interpreting Glassy Dynamics

The free volume model, proposed by Turnbull and Cohen [34], describes quite properly the viscosity dependence on temperature and it was further developed to explain the glass transformation phenomenon [35,36]. The model uses free volume as a parameter to describe the change of physical properties, with the particular achievement of predicting the equivalence of viscosity changes due to temperature or pressure variations in the supercooled liquid state. According to this theory, physical properties are related to the density of the system. The volume in the liquid could be classified into two types; type I is the volume of the elemental unit, type II is the volume where the elemental unit can move freely. Type II is called free volume. It is a small part in the whole volume and it is shared by the elemental units. When the system cools down, both volumes decrease, and when the free volume drops below a certain value, the elemental units can no longer move and thus form glass. Free volume theory is quite useful and highly accepted in the metallic glass community to explain the glass transition and aging phenomena. The weak points of the theory is that free volume is difficult to measure directly by experiment and that a single parameter model is not enough to describe the properties of a glassy state [4,37].

The potential energy landscape model is often used to interpret the relaxation dynamics [38]. According to Johari and Goldstein [7], the atomic and molecular configurations in liquids and glasses change according to motions classified as primary and secondary relaxations. Primary relaxations describe the major large scale irreversible rearrangements responsible for viscous flow. On cooling, the glass transition is reached when the decreasing mobility stifles these rearrangements. On the contrary, secondary relaxation could be viewed as a locally initiated and reversible process. Measurements of the dielectric loss factor in many rigid molecular glasses as well as amorphous polymers show secondary relaxations. According to this evidence, Johari and Goldstein suggested that secondary relaxation could be a near universal feature of the glassy state [7]. From a potential energy landscape perspective, Debenedetti and Stillinger [2,3] have identified these β-transitions as stochastically activated hopping events across sub-basins confined within the inherent mega-basin and the α-transitions as irreversible hopping events extending across different landscape mega-basins.

The mode coupling theory (MCT) is able to explain the experimental evidence that the α-relaxation time diverges from β-relaxation at some critical temperature [39]. However, the MCT theory primordially explores the microscopic dynamics of the supercooled liquids at higher temperatures than the ones probed by mechanical spectroscopy. By characterizing and classifying the secondary relaxations in many glass formers, Ngai and Paluch identified the class of secondary relaxation that bear a strong connection or correlation with the primary one in all the dynamic properties and called it Johari-Goldstein (JG) β-relaxation [40]. This link between α and β relaxations was initially found in polymers, but at present it is assumed to be universal. According to their Coupling model, the decoupling temperature and the expected effects at much lower temperatures can be calculated. Based on this Ngai suggested the excess wing manifested in mechanical spectroscopy of metallic glasses comes from a JG-relaxation [41]. A detailed review of the theoretical models proposed for glass relaxation can be found in Angell *et al.* [5].

3. Mechanical Relaxation of Glasses

3.1. Introduction to Mechanical Relaxation

In general, the self-adjustment with time of a thermodynamic system towards a new equilibrium state in response to a change in an external variable is termed relaxation. When the external variable is mechanical, the phenomenon is known as mechanical relaxation. The measurement of internal friction by dynamic mechanical analysis (DMA), also known as mechanical spectroscopy, is widely used in solid state physics, physical metallurgy and materials science to study structural defects and their mobility, as well as transport phenomena and solid-solid phase transformations. From the mechanical engineering point of view, the internal friction is responsible for the damping behavior of materials, having implications in vibration and noise reduction as well as in low-damping applications. In metallic glasses, normally, the internal friction behavior is empirically characterized and interpreted as a manifestation of internal relaxation processes, ignoring the details of their physical origins or atomistic mechanisms which are difficult to describe due to the complex structure of glasses.

The relationship between stress, σ, and strain, ε, within the elastic region is given by the modulus of elasticity M

$$\sigma = M\varepsilon \tag{6}$$

For an arbitrary deformation, the stress and strain are second order tensors and Hooke's law is a set of linear equations expressing each component of the stress tensor in terms of all the components of the strain tensor. However, considering an isotropic material and the usual modes of pure shear, uniaxial and hydrostatic loading, M corresponds to shear (G), Young's (E) and bulk (B) modulus respectively. Results of mechanical spectroscopy in metallic glasses are obtained in both shear and uniaxial modes, the latter usually adopted when only thin ribbon-shape samples are available due to a low glass-forming ability (GFA) of the alloys.

The ideal elastic behavior has three conditions to be fulfilled, namely: (1) The strain response to each level of applied stress has a unique equilibrium value; (2) The equilibrium response is achieved instantaneously and (3) The response is linear. In a solid material exposed to a time dependent load, besides the elasticity part there might be also a time dependent part generating internal friction. According to the conditions obeyed by the stress-strain relationship, mechanical responses can be classified into the following types detailed in Table 1 [42].

So, the internal friction behavior could be the result of all these effects. The work by Nowick and Berry [42] explains in detail the different behaviors observed in the relaxation of solids, focusing mainly in the linear anelastic one. The terminology of these phenomena changes within the scientific literature; the recoverable/non-recoverable phenomena may be termed as anelastic/viscoelastic, viscoelastic/viscoplastic, solid-viscoelasticy/liquid-viscoelasticity, *etc.* Here we will term the recoverable and non-recoverable phenomena as anelastic and viscoplastic relaxations respectively, while the term "viscoelasticity" will include both types of effects. Experimentally, mechanical relaxation is observable by recording the stress (or strain) change with time when strain (or stress) is modified externally. It can be measured as quasi-static measurements in terms of creep or stress relaxation. Quasi-static experiments are used to obtain information on the behavior of materials over periods of several seconds and longer. For information about the behavior of a material in a shorter timescale, dynamic experiments are more appropriate. In these experiments a stress periodic in time, $\sigma = \sigma_0 \, e^{i\omega t}$, is imposed on the system, and the phase lag ϕ of the strain, $\varepsilon = \varepsilon_0 \, e^{i(\omega t - \phi)}$, behind the stress is determined. For ideal elasticity, $\phi = 0$, the ratio ε/σ gives the elastic compliance of the material J. In the case of anelastic or viscoplastic contributions, ϕ is not null, and so the ratio ε/σ is a complex quantity called complex compliance, $J(\omega)$, which is a function of the applied frequency ω

$$J(\omega) = \frac{\varepsilon}{\sigma} = J'(\omega) - iJ''(\omega) \tag{7}$$

where $J'(\omega)$, the real part, is called the storage compliance and $J''(\omega)$, the imaginary part, is called the loss compliance. In a similar way, we could have regarded the periodic strain as given, and the stress as leading the strain by a phase angle ϕ. The complex modulus $\sigma/\varepsilon = M(\omega) = M'(\omega) + iM''(\omega)$ could be then defined in a similar way. It should be noted that $J(0) = J'(0)$, and at very high frequencies, $J(\infty) = J'(\infty)$, it follows that $J''(0) = J''(\infty) = 0$.

Table 1. Classification of mechanical behaviors.

Behavior	Complete Recoverability	Instantaneous	Linear
Ideal elasticity	Yes	Yes	Yes
Nonlinear elasticity	Yes	Yes	No
Instantaneous plasticity	No	Yes	No
Anelasticity	Yes	No	Not necessary
Viscoplasticity	No	No	Not necessary

The characterization of the internal friction of materials is commonly done by the parameter

$$Q^{-1} = \tan\phi = \frac{M''}{M'} \qquad (8)$$

which is proportional to the mechanical energy dissipated by the system. Q^{-1} has the advantage of being not influenced by uncertainties of the sample sizes and it is widely used in thermal analysis of substances, for instance in the characterization of T_g in polymeric materials, and in the determination of internal friction at high frequencies by ultrasound spectroscopy. On the other hand, as we will discuss later, the loss modulus peak is more directly related to the frequency spectrum of the mechanical relaxation.

Within the scope of linearity, the mechanical response satisfies the Boltzmann superposition principle: the response of the material to an applied stress is independent on other applied stresses. This means that each response function constitutes a complete representation of the inherent viscoelastic properties of the solid. The classical analysis of mechanical relaxation data uses mechanical models composed of springs ($\sigma = K\varepsilon$, where K is the elastic constant of the spring) and Newtonian dashpots ($\sigma = \eta\,d\varepsilon/dt$, where η is the viscosity of the dashpot) arranged in different configurations. This allows the derivation of the response of the system from the solution of the differential equations coming from the model. The two elements combined in parallel and series give rise to the Voigt and Maxwell units respectively. The standard linear viscoelastic solid is a three parameter model which can either contain a Voigt or a Maxwell unit. Different arrangements of these units allow one to model both recoverable and non-recoverable linear processes.

The general behavior of a standard viscoelastic solid shows a Debye peak in the loss modulus with the form

$$M'(\omega, T) = M_u - \frac{\Delta M}{1+\omega^2\tau^2(T)}$$
$$M''(\omega, T) = \frac{\Delta M\omega\tau}{1+\omega^2\tau^2(T)} \qquad (9)$$

where $M_u = M(\infty)$ is the modulus dictating the pure elastic response, and ΔM is the intensity of the relaxation process (*i.e.*, the decay of storage modulus observed between external forces being applied faster or slower than the characteristic time of the process). The peak in the loss modulus is observed when $\omega\tau = 1$ and it can be traced either by scanning in ω or in temperature, as $\tau(T)$ is temperature dependent. Glass relaxation involves cooperative movements of atoms in a non-regular structure, and it is far more complex than the simple standard viscoelastic model. The mathematical functions most commonly used to characterize the mechanical responses measured in both quasi-static and dynamic experiments are detailed in the following section.

3.2. Time and Frequency Domain Response Functions

Based on the observation of relaxation phenomena, the time dependent properties measured in dielectric or mechanical relaxation of glasses can be usually well-described by a KWW equation, also called stretched exponential

$$\varphi(t) = \exp\left[-(t/\tau)^{\beta_{KWW}}\right] \qquad (10)$$

$\varphi(t)$ being the time correlation function describing how the system losses memory and returns to equilibrium after being excited by an external stimulus. This expression with only two parameters is widely used to describe time dependent properties in both microscopic and macroscopic scales. Actually, the stretched exponential is known as a complementary cumulative Weibull distribution. The coverage of this equation is not only limited to models based on distributions of relaxation times but also complex correlated processes. The interpretation of the parameter β_{KWW} is of great interest as discussed by Ngai [43]. When $\beta_{KWW} = 1$ the function represents a simple exponential decay characteristic of a Debye relaxation with a characteristic time constant, while if $0 < \beta_{KWW} < 1$ the expression can be regarded as the result of a distribution of individual events with different relaxation times.

In creep-recovery experiments, the strain evolution $\varepsilon(t)$ during the recovery is

$$\varepsilon(t) = \varepsilon_{pl} + \sigma c \varphi(t) \tag{11}$$

where ε_{pl} is the residual irreversible plastic deformation, σ is the stress applied during the previous creep period and $\sigma c + \varepsilon_{pl}$ corresponds to the strain at the beginning of the recovery period [44]. On the other hand, in quasi-static stress relaxation experiments, a sudden deformation is applied to the system generating an initial stress σ_0 that decays following

$$\sigma(t) = (\sigma_0 - \sigma_R)\varphi(t) + \sigma_R \tag{12}$$

where σ_R is the residual elastic contribution. Both creep recovery and stress relaxation probe the relaxation response function of the system. However, the $\varphi(t)$ obtained from creep recovery only has contributions from anelastic processes while the $\varphi(t)$ from stress relaxation may include both anelastic and viscoplastic relaxations. An example of the expected strain or stress time evolution in these experiments is shown in Figure 2.

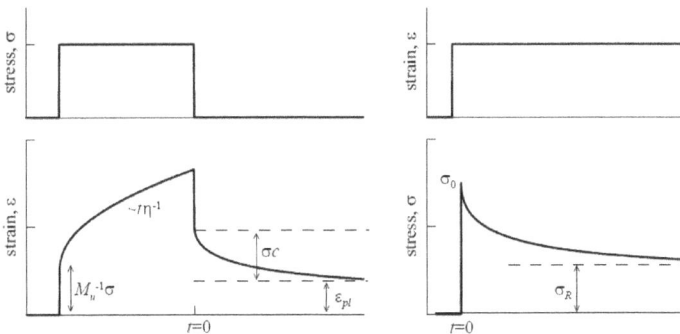

Figure 2. Expected behavior for creep-recovery (**left**) and stress-relaxation (**right**) quasi-static experiments.

The description with a two parameter stretched exponential is forcedly a simplification of the mechanical response, assuming implicitly a unimodal distribution of relaxation times. In order to gain further insight, the time response may be analyzed in terms of a distribution of relaxation times

$$\varphi(t) = \sum A_i \exp(-t/\tau_i)$$
$$\text{or} \tag{13}$$
$$\varphi(t) = \int_0^\infty A(\tau') \exp(-t/\tau')d\tau'$$

with factors $A(\tau')$ determining the contribution of each relaxation time to the whole relaxation process. This analysis is able to fit complex experimental responses that may be not well fitted by the two-parameter function. However, $\varphi(t)$ is just the Laplace transform of the intensity of the respective relaxation process and, as it is well known, the computation of the inverse Laplace transform is a very complex mathematical problem where experimental noise or inaccuracies may derive in fuzzy results. In the present case, it might give rise to fictitious relaxation time distributions, so experimental data must be cautiously analyzed.

Jiao [45] analyzed the stress relaxation of a metallic glass by assuming that the relaxation time spectrum had a log normal distribution with the form

$$A\left(\ln \tau'\right) = k \exp\left[-\left(\ln \tau' - \ln \tau\right)^2 / s^2\right] \tag{14}$$

where τ is the most probable relaxation time, s is the width of the τ' spectrum, and k is a normalizing factor. This model fitted properly the stress relaxation of the $Pd_{40}Ni_{10}Cu_{30}P_{20}$ metallic glass. Besides log-normal distribution, other distribution shapes like box or wedge-like could be also used for the spectrum to fit with experimental data [46]. The activation energy spectrum could be also determined with some approximation by using the data from calorimetry or mechanical relaxation experiments [17].

The correlation function $\varphi(t)$ is related to a complex susceptibility by Fourier transform

$$\chi(\omega) = \chi'(\omega) - i\chi''(\omega) = \int_0^\infty \left[-\frac{d\varphi(t)}{dt}\right] e^{-i\omega t} dt \tag{15}$$

which can be probed by dynamical experiments. In case of DMA experiments, the measured complex elastic modulus is

$$M(\omega) = M'(\omega) + iM''(\omega) = M_u - \Delta M \chi(\omega) \tag{16}$$

with M_u and ΔM already defined in Equation (9). It should be noticed that there is no analytical expression for the Fourier transform of the KWW function with a general value of the β_{KWW} exponent and numerical methods have to be employed in order to translate the experimental data from time to frequency domains or inversely. Furthermore, computing of Fourier transform poses numerical problems originating from cutoff effects which yield unwanted oscillations, especially when treating real data with experimental error and noise.

Characterization of the relaxation processes is generally obtained from the analysis of the loss modulus $M''(\omega)$. The $M''(\omega)$ peak related to a given relaxation process is defined by four main characteristics: The intensity of the peak (ΔM), the bluntness of the peak, and the power laws defining both the low and high-frequency tails

$$\begin{aligned} \chi''(\omega) &\propto (\omega\tau)^a & \omega \ll \tau \\ \chi''(\omega) &\propto (\omega\tau)^{-b} & \omega \gg \tau \end{aligned} \tag{17}$$

where $0 < a, b < 1$. In the case of a Debye process $a = b = 1$, and for a time-domain response defined by a KWW function (Equation (10)) $a = 1$ and $b = \beta_{KWW}$. The real loss peak found experimentally shows different degrees of asymmetry. An empirical function widely used for characterizing it is the Havriliak-Negami (HN) function

$$\chi(\omega) = \frac{1}{\left[1 + (i\omega\tau)^\alpha\right]^\gamma} \tag{18}$$

where the exponents α and γ define the broadness and the asymmetry of the peak respectively, and they produce power-laws of the tails given by $a = \alpha$ and $b = \gamma\alpha$. The Cole-Davidson (CD) function and the Cole-Cole (CC) functions, which are also commonly used in relaxation studies, correspond to the HN-function with $\alpha = 1$ or $\gamma = 1$ respectively. The CD-function, with peak shape dictated only by the γ exponent, shows an asymmetric peak very similar to the one given by the Fourier transform of the KWW-function [47]. On the other hand, the CC-function corresponds to a symmetric peak with broadening given by the exponent α. Many results show that secondary relaxations of glasses can be generally fitted by using the Cole-Cole equation; the parameter α gives information about how distributed are the relaxation times and normally it increases with the temperature. On the contrary, classical anelastic relaxations in crystalline metals are restricted to a small volume, namely defects, dislocations and grain boundaries. Accordingly they have a much smaller magnitude and show shapes very close to a Debye relaxation with α and γ ~1. It should be recalled here that the $\tau = \tau(T)$ in the previous equation is the average relaxation time of the process at a given temperature.

Other functions, either empirical or coming from physical models, are used to characterize the experimental loss modulus of glasses [48,49]; although here we only describe the most common ones, some of these other models will appear in the next section when discussing the mechanical spectroscopy results in metallic glasses. In any case, however, all the functions have to fulfill similar properties as the ones detailed here for the HN and related functions.

3.3. Thermally Activated Models

As previously stated, internal friction can be interpreted as a combination of recoverable and non-recoverable relaxation events. Previous works on crystalline metals show that anelastic relaxation can be well explained by mobility of defects in the crystalline lattice. These models consider interface relaxation (including grain boundary, twin boundary and nano-crystalline structure), dislocation, and point defect relaxation known as Snoek and Zener relaxation [42,50]. That is, most of the known mechanisms of mechanical relaxation in metals have their origin in the thermally activated motion of various kinds of defects.

The amorphous nature of metallic glasses prevents the description of internal friction in terms of these mechanisms. The only mechanism which can be easily extrapolated from crystalline to amorphous structures is that of atomic and defect migrations, directly related to the movement of single atoms inside the structure. The jump of an atom or point defect from one site to another in a crystal lattice is a simple example of a rate process. The corresponding relaxation time follows a reciprocal Arrhenius equation

$$\tau = \tau_0 \exp\left(\frac{E}{k_B T}\right) \tag{19}$$

valid when the rate limiting step of the relaxation process is the movement over an energy barrier.

From the position of the loss peak at a given temperature $\omega\tau(T) = 1$, obtained from dynamic experiments, the activation energy is calculated as

$$\ln\left(\omega_{peak}\tau\right) = 0 = \ln\left(\omega_{peak}\tau_0\right) + \left(\frac{E}{k_B}\right)\left(\frac{1}{T}\right)$$
$$\ln\left(\frac{\omega_{peak2}}{\omega_{peak1}}\right) = \left(\frac{E}{k_B}\right)\left(\frac{1}{T_1} - \frac{1}{T_2}\right) \tag{20}$$

In the case of no observable peak, the activation energy could still be calculated using the temperature dependence of a fixed value of the loss modulus as a function of frequency. In a more general way, the temperature behavior of the average relaxation time $\tau(T)$ may be obtained by application of the temperature-time-superposition (TTS) analysis of the mechanical spectroscopy curves also for non-Arrhenius behaviors [51].

The relaxation processes of glasses may involve cooperative movements much more complex than the defect migration scheme. Besides, even for a well-defined process of atomic or defect migration, the inhomogeneous structure of glasses would generate a broader distribution of activation energies than in a crystalline material. In spite of this, glass relaxations are usually interpreted in terms of the temperature dependence of a main characteristic time $\tau(T)$, which is the average value of the relaxation times distribution, and can be determined by TTS analysis. Of course, if the DMA curves involve the overlapping of different processes with quite different activation energies or $\tau(T)$ behaviors, the TTS analysis will not be applicable.

4. Mechanical Spectroscopy of Metallic Glasses

DMA can be performed both in isothermal (scanning frequency at fixed temperature) or isochronal (scanning temperature at fixed frequency) modes. In every solid, there exists a fundamental thermoelastic coupling between the thermal and mechanical states with the thermal expansion coefficient as the coupling constant. The thermoelastic damping contributes to the background of the loss modulus and Q^{-1} isochronal curves; differences between high and low frequency tests may originate from this effect. A detailed discussion on the thermoelastic background could be found in Nowick's book [42]. Other effects may also contribute to the DMA background, which increases in less compact structures and it is then more important for glassy states with higher free-volume. Castellero [52] used the change in the intensity of the Q^{-1} background in order to follow the room temperature aging of Mg-Cu-Y glasses.

In addition to the background, the basic features of isochronal DMA curves of metallic glasses are observed in Figure 3. At temperatures below T_g, a slight and constant decrease of the storage modulus is expected as temperature increases due to thermal expansion of the structure [53]. In this region, many metallic glasses also show a secondary relaxation peak in the loss modulus and the corresponding partial step-like decay of the storage modulus. Increasing the temperature, the dynamic glass transition is clearly visualized by the α-relaxation peak of M'' and a complete decay of M' once in the liquid state. At higher temperatures, crystallization returns the system to the solid state increasing again the storage modulus. At even higher temperatures, thermal expansion and softening of the solid reduce again the storage modulus and increases the internal friction.

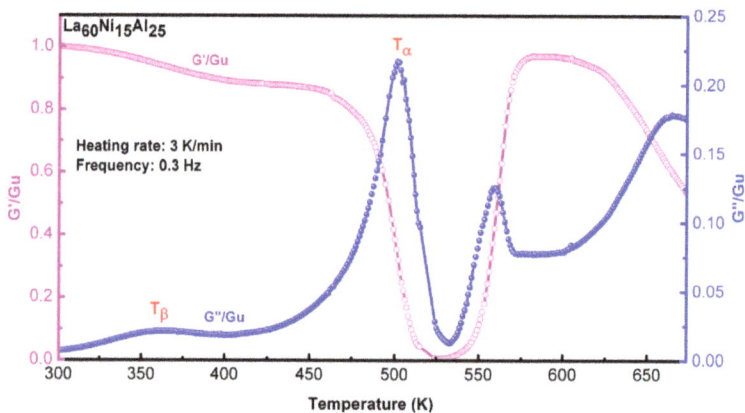

Figure 3. Normalized storage shear modulus G' and loss modulus G'' *vs.* temperature in La$_{60}$Ni$_{15}$Al$_{25}$ bulk metallic glass, G_u is the unrelaxed shear modulus. Reprinted from Reference [54] with permission from American Institute of Physics (AIP).

DMA measurements are usually performed with heating rates of 1–5 K/min and frequencies from 0.01 to 100 Hz. For this range of heating rates and frequencies, the maximum of the α-peak is found in the liquid temperature-region, and the measured temperature dependence of τ_α is in good agreement with the viscosity behavior described by Equation (1) [55]. In many glassy alloys, however, crystallization is very close or even overlapped with glass transition. The decay of the storage modulus is then stopped before reaching a zero value and the α-peak may be cut on its high-temperature side. In this case, the apparent maximum of the peak may not correspond to the real α-relaxation peak maximum. In spite of possible deviations due to crystallization or aging, the α-relaxation process observed by DMA is generally well-understood and it can be characterized by an HN-function or the Fourier transform of a KWW function with stretched exponent values

β_{KWW} ~0.5, as seen in Figure 4. Wang *et al.* [56] also found values of β_{KWW} between 0.4 and 0.5 for some of the most representative metallic glasses (Pd-Ni-Cu-P, Ce-Al-Cu and Vitreloy), while Meyer's work on $Pd_{40}Ni_{10}Cu_{30}P_{20}$ in the equilibrium state [57] found that the α-relaxation followed the stretched exponential function with $\beta_{KWW} = 0.76$. On the other hand, secondary relaxations show very diverse characteristics in different metallic glasses and their origin is less clear. Following we will review different experimental studies focusing on secondary relaxations.

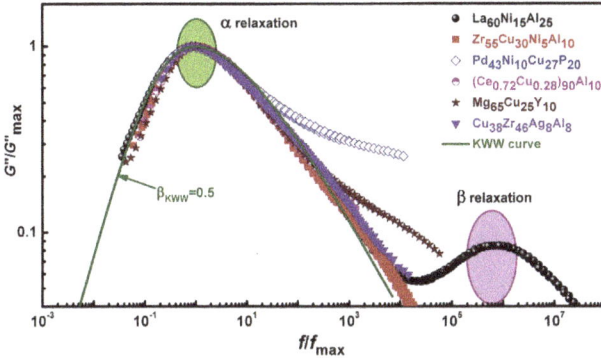

Figure 4. Dependence of the normalized loss modulus *vs.* the normalized frequency in typical metallic glasses. The solid line is fitted by the Kohlrausch-Williams-Watts (KWW) model. Reprinted from Reference [58] with permission from Elsevier. Copyright 2014, Elsevier.

4.1. Secondary Relaxations

By surveying the relaxation dynamics in organic molecular liquids and fused salts, Johari and Goldstein suggested that β relaxation was a universal feature of glassy systems [7]. In some polymers, as shown in Figure 5 from the work by Casalini [59], there is a clear picture. On one hand the α-relaxation time becomes arrested in an Arrhenius behavior once in the glass state, it controls the aging and it is coincident with the calculations from the mode coupling model. On the other hand, the β-relaxation times are similar to the primitive τ_α in agreement with the interpretation of a Johari-Goldstein relaxation as a precursor of the structural α-relaxation.

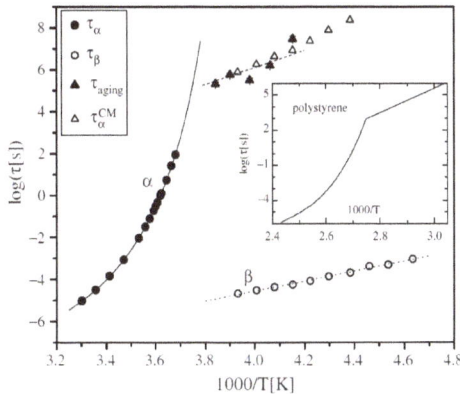

Figure 5. Relaxation time for the α and β processes, along with the aging decay time τ_{aging} and the τ_α calculated from coupling model. Reprinted from Reference [59] with permission from American Physical Society (APS).

Nowadays, it is becoming popular a description of metallic glass dynamics in terms of α and β relaxations. All the peaks below the glass transition temperature are referred as β-relaxation although their origin may not be the same. We will follow this terminology here. However, it should be noted that some β-relaxations detected by mechanical spectroscopy are not Johari-Goldstein relaxations or the ones envisioned from the potential energy landscape model but may come from different origins.

Indeed, this kind of anelastic events in metallic glasses can be dated back to the discovery by Berry that in Nb_3Ge metallic glasses thermal activated anelastic events manifest on an internal friction peak around 250 K [60]. This peak was interpreted as stress induced ordering of a similar nature to the point defect relaxations known in crystalline solids. Although it is not a Debye peak and it shows an asymmetric distribution of activation energies, the typical magnitude of the relaxation time corresponded to a single atomic jump and the intensity of the peak decreases with aging. Actually, later work suggested that in the low temperature region there might be contributions from hydrogen absorption, which exists in quite large range of metallic glasses [61]. Yoon [62] also found these peaks located around 250 K with activation energies of $E \sim 100$ kJ/mol (1.0 eV) in $Fe_{40}Ni_{40}P_{14}B_6$ and $Fe_{32}Ni_{36}Cr_{14}P_{12}B_6$ metallic glasses and they ascribed them to the movement of B atoms. Fukuhara [63] interpreted the low-temperature (150 K) relaxation peak found in $Zr_{55}Cu_{30}Al_{10}Ni_5$ as related to a topological transition or a vacancy-like defect rearrangement.

However, based on the finding of this peak in $Cu_{50}Zr_{50}$, $Co_{35}Y_{65}$ as well as $Co_{35}Dy_{65}$ metallic glass, and after excluding factors like hydrogen or oxygen absorption, Kunzi [64] suggested that the relaxation peak is due to the existence of intrinsic degrees of freedom in the amorphous structure as well as in other glasses such as oxides glasses [65]. It is also observed that cold work might lead to the observation of peaks occurring at temperatures between 100 to 300 K [66]. Actually, plastic flow both on cold rolling and hydrogenation occurs via formation and motion of dislocation-like defects which are the reason of the observed anelastic anomalies. It is suggested by Khonik [66] that low temperature internal friction peaks described in the literature for as cast, cold deformed and hydrogenated samples have common origin. Nevertheless, the characterization of local defects in amorphous structures is a complex, long-standing topic still not fulfilled in spite of many efforts since the early works of Egami [67].

These thermal activated anelastic events might also happen in a bulk metallic glass as first described by Okumura [68] in the analysis of the viscoelastic behavior of $La_{55}Al_{25}Ni_{20}$ metallic glass. As shown in Figure 6, besides the glass transition temperature region, a β-relaxation gets activated at around 400 K. In this case, the glass transition region also shows a double peak structure which was later associated by TEM analysis to glass phase separation and a corresponding two glass transitions. Further work [69] showed that aging reduces the magnitude of the relaxation peak but has little effect on the β-relaxation peak position. However, this fact was questioned by Qiao's work, which showed that the β-peak moves to higher temperature after physical aging [70]. The activation energy of the β-relaxation obtained by the time temperature superposition (TTS) shift factor method is $E_β \sim 100$ kJ/mol (1.0 eV). In calorimetric measurements, the extrapolation of the intensity of β-relaxation associated to enthalpy release when the aging process is completed shows a non-zero intercept, which suggests that the β-relaxation would still remain in the fully relaxed state. Combined with Qiao's result on partially crystallized samples where this peak remains, it seems consistent that β relaxation might be caused by short range atomic relaxation, somewhat similar to Snoek or Zener type processes, as suggest by Okumura. On the other hand, as already stated above, some works have found that the characteristics of the secondary relaxation in some metallic glasses are well in agreement with the expected JG-relaxation [41]. The debate on the origin of β-relaxation remains open.

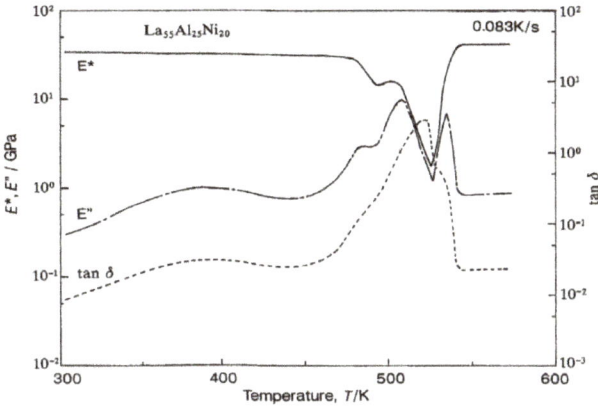

Figure 6. Temperature dependence of Dynamo-Mechanical-Analysis (DMA) behavior of $La_{55}Al_{25}Ni_{20}$ metallic glass. Reprinted from Reference [68] with permission from The Japan Institute of Metals and Materials (JIM).

Similar behavior of the loss modulus is also observed in $La_{70}Al_{15}Co_{15}$. Wang's work on La-based BMGs shows that the β-relaxation behavior could be tuned by modification of the chemical composition and could also manifest on different fragility parameter [71]. Not only the intensity, but also the temperature is strongly influenced by the composition; the loss modulus dependence on temperature of $La_{70}M_{15}Al_{15}$ with M = Ni, Co or Cu is strongly related to the composition as shown in Figure 7. In the case of Ni and Co, there are distinguishable β relaxation peaks, but in the case of Cu the onset of β relaxation is at higher temperature and overlaps with the contribution of the main relaxation, leading to a shoulder or excess wing. This is further explored by Yu [72], affirming that β-relaxation appears if all the atomic pairs have large similar negative values of enthalpy of mixing, while positive or significant fluctuations in enthalpy of mixing suppress β-relaxation. Their conclusion is based on the fact that by substituting Ni by Cu in $La_{70}Ni_{15}Al_{15}$ the loss modulus change from a separate β-relaxation peak to an excess wing behavior. The enthalpy of mixing is also used to explain the experimental observation that partially substituting Ni with Cu in $Pd_{40}Ni_{40}P_{20}$ increases the glass transition temperature while lowers the starting temperature of β-relaxation. Furthermore, they suggest that strong and comparable interactions among all the constituting atoms generate string-like atomic configurations, whose excitation emerges as the β-relaxation events.

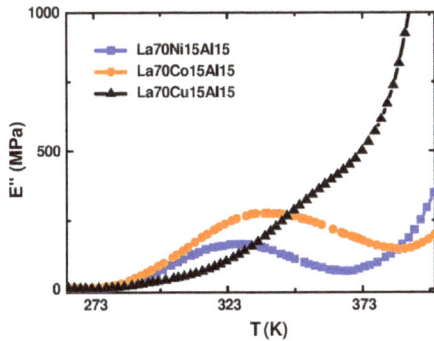

Figure 7. The *T* dependence of loss modulus of $La_{70}M_{15}Al_{15}$ with M = Ni, Co, Cu. Reprinted from Reference [71] with permission from IOP Publishing.

In systems like $Pd_{77.5}Si_{16.5}Cu_6$, $Pd_{48}Ni_{32}P_{20}$, $Pt_{58.4}Ni_{14.6}P_{27}$ and $Au_{49}Cu_{26.9}Si_{16.3}Ag_{5.5}Pd_{2.3}$ the sub-T_g relaxation is detected as a shoulder of the α-peak [73,74] and experiences important changes upon annealing due to aging. Chen pointed out [73] that this sub-T_g relaxation has different features from the JG-relaxation of polymeric and molecular glasses which shows a distinct peak at $T_m < 0.6$ T_g (at a frequency of 1 Hz) and small effect on the intensity due to thermal stabilization near T_g. In $Zr_{55}Cu_{30}Al_{10}Ni_5$ alloys or $La_{55}Al_{25}Ni_{20}$, the $M''(T)$ behavior is more similar to a double α-peak than a peak with a shoulder. These results have been interpreted in terms of double glass transitions related to phase separation in the glass [75] or because of double-stage unfreezing of the mobility of the different species during heating [76].

Cohen [77] simulated the loss modulus of a binary Lennard-Jones potential by molecular dynamics by introducing oscillatory stress. The simulation results showed that the β wing could appear on the loss modulus as a function of temperature. Based on simulated DMA curves performed with different fractions of pinned particles, β process was attributed to cooperative movements different from α relaxation. Yu [78] suggested that cooperative string-like atomic motion might be more appropriate to express β process in metallic glasses since it can explain the diffusion of the smallest atom species. Although with a nature of cooperative movement, they involve only small part of all the atoms in the system. Liu [79] measured the activation energy E_β in ultra quenched MGs, the relationship $E_\beta = 26RT_g$ suggested that it is a JG-relaxation. X-ray diffraction combined with EXAFS results showed that relaxation originated from short range collective rearrangements of large solvent atoms which could be realized by local cooperative bonding switch. In general, the microscopic mechanisms considered for secondary relaxation are also associated to aging. A short revision of various microscopic models suggested to be responsible of physical aging is already given in Section 2.2 above.

4.2. Influence of Aging

Physical aging makes the structure denser and induces changes in the mechanical, electrical, magnetic, thermal and transport properties. The oldest and widely adopted concept for interpreting aging is that of free volume being progressively reduced. Alternative concepts describe aging as annihilation of various kind of "defects" of the amorphous structure, comprising interstitial-like, stress inhomogeneities, local high free-volume zones or other microscopic motives. Some of these local motives are potential shear transformation zones (STZs), that become activated once external stress is applied. In metallic glasses aging is usually referred as irreversible structural relaxation and has long been noticed as a strong effect existing even at room temperature. Early experiments [80,81] found that when an as-quenched sample is heated cyclically at a constant rate to successively increasing temperatures, the internal friction in each heating run is reduced. This can be described as if the relaxation spectrum is reduced in its faster part by physical aging. When heating during a DMA isochronal test, physical aging may occur *in situ* and the relaxation spectrum would not correspond to a single isoconfigurational state [82]. On the other hand, if the sample has been previously properly annealed it may not suffer significant aging during the test and the results become reproducible in consecutive heating-cooling-heating cycles.

The nature of individual movement of small areas is supported by room temperature creep behavior using nanoindentation techniques by Castellero [52]. The creep behavior is viscoelastic and could be fitted by two typical relaxation times, which were found to be around 4 s and 36 s for $Mg_{65}Cu_{25}Y_{10}$ and 2.5 s and 25 s for $Mg_{85}Cu_5Y_{10}$. After aging, the relaxation time of the slow process increases. Comparing with the relaxation time obtained by positron annihilation spectroscopy, Castellero *et al.* suggested that there are small and large traps where positrons can be annihilated. Smaller defects could be intrinsic open volume regions similar to Bernal interstitial sites, while larger defects are unstable and get annihilated as a consequence of aging. The reduction of these defects, responsible for shear transformations, lead to an abrupt loss of plasticity and a continuous decrease in the creep deformation rate.

Metals **2015**, 5, 1073–1111

Kiss investigated the influence of aging on internal friction of FeB and NiP amorphous alloys [83]. Their results on $Ni_{80}P_{20}$ show that annealing decreases the internal friction and increases the storage modulus of MGs. Hettwer's work [84] on influence of heat treatment on the internal friction of $Fe_{32}Ni_{36}Cr_{14}P_{12}B_6$ shows that besides the peak observed around 665 K, there is another small peak in the range between 360 K and 400 K which nowadays could be classified as β-relaxation as shown in Figure 8. The intensity of such β-relaxation becomes reduced and shifted to higher temperatures by heat treatment. Using this β-relaxation as a probe, they investigated the aging dynamics by considering that the reduction of the damping is influenced by both temperature and annealing time as $Q^{-1} = alog t_a + Q^{-1}{}_0$ where parameters a and $Q^{-1}{}_0$ are functions of the temperature.

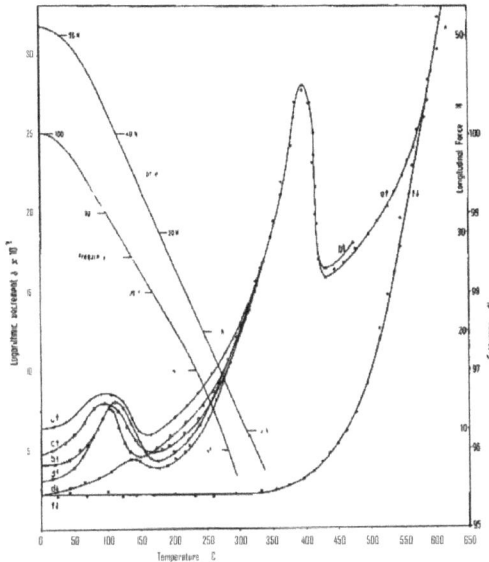

Figure 8. Internal friction of a Fe-based metallic glass for different degrees of physical aging. Reprinted from Reference [84] with permission from Elsevier.

Tests at higher frequency (280 Hz) on the same composition by Haush [85] show a similar behavior, a strong secondary relaxation peak shift to higher temperature. However, Morito [86,87] observed that the β-relaxation reported by Hettwer is not always reproducible, and he suggested that it might come from inappropriate loading. Morito and Egami's work [80,87] on the same composition shows the influence of aging on internal friction. After an extended period of annealing, the glass reaches an internal pseudo-equilibrium state revealed on the internal friction. The decay kinetics can be expressed by first order kinetics with a log normal distribution of time constants, and the pseudo-equilibrium state is a function of the annealing temperature. A change in the annealing temperature results in a reversible change from one such state to another.

The work by Deng and Argon on $Cu_{59}Zr_{41}$ and $Fe_{80}B_{20}$ shows that besides the α-relaxation, there is another relaxation process which gets activated at lower temperatures [88,89], the peak position of the β-relaxation is near 500 K. This peak shifts progressively to higher temperatures as aging continues and is used as a probe to study the aging process. Unlike the $Fe_{32}Ni_{36}Cr_{14}P_{12}B_6$, where quasi-equilibrium structures can be achieved and altered reversibly by annealing at different temperatures, in the $Cu_{59}Zr_{41}$ metallic glass these quantities continued to change until the onset of crystallization. By fitting the peak temperature at different frequencies, the activation energy for the sample aged at 573 K for 34 h is 46 kJ/mol (0.48 eV), with a frequency factor of 1.9×10^5 s^{-1}. Considering the connection between activation energy of shear transformations and the level of free volume at the transforming cluster site, they affirm that the aging related shifting to higher temperatures without change in height is a result of reduction of free volume in a specific local atomic environment existing in this composition.

The activation energy spectrum of the change of internal friction associated with aging can be obtained by subtracting the internal friction curve of the fully relaxed material from that of the as-quenched one, in a similar way as the data obtained by calorimetry measurements. It is important to keep in mind that the activation energy of internal friction is different than the activation energy of irreversible structural relaxation or aging. A well-known fact is that the relaxation time from internal friction tests is frequency dependent. However, for example in $Fe_{32}Ni_{36}Cr_{14}P_{12}B_6$, the aging characteristic time at 473 K is around 135 min and almost the same for 573 K [86].

4.3. Modeling of the Mechanical Relaxation Spectrum

The temperature dependence of internal friction or loss modulus can be modeled with the methodologies described in Section 3. Debye relaxation is normally used to describe the anelastic behavior, and the distribution of relaxation times can be related to a spectrum of activation energy. Ignoring the microscopic origin of the E distribution, the time-temperature relaxation spectrum $M''(\omega,T)$ can be modeled by combining a frequency response function (HN, CD, CC or other) with a temperature dependence of the main relaxation time $\tau(T)$, in what is called time temperature superposition (TTS) method [90]. In this approach, the shape of the response function describes the effect of the relaxation time spectrum, *i.e.*, the deviation from a Debye process. In the case of HN, CD or CC functions, this shape is determined by the exponents α and γ of Equation (18) with values obtained from fitting the experimental data. Therefore, $\tau(T)$ describes the temperature dependence of the average or main relaxation time of the process and is commonly found to follow an Arrhenius-like behavior for $T < T_g$.

If the system shows various relaxation processes well differentiated in the time scale, each one of these processes can be modeled by the corresponding response function χ_i and intensity ΔM_i as

$$M(\omega, T) = M_u - \Delta M_1 \chi_1(\omega, T) - \Delta M_2 \chi_2(\omega, T) - \ldots \tag{21}$$

where the temperature dependence of χ_i is given by the corresponding $\tau_i(T)$. Of course, if the activation energy spectrum of one of these processes is very broad, a $\tau(T)$ defined by a single activation energy and a $\chi(\omega)$ function with constant shape will not be able to reproduce the whole time-temperature spectrum and the modeling will have to take into account the explicit distribution of activation energies, computing the frequency-domain response function by numerical calculation of Equations (13) and (15). Finally, it should be taken into account that M_u, and sometimes ΔM_i, usually shows a slight temperature dependence [53] that may has a significant effect if the modeling expands over a large temperature window.

The master curve analysis is often used in the interpretation of DMA data using TTS principle; the master curve is constructed using isothermal multi frequency DMA data. Within this methodology, the temperature dependence of the shift factor follows an Arrhenius relationship with different activation energies below and above T_g. Pelletier [91] investigated the apparent activation energy in PdNiCuP using this method and obtained $E_\beta = 1.1$ eV and $E_\alpha = 3.4$ eV respectively. Jeong [92,93] analyzed the mechanical relaxation of $Mn_{55}Al_{25}Ni_{10}Cu_{10}$ and $Zr_{36}Ti_{24}Be_{40}$. For $Mn_{55}Al_{25}Ni_{10}Cu_{10}$ glass, the activation energy of the alpha relaxation was found $E_\alpha = 78$ kJ/mol (0.81 eV) and $E_\alpha = 323$ kJ/mol (3.3 eV) respectively below and above T_g. For $Zr_{36}Ti_{24}Be_{40}$, the activation energies were $E_\alpha(T < T_g) = 93$ kJ/mol (0.96 eV) and $E_\alpha(T > T_g) = 392$ kJ/mol (4.1 eV). Guo's work [94] on mechanical relaxation studies of α and slow β processes show that $Nd_{65}Fe_{15}Co_{10}Al_{10}$ have a distinct β-relaxation in the temperature region between 320 K and 420 K. The activation energy is found $E_\beta = 98$ kJ/mol (1.0 eV) with the $\tau_{\beta 0} = 10^{-14.5}$. Since there is a relationship $E_\beta = 24RT_g$ which is close to the suggested by mode coupling theory [95], they claim that β relaxation is intrinsic in metallic glasses. Activation energy data of α and β relaxations of many metallic glass systems can be found in Wang's work [96].

In a narrow range above T_g, the VFT behavior of $\tau(T)$ can be approximated to an Arrhenius law with an apparent activation energy of the liquid

$$E_{\alpha,\text{liquid}} = mRT_g \ln(10) \tag{22}$$

This gives values between 200 and 600 kJ/mol depending on the fragility and the T_g of the system. On the other hand, the activation energies of both α and β relaxations at $T < T_g$ are usually found between 80 and 160 kJ/mol. These E values of the mechanical relaxation processes below but not far from T_g coincide with the activation energy commonly found for physical aging in this temperature region, as already stated above, the same microscopic origins are expected for both processes.

Here it is interesting to note that $E_\beta \sim 26\, R\, T_g$ and E_α given by the AGV approach (Equation (3)) give very similar values. For instance, considering typical values for metallic glasses of $T_g = T_f = 600$ K, $T_0 = 450$ and $B = D^* T_0 = 4500$, Equations (1)–(3) and (22) give $E_\alpha(T > T_g) = 440$ kJ/mol, $m = 38$, $E_\alpha(T < T_g) = 128$ kJ/mol while $E_\beta = 26RT_g = 130$ kJ/mol. Therefore, the expected values of the average activation energies controlling both primary and secondary relaxations in the glassy phase are very similar for metallic glasses. This poses difficulty in interpreting the two phenomena as mega-basin and sub-basin transitions within the potential energy landscape picture.

Concerning the shape of the relaxation function, Liu and Wang [97,98] fitted the DMA behavior of Ce-based and Zr-Ti-Cu-Ni-Be glasses assuming that $\tau(T)$ follows a VFT behavior and relaxation can be described by the KWW function. The loss modulus was computed by Fourier transform finding that in the temperature region higher than T_g the experimental data was well reproduced; however, in the lower temperature region, the fitting was poorer. In $Ce_{70}Al_{10}Cu_{20}$ and Zr-Ti-Cu-Ni-Be glasses the excess wing was fitted by considering α and β relaxations. They suggested that β relaxations arise from the small scale translational motions of atoms which are hindered in its metastable atomic positions by solid-like islands.

In some metallic glass compositions, mechanical relaxation below T_g is only perceived as an excess wing of the main α-peak (see Figure 9). In other cases, a shoulder or a secondary peak is detected in as-quenched samples but vanishes after thermal cycling in more stable glassy states. Description of $M''(\omega,T)$ along the whole temperature range by consistent relaxation functions and $\tau(T)$ behaviors is maybe the main tool in order to discern if a secondary relaxation is present and what are its main characteristics.

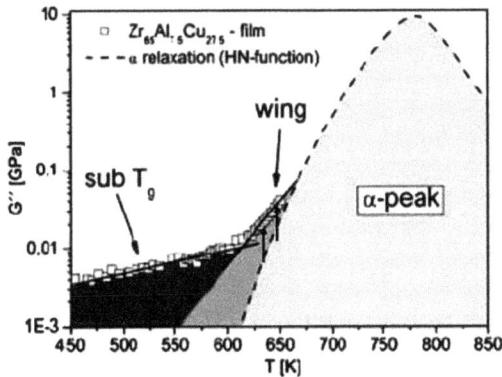

Figure 9. Low temperature side of the loss modulus peak of ZrAlCu glass. Reprinted from Reference [99] with permission from IOP Publishing.

145

Using a CC-function, Hachenberg *et al.* [100,101] showed that the α-peak of $Zr_{65}Cu_{27.5}Al_{7.5}$ and $Pd_{77}Cu_6Si_{17}$ is well described by the VFT equation and the excess wing is better fitted when taken in consideration a β-relaxation. By observing the heating rate dependence of the onset and turning point of storage modulus dependence on temperature, Hachenberg determined E_β to be 0.67 ± 0.11 eV and 0.59 ± 0.39 eV for $Pd_{77}Cu_6Si_{17}$ and $Zr_{65}Al_{17.5}Cu_{27.5}$ respectively. He ascribed this change on the storage modulus as a result of aging driven by a β-relaxation with a cooperative nature. Since these two different glassy systems have quite different strong-fragile liquid behavior (m = 52.8–77 for PdCuSi and m = 36.4–38.4 for Zr-Al-Cu) they suggested β-relaxation might be a universal feature of metallic glass dynamics. Combining the dependence on the heating rate of α-peak and MCT predictions they explained the merging of α and β relaxations.

The excess-wing of $M''(\omega,T)$ found in isochronal DMA of $Cu_{46}Zr_{46}Al_8$ was interpreted by Liu *et al.* [102] as the high-frequency tail of the α-peak once in the AGV ($T < T_g$) dynamics as shown in Figure 10. The Arrhenius/VFT transition is quite often observed in glass systems [103]. Liu showed that the DMA behavior of $Cu_{46}Zr_{46}Al_{10}$ described with a CC-function and $\tau_\alpha(T)$ showing VFT/AGV transition at $T = T_g$ was compatible with the relaxation times obtained from quasi-static stress relaxation experiments following KWW equation (see Figure 11). The broadening parameter of the CC-function and the stretching exponent of the KWW were found α $\sim\beta_{KWW}$ ~0.4. The as-quenched samples, did not follow AGV dynamics below T_g, as they suffered *in situ* aging during the heating and the measured $\tau_\alpha(T, T_f(t))$ was interpreted as the system crossing different T_f states as it undergoes simultaneously physical aging.

The same approach was previously applied to the analysis of $Mg_{65}Cu_{25}Y_{10}$ glass [82]. In this case, the deviation from VFT behavior combined with the *in situ* aging manifested a shoulder on the loss modulus as shown in Figure 12. In this case, the CC-function used for fitting the relaxation spectrum showed a significant change of the broadening parameter due to aging. The study of room temperature aging of the same system [29] shows an average activation energy coherent with the $\tau_\alpha(T < T_g)$ behavior found from DMA, implying that in this system aging is driven by molecular movements belonging to the high-frequency tail of a broad α-peak.

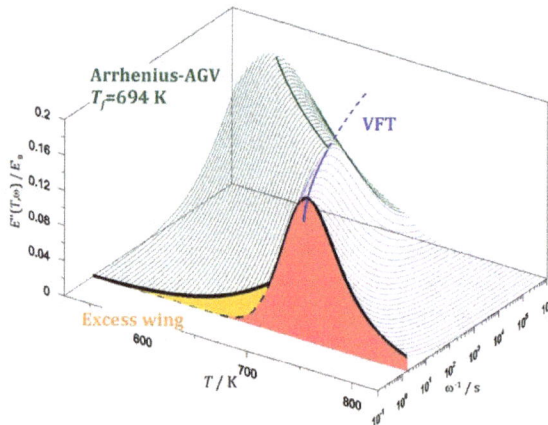

Figure 10. Time-temperature map $E''(\omega,T)$ (E Young's modulus) for $Cu_{46}Zr_{46}Al_8$ glass calculated considering a Cole-Cole function and $\tau(T)$ described by Vogel-Fulcher-Tammann (VFT)/Adams-Gibbs-Vogel (AGV) transition.

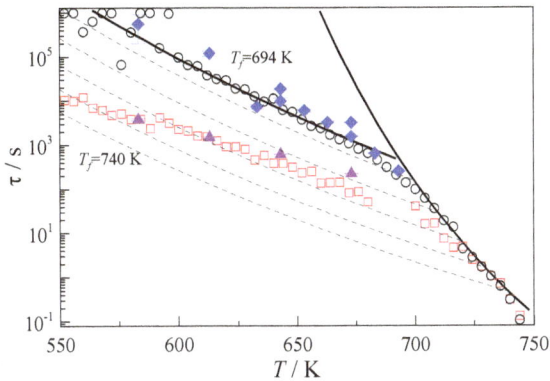

Figure 11. $\tau_\alpha(T)$ of $Cu_{46}Zr_{46}Al_8$ obtained from Cole-Cole fitting of Dynamo-Mechanical-Analysis (DMA) measurements (open symbols) and from quasi-static stress-relaxation tests (filled symbols). Diamonds and circles correspond to annealed samples while triangles and squares to as-quenched ones. Reprinted from Reference [102] with permission from Elsevier.

In some metallic glass compositions, especially Pd and La-based ones, the secondary relaxation appears as a prominent peak well-separated from the α-peak and present also for well-aged samples. Based on Cavaille's work on rheology of glasses and polymers [104], Pelletier analyzed the dynamic mechanical behavior of $Pd_{43}Ni_{10}Cu_{27}P_{20}$ in a hierarchical correlation concept [91]. Following Gauthier's [105] work on quasi-point defects, three different contributions exist in the mechanical response as elastic, anelastic and viscoplastic parts. Qiao [106] analyzed and fit the temperature dependent internal friction behavior of $Zr_{55}Cu_{30}Ni_5Al_{10}$ using the same model. In this model, the important parameter χ is a correlation factor between 0 and 1 linked to the quasi point defect concentration. $\chi = 0$ corresponds to a maximum order, when any movement of a structural unit requires the motion of all other units, while $\chi = 1$ represent maximum disorder when all the movements are independent of each other. With this methodology, in the low temperature range, when the χ is constant (~0.38 in the case of $Zr_{55}Cu_{30}Ni_5Al_{10}$), the loss factor can be easily fitted by a simple Arrhenius equation. At higher temperatures, the parameter χ is a function of temperature, and it was found that it could be fitted with a parabolic function (see Figure 13). In the point defect model, the key-question is how the order parameter χ changes with temperature. The behavior of χ is related to the viscosity change and Qiao's work shows that the quality of the fitting depends on an appropriate description of the viscosity behavior. However, due to the many orders of magnitude change within a relatively narrow temperature region, the description of viscosity behavior is still an open problem [4,107,108].

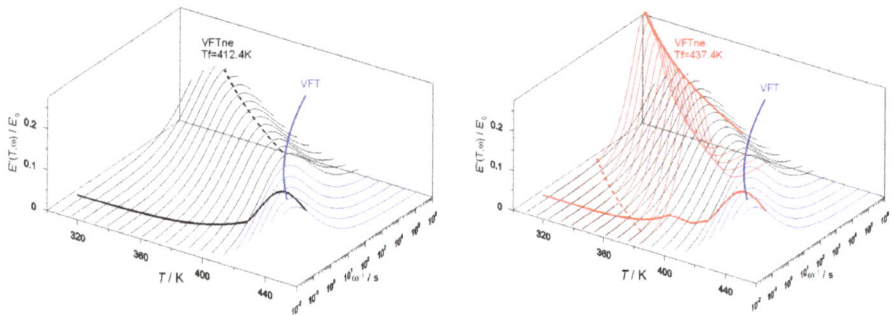

Figure 12. Thick solid line is the expected isochronal DMA behavior of $Mg_{65}Cu_{25}Y_{10}$ for different fictive temperatures. Reprinted from Reference [82] with permission from Elsevier.

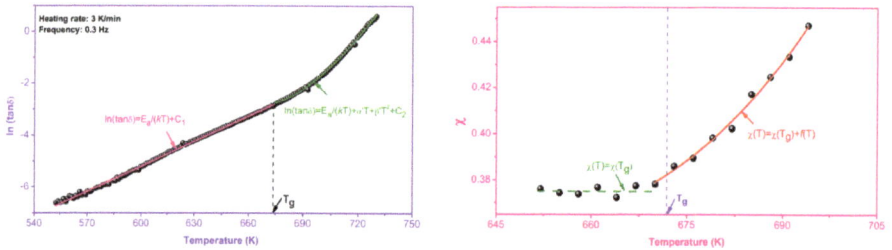

Figure 13. Internal friction modeled by the quasi-point defect theory and χ dependence on temperature. Reprinted from Reference [106] with permission from AIP.

Wang [109] showed that the DMA behavior can be fitted in the whole temperature range by coupling two KWW equations in Fourier transforms. The temperature dependence of τ_α shows a VFT equation while τ_β has an Arrhenius-like dependence. For $La_{70}Ni_{15}Al_{15}$, the pre-factors $\tau_{\alpha,0}$ and $\tau_{\beta,0}$ are 10^{-13} s and 10^{-15} s respectively and $\beta_{KWW} = 0.42$. Qiao [70] fitted the relaxation dynamics of $Pd_{40}Ni_{10}Cu_{30}P_{20}$ as well as $La_{60}Ni_{15}Al_{25}$ by combining the Fourier transform of the KWW function for the α-relaxation and the CC-function for the β. From a microscopic point of view, α-relaxation could be interpreted as collective movement of all the atoms, while β relaxation could be understood by the quasi-point defect theory which relates relaxation to thermally activated jumps of a structural unit [54]. Later on, they described the β process using a coupling model in a very similar form [70].

Mechanical spectroscopy data can also unveil the underlying distribution of relaxation times. This means obtaining the distribution of relaxation times $A(\tau')$ defined in Equation (13). Kursumovic [44,110] and Ocelik [111] analyzed creep recovery and found a trimodal distribution of τ' with maximums of the distribution peaks around 10 s, 100 s and 1000 s at temperature 50–100 K below T_g (Figure 14). The details of the $A(\tau')$ allowed them to propose different TSRO and CSRO corresponding to each mode of the distribution, the slowest one corresponding to annihilation of free-volume by cooperative motions. Ju and Atzmon applied direct spectrum analysis to strain relaxation data on $Al_{86.8}Ni_{3.7}Y_{9.5}$ at room temperature [112] and later to DMA isothermal curves of $Zr_{46.8}Ti_{8.2}Cu_{7.5}Ni_{10}Be_{27.5}$ near T_g [113]. In both cases they obtained a multimodal distribution of times, and interpreted it as associated to the activation of shear transformation zones (STZs) involving different number of atoms. The direct time spectrum analysis of mechanical spectroscopy permits to unveil more details about the microscopic movements involved in the relaxation process.

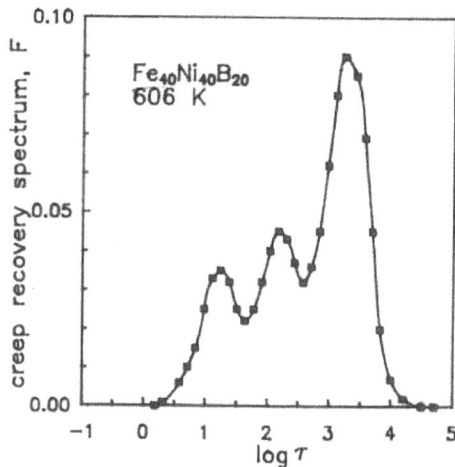

Figure 14. Time spectrum of anelastic relaxations of $Fe_{40}Ni_{40}B_{20}$ glass at $T < T_g$. Reprinted from Reference [110] with permission from Elsevier.

In addition to get insight to the microscopic origin of glassy dynamics, the determination of the relaxation spectrum $M''(\omega,T)$ by appropriate response functions and average $\tau(T)$ dependences is, *per se*, an important characterization of metallic glasses due to its consequences on the mechanical properties. The relationship between mechanical relaxation processes and mechanical properties will be briefly introduced in the following section.

5. Relationship between the Relaxation Spectrum and the Mechanical Properties

On the macroscopic scale, bulk metallic glasses can show plasticity depending on the temperature and the strain rate. At room temperature, depending on the specific system, the length scale of the plastic process zone ranges from 100 nm to 100 μm. Xi [114] determined the plastic zone size of metallic glasses and, based on the relationship between the plastic zone and the stress intensity factor K_{IC}, they suggested that fracture of metallic glasses can be regarded as a flow process at different length scales. As reviewed by Schuh [115], physical aging affects all mechanical properties, from Young's modulus to impact toughness. This is often explained in the framework of the free volume theory; the free volume decreases during annealing, the shear to bulk moduli ratio increases and the glass becomes more brittle. In general, the mechanical behavior of metallic glasses is interpreted in terms of shear transformation zones (STZs) or of the more recently developed cooperative shearing model (CSM) as described by Chen [116]. As discussed above, the β-relaxation measured by mechanical spectroscopy is interpreted as micro-events activated at temperature lower than T_g. The main point here is to describe the relationship between these events and the mechanical properties.

Kahl investigated [117] the aging paths below T_g of $Pd_{40}Ni_{40}P_{20}$ glass via ultrasonic measurements. Figure 15 shows the changes in shear modulus due to decrease in free volume after various annealing treatments. The structural changes causing the process have been attributed to JG-β relaxations. In a similar material ($Pd_{43}Ni_{10}Cu_{27}P_{20}$), Harmon [118] identifies these secondary β-relaxation events with reversible anelastic excitations within the elastic matrix confinement, while the α-relaxation event was identified with the collapse of the matrix confinement and the breakdown of elasticity.

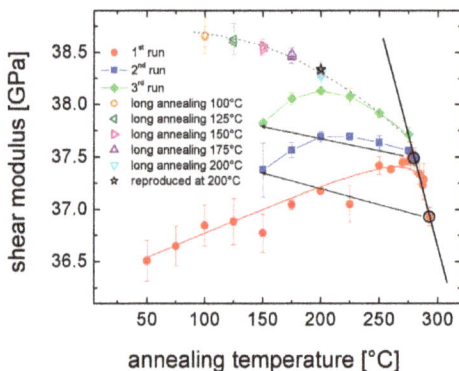

Figure 15. Shear modulus *vs.* annealing temperature of a freshly prepared $Pd_{40}Ni_{40}P_{20}$ MG and subsequent annealing procedures. Reprinted from Reference [117] with permission from AIP.

Okumura [68] investigated the mechanical behavior of $La_{55}Al_{25}Ni_{20}$ metallic glass at different temperatures. As can be seem from Figure 16, there is an increase of maximum elongation around 385 K that corresponds to the activation of β-relaxation in Figure 6. In the same work, $La_{55}Al_{25}Cu_{20}$ was also investigated showing a similar increase of elongation at the temperatures where an obvious shoulder of the loss modulus was observed. It also exhibited an obvious shoulder behavior in the mechanical spectroscopy measurements. As pointed out by Spaepen [119,120], stress or thermal activation in metallic glasses transforms nanoscale soft regions (regions with higher free volume content) into flow units able to accommodate deformation. Below the yield stress, the resulting atomic rearrangement is reversible. Above the yield stress the flow units overcome a certain energy barrier and the atomic reconfiguration becomes irreversible. Macroscopic plastic deformation is thus the result of simultaneous irreversible microscopic shearing events. Under this approach, shear banding is a consequence of a localized high density of flow units. Single flow units promote the activation of near flow units, in a cooperative mechanism which eventually results in the nucleation of a shear band.

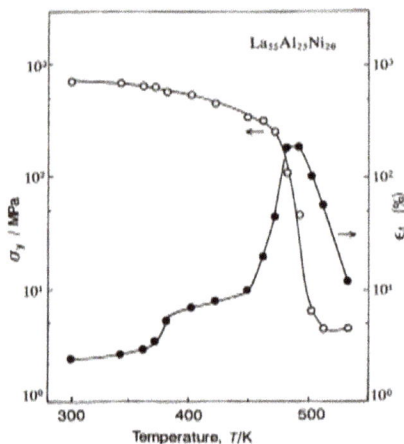

Figure 16. Changes in the yield stress and fracture elongation with testing temperature for $La_{55}Al_{25}Ni_{20}$ glass. The observed decrease in length above ~490 K is attributed to crystallization. Reprinted from Reference [68] with permission from JIM.

150

Based on potential energy landscape and the theory of shear strength in dislocation free solids, Johnson [121] proposed the CSM with the aim of understanding the rheological mechanisms and mechanical properties of metallic glasses. According to it, the volume of STZs, Ω, is proportional to their activation energy, W^*, and the number of atoms participating in the flow unit can be estimated from the model. Using this model and treating the observed shoulder or β-peak of $M''(\omega,T)$ as a thermal activated process, the activation energy of the process can be determined by DMA. Zhao *et al.* [122] obtained the E_β of several different metallic glasses. They obtained a relationship of $E_\beta = 27.5RT_g$ which is close to $24RT_g$ accepted for nonmetallic glass formers. They ascribe this difference to the different type of bonding and suggest that this is the Johari-Goldstein β relaxation in metallic glasses. Using the same methodology, Yu [123] determined the activation energy of more metallic glass alloys and found $E_\beta = 26RT_g$. By an appropriate choice of parameters and using the CSM model they found that the activation energy of β-relaxations and the potential energy barriers of STZs are the same. Liu [124] determined the activation energy of the β relaxation in La-based bulk metallic glasses and assuming that this was the activation energy of STZs they obtained $\Omega = 5.5(0.1)$ nm^3 and the number of atoms involved in an STZ, $n = 178(10)$, for La$_{60}$Al$_{25}$Ni$_{15}$. By compiling data of E_β, they found that the flow unit volume of various MGs range from 2.36 to 6.18 nm^3 and n goes from 170 to 250. These values are in agreement with Pan's [125] estimation based on nanoindentation experiments.

The importance of the Poisson's ratio, ν, on the design of modern materials is highlighted by Greaves [126]. Besides, it is generally accepted that Possion's ratio is a good indicator of the ductility of MGs. With small deviations on the exact value, it is widely accepted in the literature that there exists a critical value which divides plasticity (higher ν values) from brittleness (lower ν values) [127]. For values of ν larger than 0.32, the shear band tip tends to extend rather than induce crack initiation, allowing formation of multiple shear bands and leading to the observed macroscopic plasticity. The exact mechanism is still obscure, but it is suggested that the ductile/brittle nature of metals (in amorphous or crystalline form) is related to the viscous time dependent properties of their liquid precursors, either constrained in metallic glass shear bands or in polycrystalline grain boundaries. The analysis on STZs suggests that the average flow units also correlates with the Possion's ratio; as the value of Ω increase from 2.36 to 6.18 nm^3, the value of Poisson's ratio drops from 0.404 to 0.304.

Unlike previous work where plasticity could only be observed in constrained conditions like bending or compression, Yu [128] found a pronounced macroscopic tensile plasticity in a La$_{68.5}$Ni$_{16}$Al$_{14}$Co$_{1.5}$ metallic glass using ribbon samples. Even at room temperature, the stress strain curve deviates from linear relationship under the strain rate of 1.6×10^{-6} s^{-1}. As shown in Figure 17, by determination of the strain rate of ductile to brittle transition (DBT) at different temperatures, the activation energy of the DBT is determined to be 103 kJ/mol. This is a similar value to the E_β determined by DMA. Furthermore, by using nuclear magnetic resonance (NMR), Yu [78] determined the temperature dependent atomic (diffusive) hopping rates of P atoms in Pd$_{40}$Ni$_{10}$Cu$_{30}$P$_{20}$ and Be atoms in Zr$_{46.75}$Ti$_{8.25}$Cu$_{7.5}$Ni$_{10}$Be$_{27.5}$. They found that their activation energies are very close to E_β. Since the P and Be are the smallest atoms in the respectively metallic glasses, it is suggested that the β relaxation and self-diffusion of the smallest atoms are closely related.

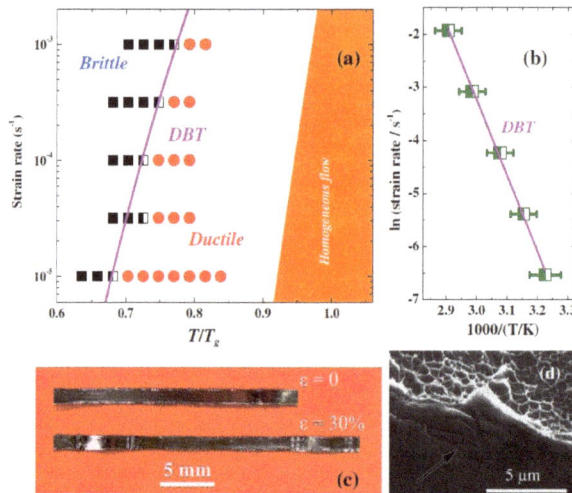

Figure 17. Map of modes of deformation of a La-based glass showing the ductile/brittle transition. Reprinted from Reference [128] with permission from APS.

It is generally accepted that the microstructural origin of the MGs plasticity can be explained by flow units or STZs. By utilizing a mandrel winding method which deform in the bend mode, Lu [129] realized homogeneous plastic deformation at room temperature for Zr, Fe, Mg, Al, and La based metallic glass ribbons. Assuming $E_\beta = 26RT_g$ and choosing metallic glasses with different T_g, they found that plastic deformation is higher for lower E_β. By annealing the sample, physical aging decreases the density of flow units and then both the β-peak of internal friction and plastic deformation get reduced. From the results, they suggest that when the loading time is longer than the relaxation time or if enough energy is applied to activate a sufficiently high density of flow units, homogeneous plastic deformation of MGs can occur at room temperature.

As discussed in the previous sections, mechanical spectroscopy is a powerful tool in order to get insight of the complex glassy dynamics and the time evolution of the system due to physical aging. Furthermore, as shown in this section, the relationships between α-relaxation and homogeneous flow, on one side, and between β-relaxation and plastic behavior at lower temperatures on the other, makes the characterization of the mechanical relaxation spectrum an essential tool for predicting the deformation behavior of MGs at given temperature/deformation rate conditions.

6. Conclusions

In this work, the present knowledge of relaxation dynamics in metallic glasses is reviewed, as well as some suitable methodologies to reveal it. It is generally accepted that primary relaxation, well described by the VFT model at $T > T_g$, reflects the homogeneous flow. On the contrary, understanding of the β process is still developing. Early work suggested that it is the result of anelastic events similar to already known processes in crystalline metals. It might be originated from diffusion processes, resembling Zener or Snoek relaxation. In the energy landscape picture, it corresponds to jumps between close energy minima separated by a low energy barrier and, nowadays, it is treated as a process related to the activation of shear transformation zones or flow units. Both low temperature anelastic events and irreversible aging are usually attributed to the presence β-relaxation, mostly due to the fact that it is very difficult to distinguish between them and the purely reversible Johari-Goldstein relaxation. Furthermore, although the relaxation times of α and β relaxations are separated orders of magnitude, the activation energies obtained for both processes when $T < T_g$ are

Metals **2015**, *5*, 1073–1111

similar. This obscures the differentiation of their microscopic origin and the effect of each relaxation in the mechanical behavior.

For as-quenched samples, obtained with high cooling rates, the relaxation times are also affected by aging and the corresponding change of fictive temperature of the glass. Dynamic mechanical analysis reveals a main α-peak combined with a well-differentiated secondary peak, a low-temperature shoulder or an excess wing depending on the composition and degree of aging of the system. The shoulder or excess wing is the signature of a relaxation time divergent from the VFT behavior which is generally associated to β-relaxation. In some systems, however, the excess wing can be explained only by the high-frequency tail of the α-peak. This picture applies both to glasses such as $Cu_{46}Zr_{46}Al_8$, with a broad distribution of relaxation energies manifested on an excess wing, and to glasses such as $Mg_{65}Cu_{25}Y_{10}$ showing a more obvious effect of physical aging reflected in an apparent shoulder of the loss modulus. In other metallic glasses like La-Ni-Al or Pd-Ni-Cu-P systems, the loss modulus behavior cannot be fitted by one single relaxation event. For La-Ni-Al β-relaxation is well-separated from the primary relaxation and shows different strengths for different compositions. In the case of Pd-Ni-Cu-P, α and β processes are close in a narrow temperature region. In this latter case, the loss modulus behavior might result from an overlap of different concurring mechanisms: thermoelastic background, β-relaxation, and the viscosity related component changing from Arrhenius to VFT behavior at the glass transition.

The link between relaxation dynamics of metallic glass and their mechanical properties is also discussed. At a given temperature, there is a ductile to brittle transition relating the applied strain rate and the β-relaxation time. Further work suggests that this transition might be originated from the notable β relaxations and its influence on the activation of shear transform zones or flow units. There might be also a connection between the β relaxations and the ability of absorbing energy which is associated with the toughness of materials. By exploring the internal friction behavior (especially the β relaxations), it is expected that we can improve our knowledge of metallic glasses, leading to obtain alloys with improved mechanical performance.

Acknowledgments: Work funded by MINECO, grant FIS2014-54734-P and Generalitat de Catalunya, grant 2014SGR00581. C. Liu is supported by Generalitat de Catalunya, FI grant 2012FI_B00237.

Author Contributions: All authors contributed equally to this work.

Conflicts of Interest: The authors declare no conflict of interest.

References

1. Angell, C.A. Relaxation in liquids, polymers and plastic crystals strong/fragile patterns and problems. *J. Non-Cryst. Solids* **1991**, *131–133*, 13–31. [CrossRef]
2. Stillinger, F.H.; Debenedetti, P.G. Glass Transition Thermodynamics and Kinetics. *Annu. Rev. Condens. Matter Phys.* **2013**, *4*, 263–285. [CrossRef]
3. Debenedetti, P.G.; Stillinger, F.H. Supercooled liquids and the glass transition. *Nature* **2001**, *410*, 259–267. [CrossRef] [PubMed]
4. Dyre, J.C. Colloquium: The glass transition and elastic models of glass-forming liquids. *Rev. Mod. Phys.* **2006**, *78*, 953–972. [CrossRef]
5. Angell, C.A.; Ngai, K.L.; McKenna, G.B.; MaMillan, P.F.; Martin, S.W. Relaxation in glassforming liquids and amorphous solids. *J. Appl. Phys.* **2000**, *88*, 3113–3157. [CrossRef]
6. Bohmer, R.; Ngai, K.L.; Angell, C.A.; Plazek, D.J. Nonexponential relaxations in strong and fragile glass formers. *J. Chem. Phys.* **1993**, *99*, 4201–4209. [CrossRef]
7. Johari, G.P.; Goldstein, M. Viscous liquids and the glass transition. II. Secondary relaxations in glasses of rigid molecules. *J. Chem. Phys.* **1970**, *53*, 2372–2388. [CrossRef]
8. Borrego, J.M.; Conde, C.F.; Conde, A. Structural relaxation processes in FeSiB-Cu(Nb, X), X = Mo, V, Zr, Nb glassy alloys. *Mater. Sci. Eng. A* **2001**, *304–306*, 491–494. [CrossRef]
9. Kumar, G.; Neibecker, P.; Liu, Y.H.; Schroers, J. Critical fictive temperature for plasticity in metallic glasses. *Nat. Commun.* **2013**, *4*. [CrossRef]

10. Hodge, I.M. Enthalpy relaxation and recovery in amorphous materials. *J. Non-Cryst. Solids* **1994**, *169*, 211–266. [CrossRef]

11. Hodge, I.M. Effects of Annealing and Prior History on Enthalpy Relaxation in Glassy-Polymers 6. Adam-Gibbs Formulation of Nonlinearity. *Macromolecules* **1987**, *20*, 2897–2908. [CrossRef]

12. Tool, A.Q. Relation Between Inelastic Deformability and Thermal Expansion of Glass in Its Annealing Range. *J. Am. Ceram. Soc.* **1946**, *29*, 240–253. [CrossRef]

13. Moynihan, C.T.; Macedo, P.B.; Montrose, C.J.; Gupta, P.K.; DeBolt, M.A.; Dill, J.F.; Dom, B.E.; Drake, P.W.; Easteal, A.J.; Elterman, P.B.; *et al.* Structural Relaxation in Vitreous Materials. *Ann. N. Y. Acad. Sci.* **1976**, *279*, 15–35. [CrossRef]

14. Narayanaswamy, O.S. A Model of Structural Relaxation in Glass. *J. Am. Ceram. Soc.* **1971**, *54*, 491–498. [CrossRef]

15. Lunkenheimer, P.; Wehn, R.; Schneider, U.; Loidl, A. Glassy aging dynamics. *Phys. Rev. Lett.* **2005**, *95*. [CrossRef]

16. Chen, H.S.; Coleman, E. Structure relaxation spectrum of metallic glasses. *Appl. Phys. Lett.* **1976**, *28*, 245–247. [CrossRef]

17. Tsyplakov, A.N.; Mitrofanov, Y.P.; Makarov, A.S.; Afonin, G.V.; Khonik, V.A. Determination of the activation energy spectrum of structural relaxation in metallic glasses using calorimetric and shear modulus relaxation data. *J. Appl. Phys.* **2014**, *116*. [CrossRef]

18. Granato, A.V.; Khonik, V.A. An interstitialcy theory of structural relaxation and related viscous flow of glasses. *Phys. Rev. Lett.* **2004**, *93*. [CrossRef]

19. Khonik, S.V.; Granato, A.V.; Joncich, D.M.; Pompe, A.; Khonik, V.A. Evidence of distributed interstitialcy-like relaxation of the shear modulus due to structural relaxation of metallic glasses. *Phys. Rev. Lett.* **2008**, *100*. [CrossRef]

20. Nagel, C.; Rätzke, K.; Schmidtke, E.; Faupel, F.; Ulfert, W. Positron-annihilation studies of free-volume changes in the bulk metallic glass $Zr_{65}Al_{17.5}Ni_{10}Cu_{17.5}$ during structural relaxation and at the glass transition. *Phys. Rev. B* **1999**, *60*, 9212–9215. [CrossRef]

21. Van den Beukel, A.; Radelaar, S. On the Kinetics of Structural Relaxation in Metallic Glasses. *Acta Mater.* **1983**, *31*, 419–427. [CrossRef]

22. Van den Beukel, A.; van der Zwaag, S.; Mulder, A.L. A semi quantitative description of the kinetics of structural relaxataion in amorphous $Fe_{40}Ni_{40}B_{20}$. *Acta Metall.* **1984**, *32*, 1895–1902. [CrossRef]

23. Gibbs, M.R.J.; Sinning, H.R. A critique of the roles of TSRO and CSRO in metallic glasses by application of the activation energy spectrum model to dilatometric data. *J. Mater. Sci.* **1985**, *20*, 2517–2525. [CrossRef]

24. Khonik, V.A.; Kosilov, A.T.; Mikhailov, V.A.; Sviridov, V.V. Isothermal creep of metallic glasses: A new approach and its experimental verification. *Acta Mater.* **1998**, *46*, 3399–3408. [CrossRef]

25. Borrego, J.M.; Blázquez, J.S.; Lozano-Pérez, S.; Kim, J.S.; Conde, C.F.; Conde, A. Structural relaxation in Fe(Co)SiAlGaPCB amorphous alloys. *J. Alloys Compd.* **2014**, *584*, 607–610. [CrossRef]

26. Khonik, V.A. The Kinetics of Irreversible Structural Relaxation and Homogeneous Plastic Flow of Metallic Glasses. *Phys. Status Solidi A* **2000**, *177*, 173–189. [CrossRef]

27. Khonik, V.A. The kinetics of irreversible structural relaxation and rheological behavior of metallic glasses under quasi-static loading. *J. Non-Cryst. Solids* **2001**, *296*, 147–157. [CrossRef]

28. Ruta, B.; Baldi, G.; Monaco, G.; Chushkin, Y. Compressed correlation functions and fast aging dynamics in metallic glasses. *J. Chem. Phys.* **2013**, *138*. [CrossRef] [PubMed]

29. Zhai, F.; Pineda, E.; Ruta, B.; Gonzalez-Silveira, M.; Crespo, D. Aging and structural relaxation of hyper-quenched $Mg_{65}Cu_{25}Y_{10}$ metallic glass. *J. Alloys Compd.* **2014**, *615*, s9–s12. [CrossRef]

30. Hu, L.; Zhou, C.; Zhang, C.; Yue, Y. Thermodynamic anomaly of the sub-T_g relaxation in hyperquenched metallic glasses. *J. Chem. Phys.* **2013**, *138*. [CrossRef] [PubMed]

31. Tsyplakov, A.N.; Mitrofanov, Y.P.; Khonik, V.A.; Kobelev, N.P.; Kaloyan, A.A. Relationship between the heat flow and relaxation of the shear modulus in bulk PdCuP metallic glass. *J. Alloys Compd.* **2015**, *618*, 449–454. [CrossRef]

32. Chen, H.S. Glass transition and secondary relaxation in metal glasses. In *Amorphous Metals and Semiconductors*; Haasen, P., Jaffee, R.I., Eds.; Pergamon Press: Coronado, CA, USA, 1985; pp. 126–150.

33. Maddin, R.; Masumoto, T. The deformation of amorphous palladium-20 at. % silicon. *Mater. Sci. Eng.* **1972**, *9*, 153–162. [CrossRef]

34. Cohen, M.H.; Turnbull, D. Molecular Transport in Liquids and Glasses. *J. Chem. Phys.* **1959**, *31*, 1164–1169. [CrossRef]

35. Turnbull, D.; Cohen, M.H. Free Volume Model of the Amorphous Phase: Glass Transition. *J. Chem. Phys.* **1961**, *34*, 120–125. [CrossRef]

36. Turnbull, D.; Cohen, M.H. On the Free-Volume Model of the Liquid-Glass Transition. *J. Chem. Phys.* **1970**, *52*, 3038–3041. [CrossRef]

37. Jackle, J. Models of the glass transition. *Rep. Prog. Phys.* **1986**, *49*, 171–231. [CrossRef]

38. Goldstein, M. Viscous Liquids and the Glass Transition: A Potential Energy Barrier Picture. *J. Chem. Phys.* **1969**, *51*, 3728–3739. [CrossRef]

39. Gotze, W.; Sjogren, L. Relaxation processes in supercooled liquids. *Rep. Prog. Phys.* **1992**, *55*, 241–370. [CrossRef]

40. Ngai, K.L.; Paluch, M. Classification of secondary relaxation in glass-formers based on dynamic properties. *J. Chem. Phys.* **2004**, *120*, 857–873. [CrossRef] [PubMed]

41. Ngai, K.L. Johari-Goldstein relaxation as the origin of the excess wing observed in metallic glasses. *J. Non-Cryst. Solids* **2006**, *352*, 404–408. [CrossRef]

42. Nowick, A.S.; Berry, B.S. *Anelastic Relaxation in Crystalline Solids*; Academic Press, Inc.: New York, NY, USA; London, UK, 1972.

43. Ngai, K.L. *Relaxation and Diffusion in Complex Systems*; Springer: New York, NY, USA, 2011.

44. Kuršumović, A.; Cantor, B. Anelastic crossover and creep recovery spectra in $Fe_{40}Ni_{40}B_{20}$ metallic glass. *Scr. Mater.* **1996**, *34*, 1655–1660. [CrossRef]

45. Jiao, W.; Wen, P.; Peng, H.L.; Bai, H.Y.; Sun, B.A.; Wang, W.H. Evolution of structural and dynamic heterogeneities and activation energy distribution of deformation units in metallic glass. *Appl. Phys. Lett.* **2013**, *102*. [CrossRef]

46. Hermida, É.B. Description of the Mechanical Properties of Viscoelastic Materials Using a Modified Anelastic Element. *Phys. Status Solidi* **1993**, *178*, 311–327.

47. Alvarez, F.; Alegria, A.; Colmenero, J. Relationship between the time domain Kohlrausch Williams Watts and frequency domain Havriliak Negami relaxation functions. *Phys. Rev. B* **1991**, *44*, 7306–7312. [CrossRef]

48. Svanberg, C. Correlation function for relaxations in disordered materials. *J. Appl. Phys.* **2003**, *94*, 4191–4197. [CrossRef]

49. Qiao, J.; Casalini, R.; Pelletier, J.-M.M.; Kato, H. Characteristics of the structural and Johari-Goldstein relaxations in Pd-based metallic glass-forming liquids. *J. Phys. Chem. B* **2014**, *118*, 3720–3730. [CrossRef] [PubMed]

50. Blanter, M.S.S.; Golovin, I.S.S.; Neuhauser, H.; Sinning, H.-R. *Internal Friction in Metallic Materials*; Springer: Heidelberg, Germany, 2007.

51. Wen, P.; Zhao, D.Q.; Pan, M.X.; Wang, W.H.; Huang, Y.P.; Guo, M.L. Relaxation of metallic $Zr_{46.75}Ti_{8.25}Cu_{7.5}Ni_{10}Be_{27.5}$ bulk glass-forming supercooled liquid. *Appl. Phys. Lett.* **2004**, *84*, 2790–2792. [CrossRef]

52. Castellero, A.; Moser, B.; Uhlenhaut, D.I.; Dalla Torre, F.H.; Löffler, J.F. Room-Temperature creep and structural relaxation of Mg–Cu–Y metallic glasses. *Acta Mater.* **2008**, *56*, 3777–3785. [CrossRef]

53. Wang, W.H. The elastic properties, elastic models and elastic perspectives of metallic glasses. *Prog. Mater. Sci.* **2011**, *57*, 487–656. [CrossRef]

54. Qiao, J.C.; Pelletier, J.M. Dynamic mechanical analysis in La-based bulk metallic glasses: Secondary (β) and main (α) relaxations. *J. Appl. Phys.* **2012**, *112*. [CrossRef]

55. Ruta, B.; Chushkin, Y.; Monaco, G.; Cipelletti, L.; Pineda, E.; Bruna, P.; Giordano, V.M.; Gonzalez-Silveira, M. Atomic-Scale relaxation dynamics and aging in a metallic glass probed by X-ray photon correlation spectroscopy. *Phys. Rev. Lett.* **2012**, *109*. [CrossRef]

56. Wang, L.-M.; Liu, R.; Wang, W.H. Relaxation time dispersions in glass forming metallic liquids and glasses. *J. Chem. Phys.* **2008**, *128*. [CrossRef] [PubMed]

57. Meyer, A.; Busch, R.; Schober, H. Time-Temperature Superposition of Structural Relaxation in a Viscous Metallic Liquid. *Phys. Rev. Lett.* **1999**, *83*, 5027–5029. [CrossRef]

58. Qiao, J.C.; Pelletier, J.M. Dynamic universal characteristic of the main (α) relaxation in bulk metallic glasses. *J. Alloys Compd.* **2014**, *589*, 263–270. [CrossRef]

59. Casalini, R.; Roland, C.M. Aging of the secondary relaxation to probe structural relaxation in the glassy state. *Phys. Rev. Lett.* **2009**, *102*. [CrossRef]

60. Berry, B.S.; Pritchet, W.C.; Tsuei, C.C. Discovery of an internal-friction peak in the metallic glass Nb_3Ge. *Phys. Rev. Lett.* **1978**, *41*, 410–413. [CrossRef]

61. Berry, B.S.; Pritchet, W.C. Hydrogen related internal friction peaks in metallic glasses. *Scr. Mater.* **1981**, *15*, 637–642. [CrossRef]

62. Yoon, H.N.; Eisenberg, A. Dynamic mechanical properties of metallic glasses. *J. Non-Cryst. Solids* **1978**, *29*, 357–364. [CrossRef]

63. Fukuhara, M.; Wang, X.; Inoue, A.; Yin, F. Low temperature dependence of elastic parameters and internal frictions for glassy alloy $Zr_{55}Cu_{30}Al_{10}Ni_5$. *Phys. Status Solidi* **2007**, *1*, 220–222.

64. Kunzi, H.U.U.; Agyeman, K.; Guntherodt, H.-J. Internal friction peaks in metallic glasses. *Solid State Commun.* **1979**, *32*, 711–714. [CrossRef]

65. Zdaniewski, W.A.; Rindone, G.E.; Day, D.E. The internal friction of glasses. *J. Mater. Sci.* **1979**, *14*, 763–775.

66. Khonik, V.A.; Spivak, L.V. On the nature of low temperature internal friction peaks in metallic glasses. *Acta Mater.* **1996**, *44*, 367–381. [CrossRef]

67. Egami, T.; Maeda, K.; Vitek, V. Structural defects in amorphous solids A computer simulation study. *Philos. Mag. A* **1980**, *41*, 883–901. [CrossRef]

68. Okumura, H.; Inoue, A.; Masumoto, T. Glass transition and viscoelastic behaviors of $La_{55}Al_{25}Ni_{20}$ and $La_{55}Al_{25}Cu_{20}$ amorphous alloys. *Mater. Trans.* **1991**, *32*, 593–598. [CrossRef]

69. Okumura, H.; Chen, H.S.; Inoue, A.; Masumoto, T. Sub-T_g mechanical relaxation of a $La_{55}Al_{25}Ni_{20}$ amorphous alloy. *J. Non-Cryst. Solids* **1991**, *130*, 304–310. [CrossRef]

70. Qiao, J.; Pelletier, J.-M.; Casalini, R. Relaxation of bulk metallic glasses studied by mechanical spectroscopy. *J. Phys. Chem. B* **2013**, *117*, 13658–13666. [CrossRef] [PubMed]

71. Wang, Z.; Yu, H.B.; Wen, P.; Bai, H.Y.; Wang, W.H. Pronounced slow beta-relaxation in La-based bulk metallic glasses. *J. Phys. Condens. Matter* **2011**, *23*. [CrossRef] [PubMed]

72. Yu, H.B.; Samwer, K.; Wang, W.H.; Bai, H.Y. Chemical influence on β-relaxations and the formation of molecule-like metallic glasses. *Nat. Commun.* **2013**, *4*. [CrossRef] [PubMed]

73. Chen, H.S.; Morito, N. Sub-T_g α' relaxation in a PdCuSi glass; internal friction measurements. *J. Non-Cryst. Solids* **1985**, *72*, 287–299. [CrossRef]

74. Evenson, Z.; Naleway, S.E.; Wei, S.; Gross, O.; Kruzic, J.J.; Gallino, I.; Possart, W.; Stommel, M.; Busch, R. β relaxation and low-temperature aging in a Au-based bulk metallic glass: From elastic properties to atomic-scale structure. *Phys. Rev. B* **2014**, *89*. [CrossRef]

75. Okumura, H.; Inoue, A.; Masumoto, T. Heating rate dependence of two glass transitions and phase separation for a $La_{55}Al_{25}Ni_{20}$ amorphous alloy. *Acta Metall. Mater.* **1993**, *41*, 915–921. [CrossRef]

76. Louzguine-Luzgin, D.V.; Seki, I.; Yamamoto, T.; Kawaji, H.; Suryanarayana, C.; Inoue, A. Double-Stage glass transition in a metallic glass. *Phys. Rev. B* **2010**, *81*. [CrossRef]

77. Cohen, Y.; Karmakar, S.; Procaccia, I.; Samwer, K. The nature of the β-peak in the loss modulus of amorphous solids. *EPL* **2012**, *100*. [CrossRef]

78. Yu, H.B.; Samwer, K.; Wu, Y.; Wang, W.H. Correlation between belta relaxation and self-diffusion of the smallest constituting atoms in metallic glasses. *Phys. Rev. Lett.* **2012**, *109*. [CrossRef]

79. Liu, Y.H.; Fujita, T.; Aji, D.P.B.; Matsuura, M.; Chen, M.W. Structural origins of Johari-Goldstein relaxation in a metallic glass. *Nat. Commun.* **2014**, *5*. [CrossRef] [PubMed]

80. Morito, N.; Egami, T. Internal friction and reversible structural relaxation in the metallic glass $Fe_{32}Ni_{36}Cr_{14}P_{12}B_6$. *Acta Metall.* **1984**, *32*, 603–613. [CrossRef]

81. Bohonyey, A.; Kiss, L.F. A quantitative study on reversible structural relaxation of metallic glasses. *J. Phys. Condens. Matter* **1999**, *3*, 4523–4531. [CrossRef]

82. Pineda, E.; Bruna, P.; Ruta, B.; Gonzalez-Silveira, M.; Crespo, D. Relaxation of rapidly quenched metallic glasses: Effect of the relaxation state on the slow low temperature dynamics. *Acta Mater.* **2013**, *61*, 3002–3011. [CrossRef]

83. Kiss, S.; Posgay, G.; Harangozo, I.Z.; Kedves, F.J. Structural relaxation and crystallization of FeB and NiP metallic glasses followed by internal friction and modulus measurements. *J. Phys.* **1981**, *42*, C5:529–C5:534. [CrossRef]

84. Hettwer, K.J.; Haessner, F. Influence of heat treatment on the internal friction of metglas $Fe_{32}Ni_{36}Cr_{14}P_{12}B_6$. *Mater. Sci. Eng.* **1982**, *52*, 147–154. [CrossRef]

85. Hausch, G. *Internal Friction and Ultrasonic Attenuation in Solids*; University of Tokyo Press: Tokyo, Japan, 1977; p. 265.

86. Morito, N.; Egami, T. Internal friction of a glassy metal $Fe_{32}Ni_{36}Cr_{14}P_{12}B_6$. *IEEE Trans. Magn.* **1983**, *5*, 1898–1900. [CrossRef]

87. Morito, N. Internal friction study on structural relaxation of a glassy metal $Fe_{32}Ni_{36}Cr_{14}P_{12}B_6$. *Mater. Sci. Eng.* **1983**, *60*, 261–268. [CrossRef]

88. Deng, D.; Argon, A.S. Structural relaxation and embrittlement of $Cu_{59}Zr_{41}$ and $Fe_{80}B_{20}$ glasses. *Acta Metall.* **1986**, *34*, 2011–2023. [CrossRef]

89. Deng, D.; Argon, A.S. Analysis of the effect of aging on distributed relaxations, hardness, and embrittlement in $Cu_{59}Zr_{41}$ and $Fe_{80}B_{20}$ glasses. *Acta Metall.* **1986**, *34*, 2025–2038. [CrossRef]

90. Olsen, N.B.; Christensen, T.; Dyre, J.C. Time-Temperature superposition in viscous liquids. *Phys. Rev. Lett.* **2001**, *86*, 1271–1274. [CrossRef] [PubMed]

91. Pelletier, J.M.; van de Moortèle, B.; Lu, I.R. Viscoelasticity and viscosity of Pd–Ni–Cu–P bulk metallic glasses. *Mater. Sci. Eng. A* **2002**, *336*, 190–195. [CrossRef]

92. Jeong, H.T.; Fleury, E.; Kim, W.T.; Kim, D.H.; Hono, K. Study on the Mechanical Relaxations of a $Zr_{36}Ti_{24}Be_{40}$ Amorphous Alloy by Time-Temperature Superposition Principle. *J. Phys. Soc. Jpn.* **2004**. [CrossRef]

93. Jeong, H.T.; Kim, J.-H.; Kim, W.T.; Kim, D.H. The mechanical relaxations of a $Mm_{55}Al_{25}Ni_{10}Cu_{10}$ amorphous alloy studied by dynamic mechanical analysis. *Mater. Sci. Eng. A* **2004**, *385*, 182–186. [CrossRef]

94. Guo, L.; Wu, X.; Zhu, Z. Mechanical relaxation studies of α and slow β processes in $Nd_{65}Fe_{15}Co_{10}Al_{10}$ bulk metallic glass. *J. Appl. Phys.* **2011**, *109*. [CrossRef]

95. Ngai, K.L.; Capaccioli, S. Relation between the activation energy of the Johari-Goldstein belta relaxation and Tg of glass formers. *Phys. Rev. E* **2004**, *69*. [CrossRef]

96. Wang, W.H. Correlation between relaxations and plastic deformation, and elastic model of flow in metallic glasses and glass-forming liquids. *J. Appl. Phys.* **2011**, *110*. [CrossRef]

97. Liu, X.F.; Zhang, B.; Wen, P.; Wang, W.H. The slow β-relaxation observed in Ce-based bulk metallic glass-forming supercooled liquid. *J. Non-Cryst. Solids* **2006**, *352*, 4013–4016. [CrossRef]

98. Wang, W.H.; Wen, P.; Liu, X.F. The excess wing of bulk metallic glass forming liquids. *J. Non-Cryst. Solids* **2006**, *352*, 5103–5109. [CrossRef]

99. Rosner, P.; Samwer, K.; Lunkenheimer, P. Indications for an excess wing in metallic glasses from the mechanical loss modulus in $Zr_{65}Al_{7.5}Cu_{27.5}$. *Europhys. Lett.* **2004**, *68*, 226–232. [CrossRef]

100. Hachenberg, J.; Bedorf, D.; Samwer, K.; Richert, R.; Kahl, A.; Demetriou, M.D.; Johnson, W.L. Merging of the alpha and beta relaxations and aging via the Johari-Goldstein modes in rapidly quenched metallic glasses. *Appl. Phys. Lett.* **2008**, *92*. [CrossRef]

101. Hachenberg, J.; Samwer, K. Indications for a slow β-relaxation in a fragile metallic glass. *J. Non-Cryst. Solids* **2006**, *352*, 5110–5113. [CrossRef]

102. Liu, C.; Pineda, E.; Crespo, D. Characterization of mechanical relaxation in a Cu–Zr–Al metallic glass. *J. Alloys Compd.* **2014**, in press.

103. Ferrari, L.; Mott, N.F.; Russo, G. A defect theory of the viscosity in glass-forming liquids. *Philos. Mag. A* **1989**, *59*, 263–272. [CrossRef]

104. Cavaille, J.Y.; Perez, J. Molecular theory for the rheology of glasses and polymers. *Phys. Rev. B* **1989**, *39*, 2411–2422. [CrossRef]

105. Gauthier, C.; Pelletier, J.M.; David, L.; Vigier, G.; Perez, J. Relaxation of non-crystalline solids under mechanical stress. *J. Non-Cryst. Solids* **2000**, *274*, 181–187. [CrossRef]

106. Qiao, J.C.; Pelletier, J.M. Mechanical relaxation in a Zr-based bulk metallic glass: Analysis based on physical models. *J. Appl. Phys.* **2012**, *112*. [CrossRef]

107. Hecksher, T.; Nielsen, A.I.; Olsen, N.B.; Dyre, J.C. Little evidence for dynamic divergences in ultraviscous molecular liquids. *Nat. Phys.* **2008**, *4*, 737–741. [CrossRef]

108. Martinez-Garcia, J.C.; Rzoska, S.J.; Drozd-Rzoska, A.; Martinez-Garcia, J. A universal description of ultraslow glass dynamics. *Nat. Commun.* **2013**, *4*. [CrossRef] [PubMed]

109. Wang, Z.; Wen, P.; Huo, L.S.; Bai, H.Y.; Wang, W.H. Signature of viscous flow units in apparent elastic regime of metallic glasses. *Appl. Phys. Lett.* **2012**, *101*. [CrossRef]

110. Kuršumović, A.; Scott, M.G.; Cahn, R.W. Creep recovery spectra in Fe$_{40}$Ni$_{40}$B$_{20}$ metallic glass. *Scr. Mater.* **1990**, *24*, 1307–1312. [CrossRef]
111. Ocelík, V.; Csach, K.; Kasardová, A.; Bengus, V.Z. Anelastic deformation processes in metallic glasses and activation energy spectrum model. *Mater. Sci. Eng. A* **1997**, *226–228*, 851–855. [CrossRef]
112. Ju, J.D.; Jang, D.; Nwankpa, A.; Atzmon, M. An atomically quantized hierarchy of shear transformation zones in a metallic glass. *J. Appl. Phys.* **2011**, *109*. [CrossRef]
113. Ju, J.D.; Atzmon, M. A comprehensive atomistic analysis of the experimental dynamic-mechanical response of a metallic glass. *Acta Mater.* **2014**, *74*, 183–188. [CrossRef]
114. Xi, X.K.; Zhao, D.Q.; Pan, M.X.; Wang, W.H.; Wu, Y.; Lewandowski, J.J. Fracture of brittle metallic glasses: Brittleness or plasticity. *Phys. Rev. Lett.* **2005**, *94*. [CrossRef]
115. Schuh, C.A.; Hufnagel, T.C.; Ramamurty, U. Mechanical behavior of amorphous alloys. *Acta Mater.* **2007**, *55*, 4067–4109. [CrossRef]
116. Chen, M. Mechanical Behavior of Metallic Glasses: Microscopic Understanding of Strength and Ductility. *Annu. Rev. Mater. Res.* **2008**, *38*, 445–469. [CrossRef]
117. Kahl, A.; Koeppe, T.; Bedorf, D.; Richert, R.; Lind, M.L.; Demetriou, M.D.; Johnson, W.L.; Arnold, W.; Samwer, K. Dynamical and quasistatic structural relaxation paths in Pd$_{40}$Ni$_{40}$P$_{20}$ glass. *Appl. Phys. Lett.* **2009**, *95*. [CrossRef]
118. Harmon, J.S.; Demetriou, M.D.; Johnson, W.L.; Samwer, K. Anelastic to plastic transition in metallic glass-forming liquids. *Phys. Rev. Lett.* **2007**, *99*. [CrossRef]
119. Spaepen, F. Homogeneous flow of metallic glasses: A free volume perspective. *Scr. Mater.* **2006**, *54*, 363–367. [CrossRef]
120. Spaepen, F. A microscopic mechanism for steady state inhomogeneous flow in metallic glasses. *Acta Metall.* **1977**, *25*, 407–415. [CrossRef]
121. Johnson, W.L.; Demetriou, M.D.; Harmon, J.S.; Lind, M.L.; Samwer, K. Rheology and Ultrasonic Properties of Metallic Glass-Forming Liquids: A Potential Energy Landscape Perspective. *MRS Bull.* **2007**, *32*, 644–650. [CrossRef]
122. Zhao, Z.F.; Wen, P.; Shek, C.H.; Wang, W.H. Measurements of slow β-relaxations in metallic glasses and supercooled liquids. *Phys. Rev. B* **2007**, *75*. [CrossRef]
123. Yu, H.B.; Wang, W.H.; Bai, H.Y.; Wu, Y.; Chen, M.W. Relating activation of shear transformation zones to β relaxations in metallic glasses. *Phys. Rev. B* **2010**, *81*. [CrossRef]
124. Liu, S.T.; Wang, Z.; Peng, H.L.; Yu, H.B.; Wang, W.H. The activation energy and volume of flow units of metallic glasses. *Scr. Mater.* **2012**, *67*, 9–12. [CrossRef]
125. Pan, D.; Inoue, A.; Sakurai, T.; Chen, M.W. Experimental characterization of shear transformation zones for plastic flow of bulk metallic glasses. *Proc. Natl. Acad. Sci. USA* **2008**, *105*, 14769–14772. [CrossRef] [PubMed]
126. Greaves, G.N.; Greer, A.L.; Lakes, R.S.; Rouxel, T. Poisson's ratio and modern materials. *Nat. Mater.* **2011**, *10*, 823–838. [CrossRef] [PubMed]
127. Lewandowski, J.J.; Wang, W.H.; Greer, A.L. Intrinsic plasticity or brittleness of metallic glasses. *Philos. Mag. Lett.* **2005**, *85*, 77–87. [CrossRef]
128. Yu, H.B.; Shen, X.; Wang, Z.; Gu, L.; Wang, W.H.; Bai, H.Y. Tensile plasticity in metallic glasses with pronounced β relaxations. *Phys. Rev. Lett.* **2012**, *108*. [CrossRef]
129. Lu, Z.; Jiao, W.; Wang, W.H.; Bai, H.Y. Flow unit perspective on room temperature homogeneous plastic deformation in metallic glasses. *Phys. Rev. Lett.* **2014**, *113*. [CrossRef]

![metals logo] *metals*

MDPI

Review

Understanding of the Structural Relaxation of Metallic Glasses within the Framework of the Interstitialcy Theory

Vitaly A. Khonik

Department of General Physics, State Pedagogical University, Lenin St. 86, Voronezh 394043, Russia;
E-Mail: khonik@vspu.ac.ru; Tel./Fax: +7-473-2390433

Academic Editors: K. C. Chan and Jordi Sort Vinas
Received: 31 January 2015 / Accepted: 19 March 2015 / Published: 25 March 2015

Abstract: A review of the new approach to the understanding of the structural relaxation of metallic glasses based on the Interstitialcy theory has been presented. The key hypothesis of this theory proposed by Granato consists of the statement that the thermodynamic properties of crystalline, liquid and glassy states are closely related to the interstitial defects in the dumbbell (split) configuration, called also interstitialcies. It has been argued that structural relaxation of metallic glasses takes place through a change of the concentration of interstitialcy defects frozen-in from the melt upon glass production. Because of a strong interstitialcy-induced shear softening, the defect concentration can be precisely monitored by measurements of the unrelaxed shear modulus. Depending on the relation between the current interstitialcy concentration c and interstitialcy concentration in the metastable equilibrium, different types of structural relaxation (decreasing or increasing c) can be observed. It has been shown that this approach leads to a correct description of the relaxation kinetics at different testing conditions, heat effects occurring upon annealing, shear softening and a number of other structural relaxation-induced phenomena in metallic glasses. An intrinsic relation of these phenomena with the anharmonicity of the interatomic interaction has been outlined. A generalized form of the interstitialcy approach has been reviewed.

Keywords: metallic glasses; structural relaxation; Interstitialcy theory; dumbbell interstitials; shear modulus; heat effects; elastic dipoles

---

1. Introduction

The non-crystallinity of glasses defines the excess Gibbs free energy and, as a result, relaxation of their structure towards the states with lower energy upon any kind of heat treatment. In metallic glasses (MGs), structural relaxation is expressed very markedly, leading to significant, sometimes even drastic changes of their physical properties. For instance, it was found long ago that the shear viscosity of a metallic glass at temperatures below the glass transition temperature T_g (far enough from the metastable equilibrium) can be increased by five orders of magnitude as a result of structural relaxation [1]. The kinetics of homogeneous plastic flow at these temperatures can be described as structural relaxation oriented by the external stress [2]. Structural relaxation is known to affect many physical properties of MGs—mechanical (elasticity, anelasticity, viscoelasticity, *etc.*), electrical, corrosion, magnetic and others [3–5]. Since the beginning of the 1980s, it has remained a subject of unabated interest.

However, in spite of decades-long investigations, the microscopic mechanism and related kinetics of structural relaxation remain a highly debated issue. Most often, structural relaxation is interpreted within the framework of the "free volume" approach, which dates back to the ideas of Doolittle [6], Turnbull and Cohen [7,8], and was later conceptually adopted by Spaepen [9] and Argon [10] to MGs.

The free volume model was numerously modified to better describe the property changes [5,11–14]. This model has both advantages (e.g., [4,15]) and drawbacks (e.g., [16,17]), which are, however, beyond the scope of the present paper. Our goal is to review a new comprehensive, versatile and verifiable approach to the understanding of the structural relaxation of metallic glasses, which provides a generic relationship between the crystal, its melt and glass produced by melt quenching. This promising approach is based on the Interstitialcy theory suggested by Granato in 1992 [18,19].

2. Background

The approach to structural relaxation of MGs based on the Interstitialcy theory starts from crystal melting. Frenkel in the 1930s suggested that melting occurs through the thermoactivated generation of point defects of a crystalline structure. There are two types of such defects known from the solid state physics—vacancies and interstitials. Frenkel chose the former and developed a vacancy model of melting [20]. The model was numerously modified [21] and gained much popularity. However, a few issues remained unanswered. One of the first doubts was formulated by Slater [22]. Assuming that the entropy of melting is determined by the generation of vacancies, he concluded that the vacancy concentration just before the melting point should be about 50%, which is clearly unrealistic. It is now known that the volume change and entropy per vacancy upon melting do not agree with the experimental values [18,22,23].

According to Frenkel, the second possibility for point-defect-mediated melting is associated with interstitials. However, available information on these defects over a long period of time was largely inadequate. In particular, it was assumed that interstitials occupy the octahedral cavity in the FCC structure (*i.e.*, in the center of the elementary cell). In 1950, Seitz [24] theoretically considered "an interstitial atom which moves by jumping into a normal lattice site and forces the atom that is there into a neighboring interstitial site" and termed it an "interstitialcy". In the beginning of the 1960s, Vineyeard *et al.* [25,26] performed detailed computer simulation of radiation damage of FCC and BCC metallic lattices and firmly concluded that interstitials "reside in a split configuration, sharing a lattice site with another atom", which is equivalent to Seitz's interstitialcy. By the middle of the 1970s, the split (dumbbell) nature of interstitials (interstitialcies) in different metals with different crystalline structures became quite evident [27]. To date, it is generally accepted that split interstitials exist in all basic crystalline structures and represent the basic state of interstitials in metals [28,29]. An interstitialcy in a molecular-dynamic model of crystalline copper [30] is shown in Figure 1 as an example.

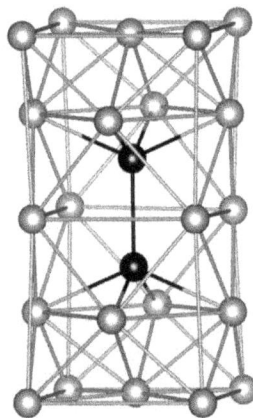

Figure 1. <100> interstitialcy in a molecular-dynamic model of copper [30].

Granato and co-workers made an important contribution to the understanding of the nature of interstitial defects in crystals. In particular, they performed a unique experiment (not repeated so far)—irradiation of a copper single crystal by thermal neutrons at $T = 4$ K with simultaneous measurement of all elastic constants [31,32]. The irradiation results in the formation of long-living (at this temperature) Frenkel pairs. It was found that all elastic moduli decrease with the concentration c of Frenkel pairs, but the shear modulus C_{44} decreases most rapidly. The extrapolation of $C_{44}(c)$-dependence led to the unexpected conclusion that the shear modulus should become zero at $c \approx 2\%$ to 3%. Zero (or very low) shear modulus is characteristic of a liquid [33]. The analysis of the magnitude and orientational dependence of this effect provided the evidence that it is conditioned by interstitial atoms in the dumbbell configuration.

Later, analyzing the thermodynamics of a crystal with interstitialcies, Granato showed that depending on temperature, it is possible to distinguish between several possibilities of stable and metastable equilibrium, which he interpreted as equilibrium/superheated crystal and equilibrium/supercooled liquid states [18,19]. Every stable and metastable state is characterized by its own interstitialcy concentration. Melting of a crystal within the framework of this approach is understood as a result of the thermoactivated generation of interstitialcies, which to a large extent define the properties of the equilibrium and supercooled melt.

This viewpoint is supported by the fundamental property of interstitialcies—the existence of low frequency resonance vibration modes in their vibration spectrum [27,34]. These modes correspond to the frequencies, which are by several times smaller than the Debye frequency. Consequently, the vibrational entropy of the defect becomes large, by several times bigger than that of the vacancy (for Cu, for instance, $S^{int}/k_B \approx 15$, while $S^{vac}/k_B \approx 2.4$ [35] for interstitialcies and vacancies, respectively, where k_B is the Boltzmann constant). This then allows explaining (contrary to vacancies) the observed entropy of melting $S_m = Q_m/T_m$ (where Q_m is the heat of melting, T_m the melting temperature). Indeed, there is a remarkable empirical Richards rule [36,37], which states that the entropy of melting per atom S_m^{at} for elemental substances is close to $1.2k_B$, with only a few exceptions [37]. Assuming that melting is connected with interstitialcy formation, one can calculate exactly this value for S_m^{at} [23]. The other major finding was that the interstitialcy formation enthalpy rapidly decreases (by several times) with their concentration at large c strongly promoting melting [30,38]. Within the framework of this approach, one can quantitatively explain the empirical Lindemann rule ($\alpha T_m = const$, where α is the crystal thermal expansion coefficient), as well as the correlation between the melting temperature and shear modulus [39].

Thus, the Interstitialcy theory implies that melting can be understood as a result of thermoactivated interstitialcy generation. This is in agreement with earlier results by Stillinger and Weber [40], who performed molecular dynamic simulation of the BCC structure and found that the elementary structural excitations are vacancy-interstitialcy pairs, which define the defect-induced softening and lead to first-order melting. The coexisting solid and liquid are close to defect free and almost maximally defective states. Several other molecular dynamic experiments noted an important role of interstitialcies in melting [38,41–43] and even in the crystallization of simple metals [44]. Recent molecular dynamics work by Betancourt *et al.* [45] makes sense of the interstitialcy in liquids, and the results seem to accord remarkably with defect concentration estimates given by Granato [18,19]. An important result was presented in [38]. The authors concluded that the "string" atoms, which were repeatedly noticed in computer simulations of supercooled liquids and glasses [46,47] and said to resemble the signatures of interstitialcies in crystals [48], have many of the same properties as interstitialcies in crystals, and these properties become even closer as the interstitialcy concentration approaches a few percent. The idea that the liquid state has an interstitialcy concentration of about a few percent [18] has led to a successful interpretation of property peculiarities for equilibrium and supercooled liquids [23,49].

If the hypothesis on interstitialcy-mediated melting has real meaning and interstitialcies indeed retain their individuality in the molten state, then one simply comes to the conclusion that the glass prepared by freezing of melt should also contain interstitialcies, as indeed suggested by computer simulations of amorphous copper [30,38,50]. Another support for this hypothesis comes from the fact that many features of low-temperature glass anomalies (low-frequency vibrations, relaxation processes and general two-level system behavior) are also observed in crystals after irradiation (which produces vacancy-interstitialcy pairs) at doses much lower than those needed for amorphization [51]. Besides that, the volume dependence of the shear elastic constant associated with radiation-induced disordering and eventual amorphization was found to be virtually identical to that associated with heating up to the melting point [52]. However, it is likely that interstitialcies in liquid and glassy states (while keeping the same or similar properties) do not have direct structural representations, as they do in the crystalline state (see Figure 1).

In any case, if glass contains interstitialcies, its structural relaxation upon heat treatment should be conditioned by a change of the interstitialcy concentration. It is this idea that is analyzed and tested below.

3. Interstitialcy-Mediated Structural Relaxation and Related Relaxation of the Shear Modulus

3.1. Relaxation Kinetics

Interstitialcies exert a pronounced impact on the high-frequency (unrelaxed) shear modulus. This is directly stated by the basic equation of the Interstitialcy theory, which suggests an exponential decrease of the shear modulus G of a crystal with the interstitialcy defect concentration c [18,19],

$$G(T,c) = G_x(T)exp(-\alpha\beta c) \tag{1}$$

where $G_x = G(c = 0)$, β is the dimensionless shear susceptibility and $\alpha = \frac{1}{G\Omega}\frac{dU}{dc}$ with U being the internal energy and Ω the volume per atom. Using a numerical fit for copper, Granato found that $\alpha \approx 1$ [18]. Since β is about 15 to 25, Equation (1) implies a strong decrease of the shear modulus with the interstitialcy defect concentration in crystal.

Following the conceptual framework described above, it is natural to assume that Equation (1) should be also valid for glass. In this case, G_x has the meaning of the shear modulus of the reference (maternal) crystal. This assumption leads to rather numerous examples of the successful interpretation of MGs' property changes upon structural relaxation, as reviewed below. The shear modulus within the framework under consideration is the key thermodynamic parameter (as being the second derivative of the free energy with respect to the shear strain [53]) of glass, and precise measurements of G provide an efficient way to monitor the defect concentration and related relaxation kinetics. It is worthy of notice in this connection that the general idea for the key role of the shear modulus in the relaxation kinetics of supercooled liquids and glasses was introduced long ago [54] and currently is gaining increasing acceptance [55–59].

The kinetics of MGs' structural relaxation is intimately related with the state of "metastable equilibrium" [5,60]. While far below the glass transition temperature T_g this state is kinetically unachievable, near T_g it can be reached from opposite sides at reasonable times, leading to different signs of property changes [60]. In particular, this is manifested in the relaxation of the shear modulus G, as illustrated in Figure 2, which gives the temperature dependence of G for bulk glassy $Pd_{40}Cu_{30}Ni_{10}P_{20}$ measured upon linear heating and cooling at the same rate together with the shear modulus G_{eq} in the metastable equilibrium determined by prolonged isothermal tests at different temperatures [61,62]. The temperature T_{pc} in this figure is interpreted as the Kauzmann pseudocritical temperature, *i.e.*, the lowest temperature at which the state of the supercooled liquid is still possible (see [61] for details). At temperatures $T < T_{pc}$, the shear modulus $G < G_{eq}$ (*i.e.*, the current defect concentration c is bigger than the defect concentration c_{eq} in the metastable equilibrium state), and therefore, structural relaxation leads to an increase of G. On the contrary, at $T > T_{pc}$, the shear modulus $G > G_{eq}$ (respectively, $c < c_{eq}$) and structural relaxation decreases it. Different signs of shear modulus relaxation are indeed experimentally observed [63].

It is interesting to notice the shear modulus behavior upon cooling (see Figure 2). Just after switching from heating to cooling, the shear modulus still continues to decrease. This behavior is soon changed into an increase of the shear modulus, but the latter remains significantly smaller than that in the course of the initial heating. Such a big hysteresis turns out to be a characteristic feature of MGs [62]. The obvious reason for it is the big underlying relaxation time [62] (see below Equation (6) and the related discussion). It should be also emphasized that the shear modulus upon cooling at temperatures $T < T_{pc}$ is quite close the metastable equilibrium (see Figure 2), confirming this state as the limit for the relaxation. Thus, structural relaxation takes place mainly during cooling, contrary to what is usually assumed. Below, we discuss this behavior in more detail (see Figure 4 and related description).

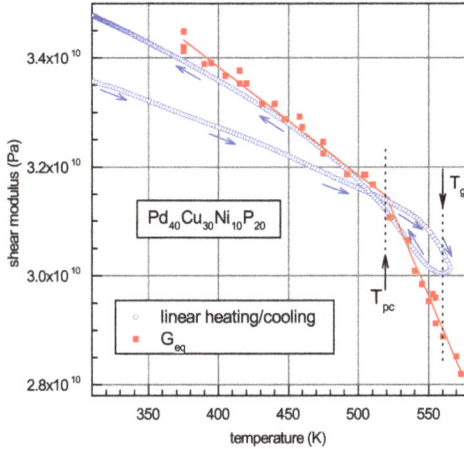

Figure 2. Temperature dependences of the shear modulus for glassy $Pd_{40}Cu_{30}Ni_{10}P_{20}$ upon linear heating and subsequent cooling at a rate of 3 K/min together with the shear modulus in the state of the metastable equilibrium [61,62]. The calorimetric glass transition temperature T_g measured at the same rate is shown by the vertical arrow. The temperature T_{pc} is ascribed to the Kauzmann pseudocritical temperature. The sequence of heating/cooling is given by the arrows.

To analyze the isothermal relaxation kinetics of as-cast glass far below T_g, one can simply consider a spontaneous decrease of the interstitialcy defect concentration, which follows the first-order kinetics. In line with numerous data (e.g., [64]), the corresponding activation energy E should be continuously distributed (e.g., because of the distribution of local shear moduli due to fluctuations in local densities, chemical bonding, *etc.*, as experimentally demonstrated in [65]). Let $N(E, T, t)$ be the temperature-/time-dependent defect concentration per unit activation energy interval. Then, the relaxation kinetics is given by $dN/N = -\nu exp(-E/kT)dt$, where ν is the attempt frequency. If $N_0(E)$ is the initial interstitialcy concentration per unit activation energy interval (*i.e.*, the initial activation energy spectrum, AES), then the time dependence of N after pre-annealing during time τ becomes [66]:

$$N(E, T, t) = N_0(E)exp\left[-\nu(\tau + t)exp(-E/kT)\right] = N_0(E)\Theta(E, T, t) \qquad (2)$$

The characteristic annealing function $\Theta(E, T, t) = exp\left[-\nu(\tau + t)exp(-E/kT)\right]$ in Equation (2) sharply increases near the characteristic activation energy $E_0 = kTln\left[\nu(\tau + t)\right]$ and, to a good precision, can be replaced by the Heaviside step function equal to zero at $E < E_0$ and unity at $E > E_0$ [66]. The total concentration c of defects available for relaxation is then given as:

$$c(T,t) = \int_{E_{min}}^{E_{max}} N(E,T,t)dE \approx \int_{kTln\nu(\tau+t)}^{E_{max}} N_0(E)dE \qquad (3)$$

where E_{min} and E_{max} are the lower and upper limits of the AES available for activation. Next, for the isothermal test, one can use the "flat spectrum" approximation, $N_0 \approx const$ [66]. Then, the concentration Equation (3) is reduced to $c(t) = N_0 E_{max} - N_0 kTln\nu(\tau + t)$. On the other hand, the basic Equation (1) of the Interstitialcy theory for small concentration changes Δc can be rewritten as $\Delta G(T,t)/G = -\beta \Delta c(T,t)$. Thus, the relaxation kinetics for the relative shear modulus change, $g(t) = G(t)/G_0 - 1$, is given by [67]:

$$g(t) = -\beta [c(t) - c_0] = \beta kTN_0 ln(1 + t/\tau) \qquad (4)$$

Equation (4) at long times $t \gg \tau$ gives the well-known "*lnt*" kinetics, which is often experimentally observed upon isothermal annealing of as-cast MGs [64,66,68]. In particular, such behavior is documented for the shear modulus change upon structural relaxation far below T_g, even for very long annealing time [67,69]. This is illustrated in Figure 3, which shows a linear growth of g with the logarithm of time after some transient for a Zr-based glass. The red curve gives the fit calculated using Equation (4). It is seen that this Equation gives a good approximation of the relaxation behavior in the range of times from tens of seconds up to about twenty-four hours.

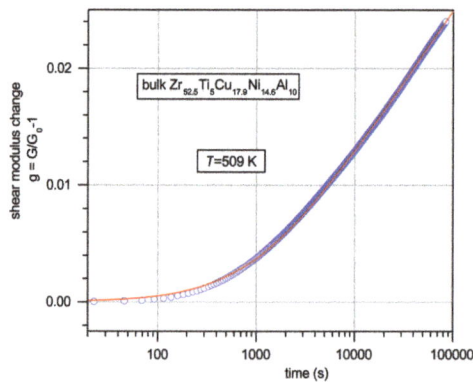

Figure 3. Time dependence of the shear modulus change upon isothermal annealing of bulk glassy $Zr_{52.5}Ti_5Cu_{17.9}Ni_{14.6}Al_{10}$ at $T = 509$ K. The solid red curve gives the fit using Equation (4). Reprinted with permission from American Physical Society, 2008 [67].

The relaxation kinetics upon linear heating can be calculated in a similar way. Since the metastable equilibrium is kinetically achievable at high temperatures, the differential equation for the relaxation should be accepted as:

$$\frac{dc}{dt} = -\frac{c - c_{eq}}{\tau} \qquad (5)$$

where c_{eq} is the interstitialcy defect concentration in the metastable equilibrium, which according to Equation (1) has the form $c_{eq} = -\beta^{-1}ln(G_{eq}/G_x)$, with G_{eq} being the shear modulus in the metastable equilibrium.

Using the "elastic" hypothesis for the activation energy of elementary atomic rearrangements [57,58], $E = GV_c$, where V_c is some characteristic volume, the underlying relaxation time can be written down as:

$$\tau = \tau_0 e^{\frac{GV_c}{k_B T}} = \tau_0 e^{\frac{G_0 V_c}{k_B T}(1+\beta\delta c)} = \tau_0 e^{\frac{G_0 V_c}{k_B T}(1+g)} \tag{6}$$

where τ_0 is of the order of the inverse Debye frequency. Then, the relaxation law (5) for heating at a rate $\dot{T}$ can be rewritten as [70]:

$$\frac{dg_{rel}}{dT} = \frac{\gamma - g_{rel}}{\dot{T}\tau_0 exp\left[\frac{G_0 V_c}{k_B T}(1 + g_{rel})\right]} \tag{7}$$

where g_{rel} is the relaxation component of relative change of the shear modulus and $\gamma = \beta c_0(1 - c_{eq}/c_0)$. Equation (7) describes the shear modulus relaxation upon heating at a given rate $\dot{T}$. It was shown that this equation gives a good description of the shear modulus relaxation behavior of bulk glassy $Pd_{40}Cu_{30}N_{10}P_{20}$ at different heating rates, both below and above T_g [70]. The same equation can be used for the interpretation of the shear modulus hysteresis, as illustrated by Figure 4. The bottom part of this figure gives the same shear modulus relaxation data that was shown in Figure 2, but replotted in terms of the relative shear modulus change g (open squares) for temperatures $T > 500$ K. The linear function g_{linf} here represents the $g(T)$-dependence at temperatures below 450 K, where any significant relaxation of the shear modulus is absent. The relaxation part of the relative shear modulus change was then calculated as $g_{rel}(T) = g(T) - g_{linf}(T)$, as given by closed circles in Figure 4. With the appropriate choice of parameters, Equation (7) can be used for the calculation of g_{rel}-dependence, as shown by the red solid/dashed curves. It is seen that, overall, there is quite acceptable correspondence between experimental and calculated shear modulus relaxation data [62], indicating the validity of the approach assumed by Equation (7). It is to be noted that the relaxation time near $T_g \approx 560$ K of glassy $Pd_{40}Cu_{30}Ni_{10}P_{20}$ calculated with Equation (6) is about 2900 s, and it is this large relaxation time that determines the shear modulus hysteresis shown in Figure 4 [62]. It should be also emphasized that the Maxwell relaxation time $\tau_m = \eta/G$ (η is the shear viscosity) at $T = T_g$ is about 20 s, and therefore, the Maxwell viscoelasticity does not constitute a proper basis for the understanding of relaxation phenomena in MGs, even near the glass transition (see [62] for more details).

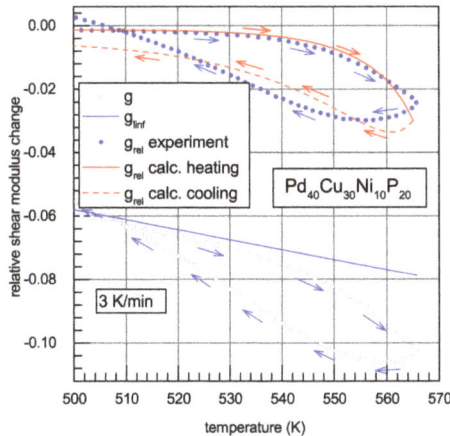

Figure 4. Temperature changes of the experimental relative shear modulus change $g(T)$ for glassy $Pd_{40}Cu_{30}Ni_{10}P_{20}$, its linear approximation $g_{linf}(T)$ for temperatures $T \leq 450$ K together with the experimental and calculated relaxation parts of the relative shear modulus change g_{rel}. The arrows give the sequence of heating/cooling [62].

3.2. Activation Energy Spectra

Upon heating at a constant rate $\dot{T}$, the characteristic activation energy (see above) linearly increases with temperature,

$$E_0 = AT \tag{8}$$

where $A \approx 3 \times 10^{-3}$ eV/K is weakly dependent on $\dot{T}$ and attempt frequency [71]. The shear modulus change then becomes $g(T) = -\beta[c(T) - c_0]$ with $c(T) = \int_{AT}^{Emax} N_0(E)dE$ and $c_0 = \int_{Emin}^{Emax} N_0(E)dE$. This gives $g(T) = \beta \int_{Emin}^{AT} N_0(E)dE$ and eventually leads to the expression for the AES [67],

$$N_0(E_0) = \beta^{-1}\partial g(E_0)/\partial E_0 \tag{9}$$

The AES for bulk glassy $Pd_{41.25}Cu_{41.25}P_{17.5}$ determined using Equation (9) is given by red triangles in Figure 5, which shows a broad pattern typical of different MGs. This AES starts from the activation energy slightly less than 1.2 eV (this corresponds to temperatures of about 400 K) and ends at activation energies answering to T_g [63,72]. Integration of the AES allows calculating the change of the concentration Δc of defect annealing out upon structural relaxation below T_g. For the AES shown in Figure 5, this gives $\Delta c = 0.00161$, close to the values determined for other MGs ($0.00112 \leq \Delta c \leq 0.00322$ [63,72,73]).

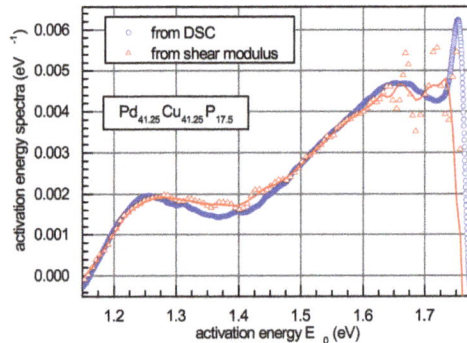

Figure 5. Activation energy spectrum of the structural relaxation of bulk glassy $Pd_{41.25}Cu_{41.25}P_{17.5}$ calculated using shear modulus and DSC data [72].

The full concentration of interstitialcy defects frozen-in upon glass formation can be determined as $c(T) = \frac{1}{\beta}ln\frac{G(T)}{G_x(T)}$ (see Equation (1)). The results of the calculation with this formula for $Pd_{40}Cu_{30}Ni_{10}P_{20}$ glass for a wide temperature range are shown in Figure 6. This figure gives the $c(T)$-dependence for both initial and relaxed (by annealing slightly above T_g) states for the heating rate of 3 K/min together with the metastable equilibrium concentration c_{eq} determined using the metastable equilibrium shear modulus G_{eq} shown in Figure 2. In the initial state, $c \approx 0.0195$ and is nearly independent of T up to temperatures slightly below T_g. At higher T, c rapidly increases, but is still smaller than the equilibrium concentration. At the beginning of the second run, c is smaller by 0.0025 compared to the initial state. This amount is quite close to the values of Δc determined by integration of the AES (see above). On the other hand, the fact that the defect concentration is significantly bigger at the beginning of the second run, but this increase is not seen during the first heating run strongly implies that structural relaxation takes place mainly upon cooling from the supercooled liquid state during the first heating cycle, as demonstrated by Figure 2. As mentioned above, the plausible reason for this is the big underlying relaxation time [62].

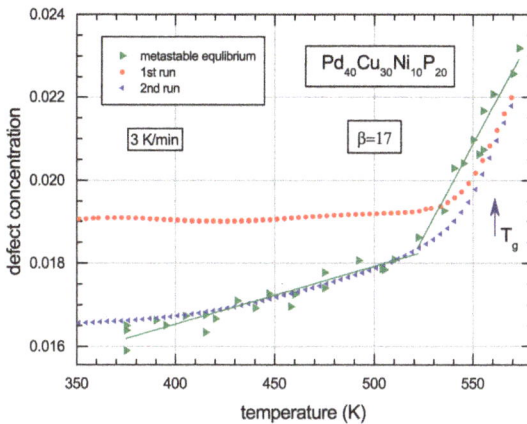

Figure 6. Temperature dependences of the interstitialcy defect concentration for bulk glassy $Pd_{40}Cu_{30}Ni_{10}P_{20}$ in the initial (first run) and relaxed (second run) states measured at 3 K/min together with the metastable equilibrium defect concentration determined from prolonged isothermal shear modulus measurements. The arrow gives the calorimetric glass transition temperature [73].

Thus, within the framework under consideration, the total interstitialcy defect concentration frozen-in upon melt quenching is about 2%, and about one tenth of this amount can be annealed out as a result of structural relaxation. This conclusion is valid for all tested MGs [72,74,75].

4. Interrelationship between the Shear Modulus of Glass, Concentration of Frozen-In Interstitialcy Defects and Shear Modulus of the Maternal Crystal

The basic Equation (1) of the Interstitialcy theory establishes a direct relationship between the shear modulus of glass, the concentration of frozen-in interstitialcy defects and the shear modulus of the maternal crystal. This relationship can be experimentally tested. For this purpose, Equation (1) can be rearranged as:

$$\frac{d}{dT} \ln \frac{G_x(T)}{G(T)} = \alpha \beta \frac{dc}{dT} \tag{10}$$

Equation (10) shows that if structural relaxation is absent, *i.e.*, if $c = const$, then the left-hand part of this equation should be zero. Then, one arrives at the equality of temperature coefficients of the shear moduli in the glassy and maternal crystalline states, *i.e.*,

$$\frac{1}{G}\frac{dG}{dT} = \frac{1}{G_x}\frac{dG_x}{dT} \tag{11}$$

Structural relaxation leading to either a decrease of the defect concentration (far below the glass transition temperature) or its increase (near T_g) should result in negative or positive values of the derivative $D = \frac{d}{dT} \ln \frac{G_x}{G}$, respectively. These predictions were tested in [73,76].

Figure 7 gives $D(T)$ for $Zr_{46}Cu_{46}Al_8$ glass in the initial and relaxed states (first run and second run on the same sample, respectively) assuming $\alpha \approx 1$. In the initial state, D is indeed very close to zero at temperatures $300 \leq T < 440$ K, reflecting the absence of structural relaxation in this range. At higher temperatures, up to $T \approx 670$ K, D is negative, which corresponds to a decrease of the defect concentration ($\frac{dc}{dT} < 0$, in line with Equation (10)) and the related increase of the shear modulus and heat release [77]. Finally, at $T > 670$ K, D becomes positive and rapidly increases with temperature due to the fast defect multiplication ($\frac{dc}{dT} > 0$), which is manifested in strong shear softening and heat absorption upon approaching the glass transition [77]. In the relaxed state, D is zero up to $T \approx 530$ K again, indicating the absence of structural relaxation. At higher temperatures, D becomes positive and rapidly increases with temperature, suggesting rapid defect multiplication near T_g, which is manifested by strong shear softening and heat absorption [77]. Similar results were obtained on glassy $Pd_{40}Cu_{30}Ni_{10}P_{20}$ [73].

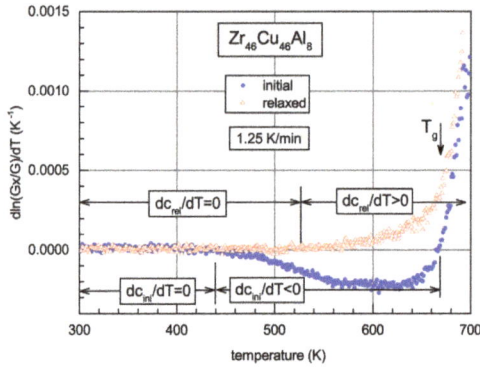

Figure 7. Temperature dependencies of the derivative $\frac{d}{dT} ln \frac{G_x(T)}{G(T)}$ for glassy $Zr_{46}Cu_{46}Al_8$ in the initial and relaxed states. The calorimetric glass transition temperature is shown by the arrow [76].

On the other hand, Equation (11) can be checked directly. Figure 8 shows the ratios of the temperature coefficients of the shear moduli in glassy (initial and relaxed) and crystalline states, $\gamma_{ini} = \left[\frac{1}{G_{ini}} \frac{dG_{ini}}{dT} \right] \left[\frac{1}{G_x} \frac{dG_x}{dT} \right]^{-1}$ and $\gamma_{rel} = \left[\frac{1}{G_{rel}} \frac{dG_{rel}}{dT} \right] \left[\frac{1}{G_x} \frac{dG_x}{dT} \right]^{-1}$, in the temperature ranges where structural relaxation is absent (see Figure 7), and therefore, Equation (11) should be valid. It is seen that both quantities, γ_{ini} and γ_{rel}, are temperature independent and close to unity, supposing a direct relationship between the shear moduli of glass and maternal crystal, as implied by Equation (11).

Thus, the experiments [73,76] described above confirm: (i) the expected relationship between the shear modulus of glass, the concentration of frozen-in defects and the shear modulus of the reference crystal; and (ii) the equality of the temperature coefficients of the shear moduli of glass and the reference crystal in the temperature range with no structural relaxation. Both conclusions validate the basic Equation (1) of the Interstitialcy theory.

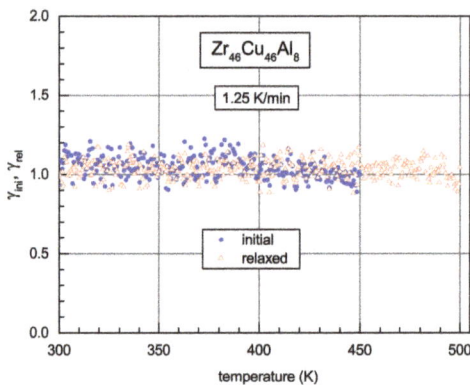

Figure 8. Ratios of the temperature coefficients of the shear moduli in glassy (initial and relaxed) and crystalline states of $Zr_{46}Cu_{46}Al_8$, γ_{ini} and γ_{rel}, for temperature ranges where structural relaxation is absent (see Figure 7). Temperature-independent $\gamma_{ini} \approx \gamma_{rel} \approx 1$ confirm Equation (11).

5. Structural Relaxation-Induced Heat Effects

Since there is certain amount of the elastic energy associated with interstitialcy defects, any change of their concentration should lead to heat effects. In particular, annihilation of interstitialcy defects should result in the release of the internal energy in the form of heat. Inversely, interstitialcy defect formation should lead to an increase of the internal energy, which can be revealed as heat absorption. These expectations can be quantified as follows. The formation enthalpy of an isolated interstitialcy is [18,39]:

$$H = \alpha \Omega G \tag{12}$$

where Ω is the volume per atom and G and α have the same sense as in Equation (1). The increment of the number of defects per mole due to an augmentation of their concentration by dc is $dN_d^{\mu} = N_A dc$, where N_A is the Avogadro number. Then, using Equation (1), the molar interstitialcy formation enthalpy becomes $H_{\mu} = \alpha \Omega N_A \int_0^c G(c)dc$, and the heat flow occurring upon heating of the molar mass m_{μ} from room temperature may be found as:

$$W = \frac{1}{m_{\mu}} \frac{dH_{\mu}}{dt} = \frac{\alpha \Omega N_A}{m_{\mu}} \frac{d}{dt} \int_{c_{RT}}^c G(c)dc \tag{13}$$

where c_{RT} is the room-temperature defect concentration. Accepting the latter to be $c_{RT} = \frac{1}{\beta} ln \left(G_x^{RT}/G^{RT} \right)$, where G^{RT} and G_x^{RT} are the shear moduli of glass and parent crystal at room temperature, respectively, substituting Equation (1) into Equation (13), one can calculate the heat heat per unit time and per unit mass flow as [78]:

$$W = \frac{\dot{T}}{\beta \rho} \left[\frac{G^{RT}}{G_x^{RT}} \frac{dG_x}{dT} - \frac{dG}{dT} \right] \tag{14}$$

with $\dot{T} = dT/dt$ being the heating rate and ρ the density. It is seen that since G^{RT} and G_x^{RT} are constants, the temperature dependence of the heat flow is simply determined by temperature derivatives of the shear moduli in the glassy and parent crystalline states. The underlying physical reason consists of the relaxation of the intrinsic defect system.

Figure 9 illustrates the calculation of the heat flow using Equation (14) with $\beta = 17$ in comparison with the experimental DSC runs for bulk glassy $Pd_{40}Ni_{40}P_{20}$ for initial and relaxed (obtained by heating into the supercooled liquid region) states [75]. One can point out a good agreement between calculated and experimental heat release below T_g and heat absorption above T_g. Similar agreement was found for other Pd- and Zr-based metallic glasses [77–79]. Below T_g, a decrease of the interstitialcy defect concentration upon heating leads to the heat release and related increase of the shear modulus. Rapid defect multiplication in the supercooled liquid region (above T_g) requires strong heat absorption and provides a significant decrease of the shear modulus. Moreover, preliminary results indicate that Equation (14) also correctly describes the crystallization-induced heat release relating it with the corresponding relaxation of the shear modulus.

169

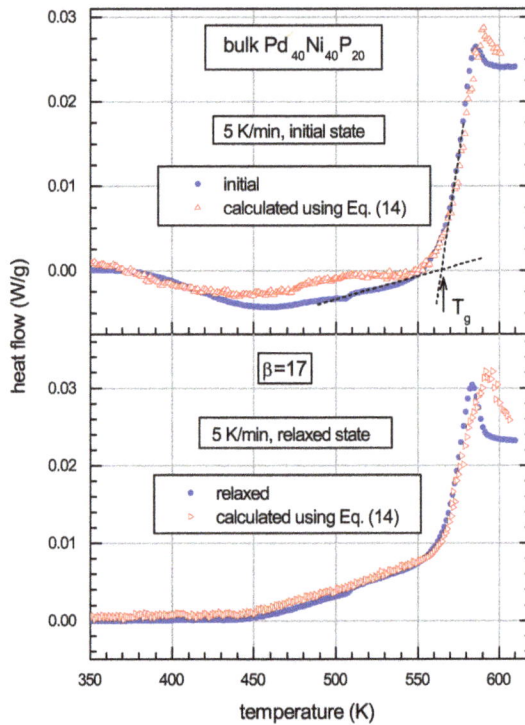

Figure 9. Experimental and calculated DSC thermograms for bulk glassy $Pd_{40}Ni_{40}P_{20}$ in the initial and relaxed states [75]. The calorimetric glass transition temperature is shown by the arrow.

For the relaxed state, the shear modulus $G = G_{rel}$, and Equation (14) determines the corresponding heat flow W_{rel}. Then, the difference between the heat flow in the relaxed and initial states, $\Delta W = W_{rel} - W$, is given by:

$$\Delta W(T) = \frac{\dot{T}}{\beta \rho} \frac{d\Delta G(T)}{dT} \tag{15}$$

where $\Delta G(T) = G(T) - G_{rel}(T)$ is the shear modulus change due to structural relaxation. As discussed above, this change is conditioned by relaxation events with distributed activation energies. Applying the approximation of the characteristic activation energy (see Equation (8) and related description) and using Equation (9), the Expression (15) can be rewritten as:

$$\Delta W(E_0) = \frac{\dot{T} G^{RT} A}{\beta \rho} \frac{d}{dE_0} \frac{\Delta G(E_0)}{G^{RT}} = \frac{\dot{T} G^{RT} A}{\rho} N_0(E_0) \tag{16}$$

From Equation (16), one obtains the formula for the AES [72],

$$N_0(E_0) = \frac{\rho}{\dot{T} A G^{RT}} \Delta W(E_0) \tag{17}$$

Equations (9) and (17) describe the same AES using the data on the shear modulus relaxation and heat flow measured by DSC, respectively. These equations should give, in principle, the same result. The AES calculated from DSC data on the structural relaxation of bulk glassy $Pd_{41.25}Cu_{41.25}P_{17.5}$ is shown in Figure 5 together with the AES determined from independent shear modulus relaxation data. A good correspondence between the two spectra is indeed seen, indicating the self-consistence of the approach under consideration. Similar results were obtained for two other Pd- and Zr-based glasses [72].

Table 1. Parameters of structural relaxation of Pd- and Zr-based glasses [72]: the change of the defect concentration $\Delta c_g = \int n_g(E_0)dE_0$ determined from G-measurements (where n_g is calculated using Equation (9)), the change of the defect concentration $\Delta c_w = \int n_w(E_0)dE_0$ determined from DSC measurements (where n_w is determined using Equation (17)), averaged concentration $\overline{\Delta c} = (\Delta c_g + \Delta c_w)/2$, molar heat of structural relaxation Q_μ (J/mole), number of defects per mole $N_\mu = \overline{\Delta c}N_A$ ($\times 10^{-20}$ mole^{-1}), heat of structural relaxation per defect $Q_d = Q_\mu/N_\mu$ (eV), room-temperature shear modulus G^{RT} (GPa), volume per atom $\Omega = \frac{m_\mu}{\rho N_A}$ ($\times 10^{-29}$ m^3) and interstitialcy formation enthalpy $H = \alpha G\Omega$ (eV).

Glass	Δc_g	Δc_w	$\overline{\Delta c}$	Q_-	N_-	Q_d	G^{RT}	Ω	H
PdCuP	0.00161	0.00165	0.00163	432	9.8	2.75	32.7	1.33	2.72
PdNiP	0.00332	0.00314	0.00323	931	19.5	2.98	38.6	1.28	3.09
ZrCuAl	0.00206	0.00220	0.00213	786	12.8	3.84	34.3	1.73	3.71

Table 1 gives interesting comparative data obtained upon analyzing the activation energy spectra for $Pd_{41.25}Cu_{41.25}P_{17.5}$, $Pd_{40}Ni_{40}P_{20}$ and $Zr_{46}Cu_{46}Al_8$ glasses [72]. First, one can calculate the change of the defect concentration during structural relaxation by integrating the spectra determined from measurements of the shear modulus and DSC (*i.e.*, using Equations (9) and (17), respectively). Table 1 shows that both methods of AES reconstruction give very close results. Second, taking the experimental molar heat of structural relaxation Q_μ and calculating the number of defects per mole, $N_\mu = \overline{\Delta c}N_A$, ($\overline{\Delta c}$ is the averaged number of defects annealed out during structural relaxation, N_A the Avogadro number), one can determine the heat of structural relaxation per defect, $Q_d = Q_\mu/N_\mu$. On the other hand, one can calculate the interstitialcy formation enthalpy using Equation (12). Table 1 illustrates a remarkable similarity between the heat of structural relaxation per defect Q_d and interstitialcy formation enthalpy H: the difference between these quantities is less than 4%. On the other hand, the obtained values of Q_d and H are quite close to the values of the interstitialcy formation enthalpy (2–3 eV) in simple close-packed crystalline metals [29,80]. These arguments constitute further evidence for the interstitialcy-mediated mechanism of structural relaxation.

6. Interstitialcies and Low Temperature Heat Capacity

A fundamental feature of the structural dynamics of glasses of different types consists of the low temperature excess heat capacity, which can be visualized as a peak in the specific heat C divided by the cube of temperature (C/T^3) in the 5 to 15 K temperature range [81]. This peak is called the Boson heat capacity peak. It is commonly accepted that this peak arises from excess vibrational states in glass, which are absent in crystalline materials [81]. There are quite a few approaches derived for the interpretation of the Boson peak (for a review, see [82]), but its physical nature still remains unclear. The Interstitialcy theory considers the Boson peak as originating from low frequency resonance vibration modes of interstitialcies, resulting mainly from excitation to their first excited state [83]. Granato calculated the height H_B of the Boson peak and showed that it is proportional to the concentration c of interstitialcy defects, $H_B = 4.6\frac{c}{0.03}\frac{f}{5}\left(\frac{\omega_D}{7\omega_R}\right)^3$, where f is the number of resonance modes per interstitialcy, ω_D is the Debye frequency and interstitialcy resonance vibration frequencies are assumed to be the same equal to ω_R [83]. The temperature of the Boson peak was calculated as $T_B = \frac{\Theta}{35}\frac{7\omega_R}{\omega_D}$. With a rough estimate, $\frac{7\omega_R}{\omega_D} \approx 1$, this leads to the Boson peak temperature $T_B \approx \frac{\Theta}{35}$, where Θ is the Debye temperature. This gives a reasonable estimate of Boson peak temperature for glasses of various types [83].

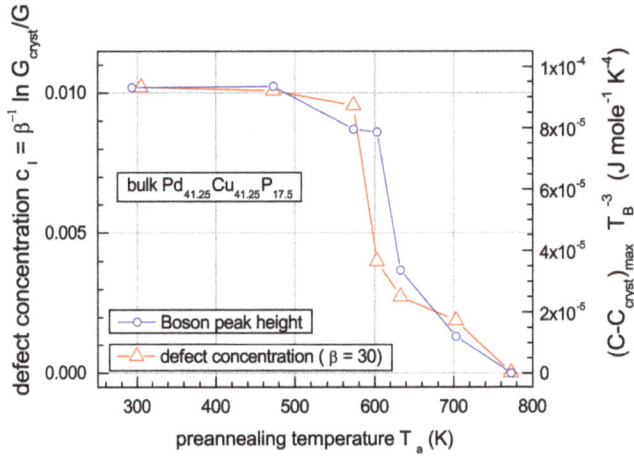

Figure 10. Interstitialcy defect concentration in bulk glassy $Pd_{41.25}Cu_{41.25}P_{17.5}$ calculated using Equation (1) together with the height of the Boson heat capacity peak as a function of the pre-annealing temperature T_a. The curves are drawn as guides for the eye. It is seen that the dependencies of the defect concentration c and Boson peak height on T_a can be superposed, indicating the direct proportionality between them and confirming, thus, the interstitialcy interpretation of the Boson peak [84].

Structural relaxation occurring after annealing at high temperatures below T_g leads to a decrease of the Boson peak height. Simultaneously, the shear modulus increases. The concentration of interstitialcy defects in glass can be estimated using its shear modulus as supposed by Equation (1). It is then possible to compare the defect concentration with the Boson peak height for different pre-annealing temperatures. This program was carried out in [84]. Figure 10 gives the defect concentration together with the Boson peak height (relatively to the crystalline state) in bulk glassy $Pd_{41.25}Cu_{41.25}P_{17.5}$ as a function of pre-annealing temperature T_a. It is seen that both quantities decrease with T_a, as one would expect. After pre-annealing at $T_a = 773$ K, the glass crystallizes, so that $c = 0$ and the Boson peak disappears. Figure 10 also shows that the dependences of the defect concentration and Boson peak height on T_a can be superposed, indicating the direct proportionality between them and confirming, thus, the understanding of the Boson heat capacity peak on the basis of the Interstitialcy theory (see [84] for more details).

7. Interstitialcies, Free Volume and Enthalpy Release

Most of metallic glasses have smaller density compared to their crystalline counterparts. Since the density is increasing upon structural relaxation of initial MGs below T_g, it is widely believed that elementary structural relaxation events take place in the regions of smaller local density, *i.e.*, in the regions containing some excess "free volume" [4,9,12,13]. In spite of the fact that the "free volume" has no clear theoretical definition, as repeatedly mentioned in the literature (e.g., [17,85]), and the application of this concept to the interpretation of experimental data sometimes leads to evident inconsistencies [16,77], the free volume-based notions still remain quite popular [4,11,15,86]. In this context, it is important to estimate what kind of volume effects could be associated with interstitialcy defects in glass and to compare these with volume effects attributed to the free volume.

For a rough estimate, it can be accepted that the free volume in MGs represents some entity similar to vacancies in crystals, while interstitialcies in MGs are analogous to those in simple metallic crystals. It is known that the insertion of vacancies (v) and interstitialcies (i) gives the volume changes $(\Delta V/\Omega)_v = 1 - \alpha_v$ and $(\Delta V/\Omega)_i = -1 + \alpha_i$, respectively, where Ω is the volume per atom, α_v and α_i are the corresponding relaxation volumes [87]. Then, the resulting relative volume change is $\Delta V/V = (\alpha_v + 1)\,c_v + (\alpha_i - 1)\,c_i$, where c_v and c_i are the concentrations of vacancies and interstitialcies. Granato [87] suggests that $\alpha_v = -0.2$ and $\alpha_i = 2.0$. The calculated values of α_v and α_i for 15 metals given in [29] after averaging are equal to -0.26 and 1.55, respectively, quite close to available experimental data [29,88]. Then, one arrives at $\Delta V/V \approx 0.74 c_v + 0.55 c_i$, and one has to conclude that the volume effects associated with vacancies and interstitialcies have the same sign and are quite comparable in the magnitude. Therefore, the observed densification of MGs below T_g cannot be solely interpreted as annealing out of the free volume, as pointed out long ago [48]. A decrease of interstitialcy concentration can almost equally lead to the volume contraction.

Isothermal structural relaxation as-cast MGs below T_g leads to a decrease of the volume, which is accompanied by a linear decrease of the molar enthalpy [15]. This dependence is usually interpreted as a result of the free volume decrease [13,15]. However, it was recently shown that this fact can be also understood within the framework of the Interstitialcy theory as a result of a decrease of the interstitialcy defect concentration. The derivative of the released molar enthalpy H_μ over the relative volume change $\Delta V/V$ then becomes [80]:

$$\frac{dH_\mu}{d\Delta V/V} = -\frac{\alpha m_\mu\,(1+g)\,G}{\rho\,(\alpha_i - 1)} \tag{18}$$

where $\alpha \approx 1$, $g = \ln\frac{G}{G_x}$ (G and G_x are the shear moduli of glass and reference crystal) and the volume per atom $\Omega = \frac{m_\mu}{\rho N_A}$, with N_A, m_μ and ρ being the Avogadro number, molar mass and density, respectively. Equation (18) shows that the derivative in the left-hand side should be nearly temperature independent and the released enthalpy should linearly increase with the relative decrease of the volume, in accordance with numerous experimental observations on different MGs [15,80,89]. The estimate of this derivative for an Au-based metallic glass using Equation (18) gives the value of 239 kJ/mol, which is quite close to the experimental value of 187 kJ/mol [80], supporting thus the notions under consideration.

8. Elastic Dipole Approach

The insertion of a defect into a crystal creates local elastic distortions. Because of these distortions, the defect interacts with the applied homogeneous elastic stress. In some sense, this interaction is similar to the interaction of an electric dipole with the applied electric field. Accordingly, the defect, which creates local elastic distortions and interacts with the external stress, is called the "elastic dipole" [90]. The necessary condition for such an interaction consists in the requirement that the symmetry of the defect must be lower than the local symmetry of the matrix structure [90]. This fully applies to dumbbell interstitials, which in fact represent a particular case of elastic dipoles.

In this case, using the conceptual framework described above, one can assume the existence of frozen-in elastic dipoles in glass and accept that they create local anisotropic elastic distortions. This approach was developed in [77,91]. It was shown that: (i) these defects lead to the elastic softening of glass with respect to the reference crystal; and (ii) their stored elastic energy, which is released as heat upon structural relaxation below and above T_g, as well as upon crystallization of glass, closely corresponds to the heat effects observed experimentally. The main points of this approach can be summarized as follows.

The expression for the internal energy U per unit mass of a deformed isotropic body taking into account third- and fourth-order expansion terms was suggested in [92,93] (the same expression was later used in [94]). It is reasonable to assume that the change of the internal energy due to strain-induced dilatation (volume change) is insignificant for MGs [77,91]. Then, the expansion for U has the form [77,91]:

$$\rho U = \rho U_0 + \mu I_2 + \frac{4}{3} v_3 I_3 + \frac{1}{2} \gamma_4 I_2^2 \tag{19}$$

where ρ is the density, U_0 the internal energy of the undeformed state, $I_2 = \varepsilon_{ij}\varepsilon_{ji}$ and $I_3 = \varepsilon_{ij}\varepsilon_{jk}\varepsilon_{ik}$ the algebraical invariants of the deformation tensor ε_{ij}, μ the second-order Lamé elastic constants, v_3 the third-order Lamé elastic constant and γ_3 and γ_4 the fourth-order Lamé elastic constants. If the strain field ε_{ij} in Equation (19) is created by randomly-oriented frozen-in elastic dipoles, then the change of the internal energy ΔU with respect to the reference crystal is:

$$\rho U - \rho U_0 = \rho \Delta U \approx \mu I_2 = \mu c \overline{\lambda_{ij}\lambda_{ji}} \tag{20}$$

where c is the concentration of frozen-in elastic dipoles, λ_{ij} the so-called λ-tensor [90], which characterizes the strain field created by an elastic dipole, and the bar denotes averaging over all elastic dipoles. The components of the λ-tensor are equal to the components of the deformation tensor per unit concentration of unidirectional elastic dipoles [90]. The change of the elastic energy given by Equation (20) is released as heat upon structural relaxation. Then, the heat flow occurring upon warming-up of glass due to the release of the internal energy associated with frozen-in elastic dipoles can be calculated as [77]:

$$W = \frac{\partial \Delta U}{\partial t} = \dot{T} \frac{\partial \Delta U}{\partial T} = \frac{\dot{T}}{\rho} \frac{\partial}{\partial T} \left(\mu c \overline{\lambda_{ij}\lambda_{ji}} \right) \tag{21}$$

With some minor further assumptions, Equation (21) leads to the expression for the heat flow occurring upon warming up of a metallic glass [77],

$$W = \frac{3\dot{T}}{\rho \beta \overline{\Omega}} \left[\frac{dG_x}{dT} - \frac{dG}{dT} \right] \tag{22}$$

where the averaged form-factor $\overline{\Omega} = 1.38$ takes into account different types of elastic dipoles involved in structural relaxation [77]; other quantities have the same meaning as above. The comparison of the heat flow given by this equation with the experimental data taken on glassy $Zr_{46}Cu_{46}Al_8$ revealed their good correspondence [77]. On the other hand, it is to be emphasized that Equation (22) is very similar, although not fully identical, to the heat flow law Equation (14) derived within the framework of the Interstitialcy theory. A detailed comparative analysis of these heat flow laws with the experimental data obtained on glassy $Pd_{41.25}Cu_{41.25}P_{17.5}$ taken as an example was reported in [95]. It was found that both equations quite correctly describe both heat release well below T_g and heat absorption near and above T_g. The elastic dipole approach Equation (22) provides a very good description of heat flow data near and above T_g, but slightly underestimates the heat flow well below T_g. The Interstitialcy theory approach Equation (14) provides a very good description of heat flow data in the whole temperature range. It is clear that the heat flow is conditioned by the relaxation of the shear modulus, as implied by both Equations (14) and (22).

Since the difference in the heat flow given by Equations (14) and (22) is small, one can derive the relationship between the shear susceptibilities entering these formulae. Designating the shear susceptibilities in Equations (14) and (22) as β_i and β_d, respectively, taking into account that the expressions in square brackets of these equations are approximately equal, one arrives at $\beta_i \approx \frac{1}{3}\overline{\Omega}\beta_d$. Fitting to the calorimetric data for $Pd_{41.25}Cu_{41.25}P_{17.5}$ glass gives $\beta_d = 38$ (just the same value as for glassy $Zr_{46}Cu_{46}Al_8$ [77]). Then, with the above relationship and $\overline{\Omega} = 1.38$, one calculates $\beta_i = 17.5$, which is quite close to $\beta_i = 20$ derived by fitting to the calorimetric data for the same glass [95].

The elastic dipole approach sketched above gives clear information on the reason for the shear softening of glass with respect to the reference crystal. The expression for the shear modulus of glass can be written down as [77,91]:

$$G = \mu + \gamma_4 \Omega_t c \overline{\lambda_{ij}\lambda_{ji}} \tag{23}$$

where Ω_t is the averaged dipole form factor for the shear deformation. If the defect concentration $c = 0$, the quantity μ equals the shear modulus G_x of the reference crystal. The only metallic glass for which the fourth-rank modulus γ_4 is so far known is $Zr_{52.5}Ti_5Cu_{17.9}Ni_{14.6}Al_{10}$ [93]. With $\gamma_4 = -171$ GPa and other estimates for this glass, $c \approx 0.039$ and $\overline{\lambda_{ij}\lambda_{ji}} = 0.92 \pm 0.12$ [91], Equation (23) gives

$\Delta G = G_x - G \approx 8.0$ GPa, fairly close to the experimental value $\Delta G \approx 9.3$ GPa [91], supporting thus the elastic dipole approach.

It is important to emphasize that the elastic constants ν_3 and γ_4 in the non-linear expansion Equation (19) for the internal energy are essentially anharmonic (in the harmonic approximation, these constants are equal to zero). Since G is determined by γ_4 (see Equation (23)), one has to conclude that the shear softening of MGs with respect to the reference crystalline state is determined by the anharmonicity of the interatomic potential.

The same conclusion immediately comes from the Interstitialcy theory. Indeed, the shear susceptibility in the main Equation (1) constitutes a major parameter, which is defined as $\beta = -\frac{1}{G}\frac{\partial^2 G}{\partial \varepsilon^2}$ (ε is the shear strain) [18]. The shear susceptibility is non-zero only if the non-linear elasticity is taken into account. A similar definition of β is considered in the elastic dipole approach, which directly gives the linear proportionality between the shear susceptibility and the absolute value of γ_4 elastic constant, $\beta = -\frac{3\gamma_4}{G}$ [77]. The understanding that non-linear elastic effects occurring due to the anharmonicity of the interatomic potential are manifested in many of MGs' properties has now been increasing [59,96–101]. In particular, the role of non-linear elastic effects in the mechanical behavior of metallic glasses at high stresses seems to be quite evident [94,102,103].

9. Summary

The Interstitialcy theory provides a new comprehensive, versatile and verifiable approach to the understanding of the structural relaxation of metallic glasses. This approach starts with the assumption that melting of simple metallic crystals takes place as a result of rapid multiplication of dumbbell (split) interstitials (= interstitialcies). The nucleus of such a defect can be interpreted as two atoms trying to occupy the same minimum of the potential energy. On the other hand, these defects create internal stresses interacting with the external stress and can be considered as elastic dipoles. In the liquid state, the defects retain their individuality, but become inherent structural elements, rather than "defects" of the structure. Rapid melt quenching partially freezes the interstitialcy defect structure in solid glass. Heat treatment of the glass leads to a change of the interstitialcy defect concentration, which can be precisely monitored by measurements of the shear modulus. The latter represents the major physical quantity controlling the relaxation kinetics through the main Equation (1) of the Interstitialcy theory.

The sign of structural relaxation monitored by measurements of the shear modulus is conditioned by current temperature and glass thermal prehistory that results in a different relations between the current interstitialcy defect concentration, c, and the defect concentration in the metastable equilibrium, c_{eq}. At temperatures far below T_g, usually $c > c_{eq}$ and the relaxation lead to a decrease of c, with a corresponding increase of the shear modulus. Near and above T_g, c can be smaller than c_{eq}, leading to a decrease of the shear modulus. It has been found that the Interstitialcy theory provides a good description of the relaxation kinetics for different metallic glasses and testing conditions.

The basic Equation (1) of the Interstitialcy theory establishes the direct relationship between the shear modulus of glass, the concentration of frozen-in interstitialcy defects and the shear modulus of the maternal crystal. It has been revealed that this relationship conforms with the available experimental data. In particular, when structural relaxation is absent, Equation (1) implies the equality of temperature coefficients of the shear moduli in the glassy and crystalline states (Equation (11)), which is proved experimentally.

Any change of the interstitialcy concentration alters the elastic energy associated with them. If c decreases upon structural relaxation, the stored elastic energy is released as heat. Increasing c requires an augmentation of the external energy, which is manifested as heat absorption. The Interstitialcy theory leads to the general heat flow law Equation (14), which directly states that the heat effects occurring upon heating are controlled by the relaxation of the shear modulus, while the underlying physical reason consists of the relaxation of the internal interstitialcy defect system. It has been found that this law correctly describes exothermal heat flow well below T_g, as well as endothermal heat reaction near and above T_g.

Metals **2015**, *5*, 504–529

The Interstitialcy theory provides two independent ways for the reconstruction of the activation energy spectrum of atomic rearrangements occurring upon structural relaxation. The first of them (Equation (9)) is based on shear modulus relaxation data, while the second one (Equation (17)) makes use of heat flow data. It has been found that both methods give nearly the same results, and the formation energy of defects responsible for structural relaxation is close to the interstitialcy formation energy implied by the Interstitialcy theory (Equation (12)).

It has been found that the height of the low temperature Boson heat capacity peak strongly correlates with the changes in the shear modulus upon high temperature annealing. The Interstitialcy theory connects this peak with low frequency resonant localized vibrations of interstitialcy defects frozen-in upon glass production. The height of this peak is proportional to the defect concentration and, together with the peak temperature, reasonably agrees with the experiment.

Since dumbbell interstitials are in fact elastic dipoles, it is possible to develop a more general approach based on the non-linear theory of elasticity. This approach gives nearly the same expression (Equation (22)) for structural relaxation-induced heat flow and makes clear evidence that the heat effects, as well as the shear softening of metallic glasses are intrinsically connected with the anharmonicity of the interatomic interaction.

Acknowledgments: The author is cordially grateful to Professor A.V. Granato for many inspiring discussions and fruitful cooperation over the years. Long-term collaboration with D.M. Joncich (University of Illinois at Urbana-Champaign, IL, USA), N.P. Kobelev (Institute for Solid State Physics, Chernogolovka, Russian Academy of Sciences), Yu.P. Mitrofanov, R.A. Konchakov, G.V. Afonin and A.S. Makarov (State Pedagogical University, Voronezh, Russia) is kindly acknowledged. The support for this work was provided by the Ministry of Education and Science of the Russian Federation (Project 3.114.2014/K).

Conflicts of Interest: The author declares no conflict of interest.

References

1. Taub, A.I.; Spaepen, F. Isoconfigurational flow of amorphous $Pd_{82}Si_{18}$. *Scr. Metall.* **1979**, *13*, 195–198.
2. Khonik, V.A. The kinetics of irreversible structural relaxation and rheological behavior of metallic glasses under quasi-static loading. *J. Non-Cryst. Sol.* **2001**, *296*, 147–157.
3. Chen, H.S. Glassy metals. *Rep. Prog. Phys.* **1980**, *43*, 353–432.
4. Schuh, C.A.; Hufnagel, T.C.; Ramamurty, U. Mechanical behavior of amorphous alloys. *Acta Mater.* **2007**, *55*, 4067–4109.
5. Greer, A.L. Metallic Glasses. In *Physical Metallurgy*; Volume I; Laughlin, D.E., Hono, K., Eds.; Elsevier: Oxford, UK, 2014; pp. 305–385.
6. Doolittle, A.K. Studies in newtonian flow. II. The dependence of the viscosity of liquids on free-space. *J. Appl. Phys.* **1951**, *22*, 1471–1475.
7. Turnbull, D.; Cohen, M.H. Free volume model of the amorphous phase: glass transition. *J. Chem. Phys.* **1961**, *34*, 120–125.
8. Turnbull, D.; Cohen, M.H. On the free volume model of the liquid-glass transition. *J. Chem. Phys.* **1970**, *52*, 3038–3041.
9. Spaepen, F. A microscopic mechanism for steady state inhomogeneous flow in metallic glasses. *Acta Metall.* **1977**, *25*, 407–415.
10. Argon, A.S. Plastic deformation in metallic glasses. *Acta Metall.* **1979**, *27*, 47–58.
11. Spaepen, F. Homogeneous flow of metallic glasses: A free volume perspective. *Scr. Mater.* **2006**, *54*, 363–367.
12. Van den Beukel, A.; Radelaar, S. On the kinetics of structural relaxation in metallic glasses. *Acta Metall.* **1990**, *31*, 419–427.
13. Van den Beukel, A.; Sietsma, J. The glass transition as a free volume related kinetic phenomenon. *Acta Metall. Mater.* **1990**, *38*, 383–389.
14. Koebrugge, G.W.; Sietsma, J.; van den Beukel, A. Structural relaxation in amorphous $Pd_{40}Ni_{40}P_{20}$. *Acta Metall. Mater.* **1992**, *40*, 753–760.
15. Slipenyuk, A.; Eckert, J. Correlation between enthalpy change and free volume reduction during structural relaxation of $Zr_{55}Cu_{30}Al_{10}Ni_5$ metallic glass. *Scr. Mater.* **2004**, *50*, 39–44.

16. Bobrov, O.P.; Khonik, V.A.; Lyakhov, S.A.; Csach, K.; Kitagawa, K.; Neuhäuser, H. Shear viscosity of bulk and ribbon glassy $Pd_{40}Cu_{30}Ni_{10}P_{20}$ well below and mear the glass transition. *J. Appl. Phys.* **2006**, *100*, 033518.

17. Cheng, Y.Q.; Ma, E. Indicators of internal structural states for metallic glasses: Local order, free volume, and configurational potential energy. *Appl. Phys. Lett.* **2008**, *93*, 051910.

18. Granato, A.V. Interstitialcy model for condensed matter states of face-centered-cubic metals. *Phys. Rev. Lett.* **1992**, *68*, 974–977.

19. Granato, A.V. Interstitialcy theory of simple condensed matter. *Eur. J. Phys.* **2014**, *87*, 18.

20. Frenkel, J. *Kinetic Theory of Liquids*; Oxford University Press: New York, NY, USA, 1946.

21. Mei, Q.S.; Lu, K. Melting and superheating of crystalline solids: From bulk to nanocrystals. *Prog. Mater. Sci.* **2007**, *52*, 1175–1262.

22. Slater, J.C. *Introduction to Chemical Physics*; McGraw-Hill Book Company: New York, NY, USA; Toronto, ON, Canada; London, UK, 1963.

23. Granato, A.V. A comparison with empirical results of the Interstitialcy theory of condensed matter. *J. Non-Cryst. Sol.* **2006**, *352*, 4821–4825.

24. Seitz, F. On the theory of diffusion in metals. *Acta Cryst.* **1950**, *3*, 355–363.

25. Gibson, J.B.; Goland, A.N.; Milgram, M.; Vineyard, G.H. Dynamics of radiation damage. *Phys. Rev.* **1960**, *120*, 1229–1253.

26. Erginsoy, C.; Vineyard, G.H.; Englert, A. Dynamics of radiation damage in a body-centered cubic lattice. *Phys. Rev.* **1964**, *133*, A595–A606.

27. Schilling, W. Self-interstitial atoms in metals. *J. Nucl. Mater.* **1978**, *69–70*, 465–489.

28. Robrock, K.H. *Mechanical Relaxation of Interstitials in Irradiated Metals*; Springer-Verlag: Berlin, Germany, 1990.

29. Wolfer, W.G. Fundamental Properties of Defects in Metals. In *Comprehensive Nuclear Materials*; Konings, R.J.M., Ed.; Elsevier: Amsterdam, The Netherlands, 2012.

30. Konchakov, R.A.; Khonik, V.A.; Kobelev, N.P. Split Interstitials in computer models of single-crystal and amorphous copper. *Phys. Sol. State (Pleiades Publishing)* **2015**, *57*, 844–852.

31. Holder, J.; Granato, A.V.; Rehn, L.E. Experimental evidence for split interstitials in copper. *Phys. Rev. Lett.* **1974**, *32*, 1054–1057.

32. Holder, J.; Rehn, L.E.; Granato, A.V. Effect of self-interstitials on the elastic constants of copper. *Phys. Rev. B* **1974**, *10*, 363–375.

33. Born, M. Thermodynamics of crystals and melting. *J. Chem. Phys.* **1939**, *7*, 591–603.

34. Dederichs, P.H.; Lehman, C.; Schober, H.R.; Scholz, A.; Zeller, R. Lattice theory of point defects. *J. Nucl. Mater.* **1978**, *69–70*, 176–199.

35. Nordlund, K.; Averback, R.S. Role of self-interstitial atoms on the high temperature properties of metals. *Phys Rev. Lett.* **1998**, *80*, 4201–4204.

36. Spaepen, F. A survey of energies in materials science. *Phil. Mag.* **2005**, *85*, 2979–2987.

37. De Podesta, M. *Understanding the Properties of Matter*, 2nd ed.; Taylor & Francis: London, UK, 2001.

38. Nordlund, K.; Ashkenazy, Y.; Averback, R.S.; Granato, A.V. Strings and interstitials in liquids, glasses and crystals. *Europhys. Lett.* **2005**, *71*, 625–631.

39. Granato, A.V.; Joncich, D.M.; Khonik, V.A. Melting, thermal expansion, and the Lindemann rule for elemental substances. *Appl. Phys. Lett.* **2010**, *97*, 171911.

40. Stillinger, F.H.; Weber, T.A. Point defects in bcc crystals: Structures, transition kinetics, and melting implications. *J. Chem. Phys.* **1984**, *81*, 5095–51034.

41. Lee, G.C.S.; Li, J.C.M. Molecular-dynamics studies of crystal defects and melting. *Phys. Rev. B* **1989**, *39*, 9302–9311.

42. Kanigel, A.; Adler, J.; Polturak, E. Influence of point defects on the shear elastic coefficients and on the melting temperature of copper. *Int. J. Mod. Phys. C* **2001**, *12*, 727–737.

43. Zhang, H.; Khalkhali, M.; Liu, Q.; Douglas, J.F. String-like cooperative motion in homogeneous melting. *J. Chem. Phys.* **2013**, *138*, 12A538.

44. Ashkenazy, Y.; Averback, R.S. Atomic mechanisms controlling crystallization behavior in metals at deep undercoolings. *Europhys. Lett.* **2007**, *79*, 26005.

45. Betancourt B.A.P.; Douglas, J.F.; Starr, F.W. String model for the dynamics of glass-forming liquids. *J. Chem. Phys.* **2014**, *140*, 204509.

46. Schober, H.R. Collectivity of motion in undercooled liquids and amorphous solids. *J. Non-Cryst. Sol.* **2002**, *307-310*, 40–49.

47. Donati, C.; Douglas, J.F.; Kob, W.; Plimpton, S.J.; Poole, P.H.; Glotzer, S.C. Stringlike cooperative motion in a supercooled liquid. *Phys. Rev. Lett.* **1998**, *80*, 2338–2341.

48. Oligschleger, C.; Schober, H.R. Collective jumps in a soft-sphere glass. *Phys. Rev. B* **1999**, *59*, 811–821.

49. Granato, A.V. The specific heat of simple liquids. *J. Non-Cryst. Sol.* **2002**, *307-310*, 376–386.

50. Konchakov, R.A.; Khonik, V.A. Effect of vacancies and interstitials in the dumbbell configuration on the shear modulus and vibrational density of states of copper. *Phys. Sol. State (Pleiades Publishing)* **2014**, *56*, 1368–1373.

51. Schober H.R. *Phonons 89*; Hunklinger, S., Ludwig, W., Weiss, G., Eds.; World Scientific: Singapore, 1989; Volume I, p. 444.

52. Okamoto, P.R.; Rehn, L.E.; Pearson, J.; Bhadra, R.; Grimsditch, M. Brillouin scattering and transmission electron microscopy studies of radiation-induced elastic softening, disordering and amorphization of metallic compounds. *J. Less-Common Met.* **1988**, *14*, 231–244.

53. Landau, L.D.; Lifshitz, E.M. *Theory of Elasticity*; Pergamon Press: Oxford, UK, 1970.

54. Nemilov, S.V. The kinetics of elementary processes in the condensed state. II. Shear relaxation and the equation of state for solids. *Russ. J. Phys. Chem.* **1968**, *42*, 726–731.

55. Dyre, J.C.; Olsen, N.B.; Christensen, T. Local elastic expansion model for viscous-flow activation energies of glass-forming molecular liquids. *Phys. Rev. B* **1996**, *53*, 2171–2174.

56. Johnson, W.L.; Samwer, K. A universal criterion for plastic yielding of metallic glasses with a $(T/T_g)^{2/3}$ temperature dependence. *Phys Rev. Lett.* **2005**, *95*, 195501.

57. Nemilov, S.V. Interrelation between shear modulus and the molecular parameters of viscous flow for glass forming liquids. *J. Non-Cryst. Sol.* **2006**, *352*, 2715–2725.

58. Dyre, J.C. The glass transition and elastic models of glass-forming liquids. *Rev. Mod. Phys.* **2006**, *78*, 953–972.

59. Wang, W.H. The elastic properties, elastic models and elastic perspertives of metallic glasses. *Prog. Mater. Sci.* **2012**, *57*, 487–656.

60. Tsao, S.S.; Spaepen, F. Structural relaxation of a metallic glass near equilibrium. *Acta Metall.* **1985**, *33*, 881–889.

61. Mitrofanov, Yu.P.; Khonik, V.A.; Granato, A.V.; Joncich, D.M.; Khonik, S.V.; Khoviv, A.M. Relaxation of a metallic glass to the metastable equilibrium: Evidence for the existence of the Kauzmann pseudocritical temperature. *Appl. Phys. Lett.* **2012**, *100*, 171901.

62. Khonik, V.A.; Mitrofanov, Y.P.; Makarov, A.S.; Konchakov, R.A.; Afonin, G.V.; Tsyplakov, A.N. Structural relaxation and shear softening of Pd- and Zr-based bulk metallic glasses near the glass transition. *J. Alloys Comp.* **2015**, *628*, 27–31.

63. Khonik, V.A.; Mitrofanov, Y.P.; Khonik, S.V.; Saltykov, S.N. Unexpectedly large relaxation time determined by in situ high-frequency shear modulus measurements near the glass transition of bulk glassy $Pd_{40}Cu_{30}Ni_{10}P_{20}$. *J. Non-Cryst. Sol.* **2010**, *356*, 1191–1193.

64. Gibbs, M.R.J.; Evetts, J. E.; Leake, J.A. Activation energy spectra and relaxation in amorphous materials. *J. Mater. Sci.* **1983**, *18*, 278–288.

65. Wagner, H.; Bedorf, D.; Küchemann, S.; Schwabe, M.; Zhang, B.; Arnold, W.; Samwer, K. Local elastic properties of a metallic glass. *Nat. Mater.* **2011**, *10*, 439–442.

66. Khonik, V.A. The kinetics of irreversible structural relaxation and homogeneous plastic flow of metallic glasses. *Phys. Status Sol. (A)* **2000**, *177*, 173–189.

67. Khonik, S.V.; Granato, A.V.; Joncich, D.M.; Pompe, A.; Khonik, V.A. Evidence of distributed interstitialcy-like relaxation of the shear modulus due to structural relaxation of metallic glasses. *Phys. Rev. Lett.* **2008**, *100*, 065501.

68. Bothe, K.; Neuhäuser, H. Relaxation of metallic glass structure measured by elastic modulus and internal friction. *J. Non-Cryst. Sol.* **1983**, *56*, 279–284.

69. Mitrofanov, Yu.P.; Khonik, V.A.; Vasil'ev, A.N. Isothermal kinetics and relaxation recovery of high-frequency shear modulus in the course of structural relaxation of $Pd_{40}Cu_{30}Ni_{10}P_{20}$ bulk glass. *J. Exp. Theor. Phys.* **2009**, *108*, 830–835.

70. Mitrofanov, Yu.P.; Khonik, V.A.; Granato, A.V.; Joncich, D.M.; Khonik, S.V. Relaxation of the shear modulus of a metallic glass near the glass transition. *J. Appl. Phys.* **2011**, *109*, 073518.

71. Khonik, V.A.; Kitagawa, K.; Morii, H. On the determination of the crystallization activation energy of metallic glasses. *J. Appl. Phys.* **2000**, *87*, 8440–8443.

72. Tsyplakov, A.N.; Mitrofanov, Yu.P.; Makarov, A.S.; Afonin, G.V.; Khonik, V.A. Determination of the activation energy spectrum of structural relaxation in metallic glasses using calorimetric and shear modulus relaxation data. *J. Appl. Phys.* **2014**, *116*, 123507.

73. Makarov, A.S. Khonik, V.A.; Mitrofanov, Yu.P.; Granato, A.V.; Joncich, D.M.; Khonik, S.V. Interrelationship between the shear modulus of a metallic glass, concentration of frozen-in defects, and shear modulus of the parental crystal. *Appl. Phys. Lett.* **2013**, *102*, 091908.

74. Khonik, V.A.; Mitrofanov, Yu.P.; Lyakhov, S.A.; Khoviv, D.A.; Konchakov, R.A. Recovery of structural relaxation in aged metallic glass as determined by high-precision in situ shear modulus measurements. *J. Appl. Phys.* **2009**, *105*, 123521.

75. Makarov, A.S.; Khonik, V.A.; Wilde, G.; Mitrofanov, Yu.P.; Khonik, S.V. "Defect"-induced heat flow and shear modulus relaxation in a metallic glass. *Intermetallics* **2014**, *44*, 106–109.

76. Makarov, A.S.; Mitrofanov, Yu.P.; Afonin, G.V.; Khonik, V.A.; Kobelev, N.P. The dependence of the shear modulus of glass on the shear modulus of crystal and kinetics of structural relaxation for $Zr_{46}Cu_{46}Al_8$ system. *Phys. Sol. State (Pleiades Publishing)* **2015**, *57*, in press.

77. Kobelev, N.P.; Khonik, V.A.; Makarov, A.S.; Afonin, G.V.; Mitrofanov, Yu.P. On the nature of heat effects and shear modulus softening in metallic glasses: A generalized approach. *J. Appl. Phys.* **2014**, *115*, 033513.

78. Mitrofanov, Yu.P.; Makarov, A.S.; Khonik, V.A.; Granato, A.V.; Joncich, D.M.; Khonik, S.V. On the nature of enthalpy relaxation below and above the glass transition of metallic glasses. *Appl. Phys. Lett.* **2012**, *101*, 191903.

79. Makarov, A.S.; Khonik, V.A.; Mitrofanov, Yu.P.; Granato, A.V.; Joncich, D.M. Determination of the susceptibility of the shear modulus to the defect concentration in a metallic glass. *J. Non-Cryst. Sol.* **2013**, *370*, 18–20.

80. Khonik, V.A.; Kobelev, N.P. Alternative understanding for the enthalpy *vs.* volume change upon structural relaxation of metallic glasses. *J. Appl. Phys.* **2014**, *115*, 093510.

81. Phillips, W.A. *Amorphous Solids: Low Temperature Properties*; Springer: Berlin, Germany, 1981.

82. Shintani, H.; Tanaka, H. Universal link between the boson peak and transverse phonons in glass. *Nat. Mater.* **2008**, *7*, 870–877.

83. Granato, A.V. Interstitial resonance modes as a source of the boson peak in glasses and liquids. *Phys. B* **1996**, *219–220*, 270–272.

84. Vasiliev, A.N.; Voloshok, T.N.; Granato, A.V.; Joncich, D.M.; Mitrofanov, Yu.P.; Khonik, V.A. Relationship between low-temperature boson heat capacity peak and high-temperature shear modulus relaxation in a metallic glass. *Phys. Rev. B* **2009**, *80*, 172102.

85. Miracle, D.B.; Egami, T.; Flores, K.M.; Kelton, K.F. Structural aspects of metallic glasses. *MRS Bull.* **2007**, *32* 629–634.

86. Trexler, M.M.; Thadhani, N.N. Mechanical properties of bulk metallic glasses. *Prog. Mater. Sci.* **2010**, *55*, 759–839.

87. Gordon, C.A.; Granato, A.V.; Simmons, R.O. Evidence for the self-interstitial model of liquid and amorphous states from lattice parameter measurements in krypton. *J. Non-Cryst. Sol.* **1996**, *205–207*, 216–220.

88. Ehrhart, P. Jung, P.; Schultz, H.; Ullmaier, H. Properties and Interactions of Atomic Defects in Metals and Alloys. In *Atomic Defects in Metals, Landolt-Börnstein New Series III*; Madelung, O., Ed.; Springer: Berlin, Germany, 1991; Volume 25, pp. 88–371.

89. Bünz, J.; Wilde, G. Direct measurement of the kinetics of volume and enthalpy relaxation of an Au-based metallic glass. *J. Appl. Phys.* **2013**, *114*, 223503.

90. Nowick, A.S.; Berry, B.S. *Anelastic Relaxation in Crystalline Solids*; Academic Press: New York, NY, USA, 1972.

91. Kobelev, N.P.; Khonik, V.A.; Afonin, G.V.; Kolyvanov, E.L. On the origin of the shear modulus change and heat release upon crystallization of metallic glasses. *J. Non-Cryst. Sol.* **2015**, *411*, 1–4.

92. Erofeyev, V.I. *Wave Processes in Solids With Microstructure*; World Scientific: Singapore, 2003.

93. Kobelev, N.P.; Kolyvanov, E.L.; Khonik, V.A. Higher order elastic moduli of the bulk metallic glass $Zr_{52.5}Ti_5Cu_{17.9}Ni_{14.6}Al_{10}$. *Phys. Sol. State (Pleiades Publishing)* **2007**, *49*, 1209–1215.

94. Nakamura, A.; Kamimura, Y.; Edagawa, K.; Takeuchi, S. Elastic and plastic characteristics of a model Cu-Zr amorphous alloy. *Mater. Sci. Eng.* **2014**, *A614*, 16–26.

95. Tsyplakov, A.N.; Mitrofanov, Yu.P.; Khonik, V.A.; Kobelev, N.P.; Kaloyan, A.A. Relationship between the heat flow and relaxation of the shear modulus in bulk PdCuP metallic glass. *J. Alloys Comp.* **2015**, *618*, 449–454.

96. Lambson, E.F.; Lambson, W.A.; Macdonald, J.E.; Gibbs, M.R.J.; Saunders, G.A.; Turnbull, D. Elastic behavior and vibrational anharmonicity of a bulk $Pd_{40}Ni_{40}P_{20}$ metallic glass. *Phys. Rev. B* **1986**, *33*, 2380–2385.

97. Wang, R.J.; Wang, W.H.; Li, F.Y.; Wang, L.M.; Zhang, Y.; Wen, P.; Wang, J.F. The Grüneisen parameter for bulk amorphous materials. *J. Phys.: Condens. Matter* **2003**, *15*, 603–608.

98. Novikov, V.N.; Sokolov, A.P. Poisson ratio and the fragility of glass-forming liquids. *Nature* **2004**, *431*, 961–963.

99. Tarumi, R.; Hirao, M.; Ichitsubo, T.; Matsubara, E.; Saida, J.; Kato, H. Low-temperature acoustic properties and quasiharmonic analysis for Cu-based bulk metallic glasses. *Phys. Rev. B* **2007**, *76*, 104206.

100. Safarik, D.J.; Schwarz, R.B. Evidence for highly anharmonic low-frequency vibrational modes in bulk amorphous $Pd_{40}Cu_{40}P_{20}$. *Phys. Rev. B* **2009**, *80*, 094109.

101. Chen, L.Y.; Li, B.Z.; Wang, X.D.; Jiang, F.; Ren, Y.; Liaw, P.K.; Jiang, J.Z. Atomic-scale mechanisms of tension–compression asymmetry in a metallic glass. *Acta Mater.* **2013**, *61*, 1843–1850.

102. Wang, H.; Li, M. Symmetry breaking and other nonlinear elastic responses of metallic glasses subject to uniaxial loading. *J. Appl. Phys.* **2013**, *113*, 213515.

103. Wang, H.; Li, M. Estimate of the maximum strength of metallic glasses from finite deformation theory. *Phys. Rev. Lett.* **2013**, *111*, 065507.

metals

MDPI

Editorial

Metallic Glasses

Kang Cheung Chan [1,*] and Jordi Sort [2,*]

[1] Advanced Manufacturing Technology Research Centre, Department of Industrial and Systems Engineering, The Hong Kong Polytechnic University, Hung Hom, Kowloon, HK

[2] Institució Catalana de Recerca i Estudis Avançats (ICREA) and Departament de Física, Universitat Autònoma de Barcelona, E-08193 Bellaterra, Spain

* Authors to whom correspondence should be addressed; kc.chan@polyu.edu.hk (K.C.C.); jordi.sort@uab.cat (J.S.); Tel.: +852-2766-4981 (K.C.C.); +34-935812085 (J.S.); Fax: +34-935812155 (J.S).

Received: 15 December 2015; Accepted: 15 December 2015; Published: 17 December 2015

Metallic glasses are a fascinating class of metallic materials that do not display long-range atomic order. Due to their amorphous character and the concomitant lack of dislocations, these materials exhibit mechanical properties that are quite different from those of other solid materials [1,2]. For example, they can be twice as strong as steels, show more elasticity and fracture toughness than ceramics and be less brittle than conventional oxide glasses. In addition to their unique mechanical properties, metallic glasses have also demonstrated interesting physical and chemical properties. For example, some metallic glasses have been found to exhibit superior soft magnetic properties [3], good magnetocaloric effects [4] and outstanding catalytic performance [5], thus having potential for a widespread range of technological applications [6].

Metallic glasses become relatively malleable when heated in the supercooled liquid region, allowing moulding and shaping with microscale precision by means of thermoplastic processing [7]. This has facilitated the development of diverse products based on these alloys, including sporting goods, medical and electronic devices and advanced defence and aerospace applications. However, in spite of their large elasticity, metallic glasses exhibit poor room-temperature macroscopic plasticity compared to polycrystalline metals [8]. This low plastic deformation, particularly evidenced when testing metallic glasses under tension, is related to the formation and rapid propagation of shear bands [2]. Therefore, novel routes to enhance plasticity of metallic glasses include procedures to hinder shear band propagation. This can be achieved, for example, by designing composite materials consisting of particles which act as second-phase reinforcements embedded in the amorphous matrix [9,10]. Other approaches towards toughening of metallic glasses have also been developed, such as the preparation of the so-called dual-phase amorphous metals [11], some specific surface treatments (e.g., laser or shot pinning) [12], or making their overall structure porous (*i.e.*, metallic glass foams) [13].

The aim of this Special Issue is to address, from both experimental and theoretical points of view, some of the challenges to improve the glass forming ability of metallic glasses, to optimize their overall mechanical performance, and to enhance their physico-chemical properties. The intricate dynamic and geometrical features of the icosahedral atomic clusters arrangements in liquid and amorphous states in Zr–Cu alloys are investigated by M. Shimono and H. Onodera using molecular dynamics simulations [14]. In spite of the lack of long-range order, the simulations reveal that such icosahedral clusters induce medium range order in these alloys. The atomic structure frozen-in during the ultra-rapid cooling processes needed to generate metallic glasses is usually not stable at room temperature. V. A. Khonik reviews how the presence of interstitialcy defects and the change in their concentration can induce structural relaxation when metallic glasses are heat treated to temperatures below or around the glass transition [15]. Such structural relaxation is important since it can induce changes in many physical properties of the glasses, particularly in the mechanical behavior (elasticity, anelasticity, viscoelasticity, *etc.* [16]), but also in the electrical, corrosion and even magnetic performances. Of course, atomic rearrangements of the glassy structure take place also during mechanical treatments. This can result in mechanically-induced structural relaxation, as reviewed by C. Liu *et al.* [17]. Dynamic mechanical measurements as a function of temperature reveal that structural relaxation is a complex phenomenon that depends on the frequency of the treatments. In fact, one can distinguish between primary and secondary relaxation effects, both of which are relevant for a full understanding of the complex and unique properties of amorphous metallic alloys.

As it might be anticipated, the properties of metallic glasses not only depend on their structure (*i.e.*, medium-range order, presence of flow defects, *etc.*) but also on the actual composition of the glasses. T. Bitoh and D. Watanabe show, for example, that incorporation of Y in the amorphous structure of Fe–Co–B–Si–Nb bulk metallic glasses induces changes in their glass forming ability and the magnetic properties of these alloys [18]. The properties of newly developed Ni-Cu-W-B metallic glasses are reported by A. Hitit *et al.* [19]. The current theories describing the links between toughness and material parameters, including elastic constants, alloy chemistry and ordering degree in the glass are reviewed by S. V. Madge [20]. As aforementioned, metallic glasses are relatively brittle when tested in their fully amorphous as-prepared state. Strategies to increase plastic deformation based partial crystallization or formation of metallic glass composites are reported by D. Wu *et al.* [21] and Ö. Balcı *et al.* [22]. Formation of metallic glass porous frameworks is described by B. Sarac *et al.* [23]. Finally, the interplay between corrosion effects and mechanical properties of Zr-based metallic glasses are surveyed by P.F. Gostin *et al.* [24].

Although the plasticity of metallic glasses in bulk form is rather limited due to the rapid propagation of single shear bands [1,2], an improved mechanical performance has been reported in miniaturized metallic glasses [25,26]. For this reason, there is a growing interest in new approaches towards the synthesis of metallic glass microwires [27] and nanowires [26], in view of their potential applications in micro/nano-electro-mechanical systems (MEMS/NEMS). From the aforementioned aspects and the ongoing works on the topic, it is likely that the interest in metallic glasses will continue to increase in the near future with the development of new compositions and novel applications, particularly in devices with micrometer and submicrometer sizes, where the full potential of these glassy materials is yet to come.

Acknowledgments: This work was supported by the "Ministerio de Economía y Competitividad" through project MAT2014-57960-C3-1-R (co-financed by the Fondo Europeo de Desarrollo Regional (FEDER)). Partial financial support from the 2014-SGR-1015 project from D.G.U. Catalunya is also acknowledged.

Conflicts of Interest: The authors declare no conflict of interest.

References

1. Schuh, C.A.; Hufnagel, T.C.; Ramamurty, U. Mechanical behavior of amorphous alloys. *Acta Mater.* **2007**, *55*, 4067–4109. [CrossRef]
2. Greer, A.L.; Cheng, Y.Q.; Ma, E. Shear bands in metallic glasses. *Mater. Sci. Eng. R* **2013**, *74*, 71–132. [CrossRef]
3. Suryanarayana, C.; Inoue, A. Iron-based bulk metallic glasses. *Int. Mater. Rev.* **2013**, *58*, 131–166. [CrossRef]
4. Xia, L.; Chan, K.C.; Tang, M.B. Enhanced glass forming ability and refrigerant capacity of a $Gd_{55}Ni_{22}Mn_3Al_{20}$ bulk metallic glass. *J. Alloys Compd.* **2011**, *509*, 6640–6643. [CrossRef]
5. Zhao, M.; Abe, K.; Yamaura, S.-I.; Yamamoto, Y.; Asao, N. Fabrication of Pd–Ni–P metallic glass nanoparticles and their application as highly durable catalysts in methanol electro-oxidation. *Chem. Mater.* **2014**, *26*, 1056–1061. [CrossRef]
6. Inoue, A.; Wang, X.M.; Zhang, W. Developments and applications of bulk metallic glasses. *Rev. Adv. Mater. Sci.* **2008**, *18*, 1–9.
7. Kumar, G.; Tang, H.X.; Schroers, J. Nanomoulding with amorphous metals. *Nature* **2009**, *457*, 868–872. [CrossRef] [PubMed]
8. Fornell, J.; Concustell, A.; Suriñach, S.; Li, W.H.; Cuadrado, N.; Gebert, A.; Baró, M.D.; Sort, J. Yielding and intrinsic plasticity of Ti–Zr–Ni–Cu–Be bulk metallic glass. *Int. J. Plast.* **2008**, *25*, 1540–1559. [CrossRef]

9. Hofmann, D.C.; Suh, J.-Y.; Wiest, A.; Duan, G.; Lind, M.-L.; Demetriou, M.D.; Johnson, W.L. Designing metallic glass matrix composites with high toughness and tensile ductility. *Nature* **2008**, *451*, 1085–1089. [CrossRef] [PubMed]

10. Wu, F.F.; Chan, K.C.; Jiang, S.S.; Chen, S.H.; Wang, G. Bulk metallic glass composite with good tensile ductility, high strength and large elastic strain limit. *Sci. Rep.* **2014**, *4*, 5302. [CrossRef] [PubMed]

11. Concustell, A.; Mattern, N.; Wendrock, H.; Kuehn, U.; Gebert, A.; Eckert, J.; Greer, A.L.; Sort, J.; Baró, M.D. Mechanical properties of a two-phase amorphous Ni–Nb–Y alloy studied by nanoindentation. *Scr. Mater.* **2007**, *56*, 85–88. [CrossRef]

12. González, S.; Fornell, J.; Pellicer, E.; Suriñach, S.; Baró, M.D.; Greer, A.L.; Belzunce, F.J.; Sort, J. Influence of the shot-peening intensity on the structure and near-surface mechanical properties of $Ti_{40}Zr_{10}Cu_{38}Pd_{12}$ bulk metallic glass. *Appl. Phys. Lett.* **2013**, *103*, 211907. [CrossRef]

13. Brothers, A.H.; Dunand, D.C. Ductile bulk metallic glass foams. *Adv. Mater.* **2005**, *17*, 484–486. [CrossRef]

14. Shimono, M.; Onodera, H. Dynamics and geometry of icosahedral order in liquid and glassy phases of metallic glasses. *Metals* **2015**, *5*, 1163–1187. [CrossRef]

15. Khonik, V.A. Understanding of the structural relaxation of metallic glasses within the framework of the interstitialcy theory. *Metals* **2015**, *5*, 504–529. [CrossRef]

16. Van Steenberge, N.; Sort, J.; Concustell, A.; Das, J.; Scudino, S.; Suriñach, S.; Eckert, J.; Baró, M.D. Dynamic softening and indentation size effect in a Zr-based bulk glass-forming alloy. *Scr. Mater.* **2007**, *56*, 605–608. [CrossRef]

17. Liu, C.; Pineda, E.; Crespo, D. Mechanical relaxation of metallic glasses: An overview of experimental data and theoretical models. *Metals* **2015**, *5*, 1073–1111. [CrossRef]

18. Bitoh, T.; Watanabe, D. Effect of yttrium addition on glass-forming ability and magnetic properties of Fe–Co–B–Si–Nb bulk metallic glass. *Metals* **2015**, *5*, 1127–1135. [CrossRef]

19. Hitit, A.; Şahin, H.; Öztürk, P.; Aşgın, A.M. A new Ni-based metallic glass with high thermal stability and hardness. *Metals* **2015**, *5*, 162–171. [CrossRef]

20. Madge, S.V. Toughness of bulk metallic glasses. *Metals* **2015**, *5*, 1279–1305. [CrossRef]

21. Wu, D.; Song, K.; Cao, C.; Li, R.; Wang, G.; Wu, Y.; Wan, F.; Ding, F.; Shi, Y.; Bai, X.; *et al.* Deformation-induced martensitic transformation in Cu-Zr-Zn bulk metallic glass composites. *Metals* **2015**, *5*, 2134–2147. [CrossRef]

22. Balcı, Ö.; Prashanth, K.G.; Scudino, S.; Ağaoğulları, D.; Duman, İ.; Öveçoğlu, M.L.; Uhlenwinkel, V.; Eckert, J. Effect of milling time and the consolidation process on the properties of Al matrix composites reinforced with Fe-based glassy particles. *Metals* **2015**, *5*, 669–685. [CrossRef]

23. Sarac, B.; Sopu, D.; Park, E.; Hufenbach, J.K.; Oswald, S.; Stoica, M.; Eckert, J. Mechanical and structural investigation of porous bulk metallic glasses. *Metals* **2015**, *5*, 920–933. [CrossRef]

24. Gostin, P.F.; Eigel, D.; Grell, D.; Uhlemann, M.; Kerscher, E.; Eckert, J.; Gebert, A. Stress-corrosion interactions in Zr-based bulk metallic glasses. *Metals* **2015**, *5*, 1262–1278. [CrossRef]

25. Ashby, M.F.; Greer, A.L. Metallic glasses as structural materials. *Scr. Mater.* **2006**, *56*, 321–326. [CrossRef]

26. Zeeshan, M.A.; Esqué-de los Ojos, D.; Castro-Hartmann, P.; Guerrero, M.; Nogués, J.; Suriñach, S.; Baró, M.D.; Nelson, B.J.; Pané, S.; Pellicer, E.; *et al.* Electrochemically synthesized amorphous and crystalline nanowires: Dissimilar nanomechanical behavior in comparison with homologous flat films. *Nanoscale* **2016**. [CrossRef] [PubMed]

27. Olofinjana, A.; Voo, N.Y. On the stability of the melt jet stream during casting of metallic glass wires. *Metals* **2015**, *5*, 1029–1044. [CrossRef]

MDPI AG

St. Alban-Anlage 66

4052 Basel, Switzerland

Tel. +41 61 683 77 34

Fax +41 61 302 89 18

http://www.mdpi.com

Metals Editorial Office

E-mail: metals@mdpi.com

http://www.mdpi.com/journal/metals